James Hutton and the History of Geology

1. James Hutton in middle age. Portrait by Henry Raeburn, ca. 1775; Hutton is approaching fifty. He is dressed in brown. Since he had published nothing as of the mid 1770s, it is all the more remarkable that he is presented here as not only an author but specifically a geological one. Beneath the symbolic quill are two apparently completed manuscripts of book length and a third in progress. The rocks to his left have been identified as a Chalk fossil (shell); two examples of mineral veins; a druse; a septarian nodule; and a breccia. All were directly relevant to the geological theory he would formally propose a decade later. This is the only seriously intended portrait of Hutton that we have. According to Playfair (1805b, 95n), it was commissioned by John Davidson of Stewartfield and "conveys a good idea of a physiognomy and character of face to which it was difficult to do complete justice." (Scottish National Portrait Gallery).

James Hutton and the History of Geology

DENNIS R. DEAN

Cornell University Press

Ithaca and London

First published 1992 by Cornell University Press.

International Standard Book Number 0-8014-2666-9
Library of Congress Catalog Card Number 91-55545

Printed in the United States of America

Librarians: Library of Congress cataloging information appears on the last page of the book.

♾ The paper in this book meets the minimum requirements of the American National Standard for Information Sciences—Permanence of Paper for Printed Library Materials, ANSI Z39.48-1984.

In Memoriam
John Playfair
(1748–1819)

Contents

Illustrations

Preface

James Hutton's theory of the earth (1788; 1795) is usually regarded as the basis—or one of them—of modern geology, but few of those who think so have actually confronted that theory at first hand. Even between 1802 and 1830, Hutton was known almost exclusively through a polished restatement of his geological ideas by John Playfair (*Illustrations of the Huttonian Theory*, 1802). After 1830, for some years, the major source became Charles Lyell's *Principles of Geology* (1830–33), which derived in part from Playfair but not at all from Hutton. We have since discovered that neither Playfair's nor Lyell's influential expositions of Huttonian theory were altogether accurate. No single work by Hutton himself, moreover, explains his full position adequately.

To be represented as he actually was, Hutton must be read comprehensively and in the original. Unfortunately, his bulky, old-fashioned prose is often tedious and difficult to follow—or even to find. Though adequate facsimiles now exist for all but one of his major geological writings, the originals are valuable and scarce. Even fine libraries rarely contain copies of his other works. This book, in part a guide to his publications, serves rather to introduce than to replace them. Thus, quoted passages have been conservatively modernized (in spelling and other accidentals only) for ease of comprehension, but advanced scholars will surely require the originals, which are fully cited in all cases. Only the major geological writings are considered in detail. Those who cannot muster sufficient time to read Hutton through, however, may rely on the present book's description of his thought.

After summarizing Hutton in Chapters 1–3, I present various responses to his geological theory and some highly diverse evaluations of its significance. This methodology then shapes the remainder of the book, permitting me to incorporate or cite most of what we have so far learned about Hutton, his theory, and its reception over the past two centuries. There are still things in all these areas that we do not know, but Huttonian studies have been notably productive in recent years and a comprehensive survey is now possible. Though some of the materials assembled here will be familiar to a few specialized historians, numerous others appear for the first time. My synthesis, moreover, is almost wholly new. Throughout my presentation I have avoided elaborate commentary, hoping that an accessible Hutton and his critical legacy will be explored, discussed, and evaluated freely, as the man himself would have wished.

Though I had no suspicions of it at the time, this book began almost thirty-five years ago, when I first read a selection from Playfair's *Illustrations* (in the White facsimile) for a class taught at Stanford by Konrad Krauskopf. My growing interest in the history of geology was then furthered by a series of Midwestern teachers and colleagues, including Robert Stauffer, Robert Dott, James Shea, Frank Egerton, and George White. I began the actual writing of this book on Christmas Day 1985, after a speaking invitation from Keith Tinkler (for 1988!) led me to realize how voluminous my Hutton files had become. At the 1988 conference, held in Ontario, I met for the first time not only Tinkler but also François Ellenberger, Mott Greene, R. P. Beckinsale, and especially Gordon Herries Davies. As history of geology editor for *Annals of Science* since the later 1970s, Davies had encouraged my Huttonian contributions for years. His fine book *The Earth in Decay* did more than any other to sophisticate my thinking about Hutton and has since influenced much of what I have written about him.

Many classics of the Huttonian tradition described in the present book are from my own library. For the others, I am indebted to a number of institutions, the University of Illinois at Urbana in particular. At the University of Oklahoma, Marcia Goodman responded to several mail queries with exemplary competence. I am also grateful to the staff of the British Library in London for courteous assistance during my several visits there and to the Fitzwilliam Museum, Cambridge, for permission to reprint the text of Hutton's "Memorial" (in two versions).

Annals of Science has permitted me to reuse material originally published by them. The Alexander Turnbull Library, Wellington, New Zealand, and other institutions have allowed me to quote from letters in their possession. For permission to reproduce illustrations, I am pleased to thank the Scottish National Portrait Gallery (Hutton portrait) and Sir

John Clerk of Penicuik. Some of the information in my captions appeared first in Craig (1978). The two maps were drawn by Donald Lintner, with thanks to the National Library of Scotland.

I am very grateful to the library staff at the University of Wisconsin–Parkside for extraordinary patience and helpfulness. In my departmental office, Trudy Rivest and Jacqueline Battersby cheerfully typed and retyped my sometimes puzzling text, which was then emended, augmented, and in other ways improved by several readers, especially Donald McIntyre, a most helpful referee (and later guide). He and I spent two weeks together in Scotland while this book was in press. I then went on to visit Werner sites in Saxony.

At Cornell University Press, Roger Haydon was a model acquisitions editor, unfailingly positive, urbane, tactful, and patient. Close readings of the text by Kay Scheuer and John Thomas were most conscientious and helpful. I very much wanted a book that would be typographically pleasing and am grateful to George Whipple for having designed one.

DENNIS R. DEAN

Evanston, Illinois

James Hutton
and the
History of Geology

[1]

The Man and His Theory

James Hutton, the "founder of modern geology," was born in Edinburgh, Scotland (then a city of perhaps thirty thousand inhabitants) on 3 June 1726. He was the only son of William Hutton, a merchant, who died three years later, in 1729. For the next eleven years James lived at home with his three sisters, being educated by his mother Sarah, whose father had been a merchant also. He then attended Edinburgh High School and in November 1740 (when fourteen and a half), Edinburgh University. Though nominally committed to the humanities, Hutton proved to be a desultory student. While privileged, for example, to attend the Newtonian lectures of Colin Maclaurin—an instructor he admired—Hutton "cultivated the mathematical sciences less than any other," as his biographer John Playfair (himself a mathematician) assures us. John Stevenson's rigorous lectures on logic and rhetoric also failed to stimulate, but by illustrating one of those lectures with an experiment Stevenson first introduced Hutton to the wonders of chemistry.[1]

In 1743, at age seventeen, Hutton left the university to accept apprenticeship under George Chalmers, an Edinburgh lawyer. Unfortunately, Chalmers repeatedly found Hutton amusing himself and his fellow apprentices with chemical experiments when he should have been copying papers or studying legal procedure; the apprenticeship, therefore, came

[1]The epithet "founder of modern geology," originally promulgated by Archibald Geikie, now appears on Hutton's gravestone in Greyfriars Churchyard, Edinburgh. For details of Hutton's life, we have Playfair 1805b, with subsequent additions by V. A. Eyles, Jean Jones, and Dennis R. Dean, among others. The publication of a letter from Joseph Black to James Watt (13 June 1797) by Robinson and McKie (1970) revealed the existence of Hutton's illegitimate son, James, who had been born about fifty years earlier.

to an abrupt end. Having meanwhile committed himself to chemistry (for which there was no separate curriculum), Hutton soon reentered Edinburgh University as a medical student. These efforts continued through 1747, when he turned twenty-one and abruptly moved to Paris, where he studied chemistry and anatomy. After nearly two years in France, Hutton returned to Britain by way of Holland. He was awarded his Doctor of Medicine degree at Leyden in September 1749, for a thesis entitled *De Sanguine et Circulatione Microcosmi* (On human blood and its circulation).

In his dissertation, Hutton first adumbrated certain deistic attitudes important to his later work. Thus, he began by asserting that nature is everywhere arranged according to physical, mechanical, and chemical laws, displaying the wisdom and perfection worthy of its creator, a benign and omnipotent God. From this point of view, a most significant feature of human anatomy is the blood's circulation, which was providentially designed to nourish and maintain the body. Blood, he stressed, is equally being used up and replenished, so that its agency is constant. Though Hutton did not specifically say so, his designation of man as microcosm indicates that he assumed there to be some equivalent process in the inanimate world. He later discovered one, for Hutton's subsequent conceptualization of the geocosm, our earth, would be remarkably similar to his view of the human body, with soil replacing blood and fertility, life. By 1749, then, assumptions destined to manifest themselves throughout Hutton's wide-ranging thought had already appeared in his work.[2]

Though Hutton had now at last qualified himself for one of the learned professions, he knew that he did not wish to practice medicine. On returning to London at the end of 1749, he wrote a number of anxious letters to Edinburgh regarding his own future. In the most important reply, his medical school friend John Davie informed Hutton that some experiments they had made together a few years earlier on the nature and production of sal ammoniac (ammonium chloride, a useful industrial chemical) had been continued in his absence; as a result, Davie and Hutton became partners in a firm that extracted ammoniac from coal soot by a relatively simple process. Hutton then returned to Edinburgh (probably to start things up) but soon left the firm's management to Davie, who ran it honestly and well. Their partnership remained profitable throughout Hutton's lifetime; its dividends, together with the remainder of a generous paternal legacy, eventually relieved him from any obligation to earn a living through regular employment.[3]

As part of his inheritance, Hutton also acquired two small farms,

[2]Ellenberger 1972b; 1973.
[3]Playfair 1805b; Clow and Clow 1947.

Slighhouses and Nether Monynut, in Berwickshire, Scotland, some forty miles southeast of Edinburgh. Having nothing more promising before him, he decided to become a farmer. This decision may have been influenced by Sir John Hall, an innovative agriculturalist whose properties were about nine miles from Hutton's. In Edinburgh, moreover, Hutton had become acquainted with Dr. John Manning of Norfolk, where the most innovative farming in Britain was going on. Hutton next went to Norfolk, living in Yarmouth as a guest of Manning's father, Thomas. From there he moved to the farmhouse of a John Dybold in Belton; remaining more than a year, Hutton learned Norfolk husbandry at first hand under Dybold's instruction. Finally, he also spent an unspecified period of time in "High Suffolk," where the soil was heavier and more like Scotland's. His studies there concentrated on plowing, dairying, and butter making. Meanwhile, tours (in part geological) through southern England and the Low Countries further broadened his knowledge of contemporary agricultural practices. After returning to Scotland in the fall of 1754, Hutton attempted to farm his own rather barren land. Having brought some newfangled two-horse plows with him from East Anglia, he found it necessary to return there for a plowman who could use them. Introducing these methods to his own farm and eventually throughout the Scottish Lowlands, Hutton remained at Slighhouses almost fourteen years—until the end of 1767, when (having lost his Norfolk plowman sometime earlier) he rented out his property and moved to Edinburgh.[4]

Now fully recovered from the national turmoil of some twenty years before, Edinburgh had in the interim become one of the liveliest and fastest growing cities in Europe, particularly characterized at this time by the vigor of its intellectual life. As the stern, repressive influence of the old kirk steadily declined, Scottish universities—and Edinburgh in particular—increasingly emphasized critical thinking, as fostered by the earlier English and Continental phases of the Enlightenment. Hutton joined such articulate, influential colleagues as David Hume, William Robertson, Adam Smith, Adam Ferguson, and Thomas Reid, all of whom would contribute significantly to philosophy and other fields. Besides Robertson, Smith, and Ferguson, his closest friends included Sir George Clerk-Maxwell, John Clerk of Eldin, Dr. James Lind, and especially Dr. Joseph Black, the physician and chemist whose name remains familiar to us for his discovery of carbon dioxide (1754) and his concepts of latent and specific heat.[5]

Obviously, then, Hutton was by no means unsociable. "He was little

[4]Jones 1985.

[5]For Edinburgh at this time, see Macgregor 1950, McElroy 1969, and Daiches 1986. Further biographical information on the major figures is widely available.

known, indeed, in general company," Playfair conceded, "and had no great relish for the enjoyment which it affords, yet he was fond of domestic society and took great delight in a few private circles, where several excellent and accomplished individuals of both sexes thought themselves happy to be reckoned in the number of his friends" (1805b, 97–98). After spending the day at home among his sisters, generally reading and writing, Hutton would often socialize at night with various male groups. Of these, his favorite was the Oyster Club, a congenial refuge founded by himself, Black, and Smith—probably in 1778, when the latter came to live in Edinburgh. Its other distinguished members (sooner or later) included William Robertson, Dr. John Hope, Clerk of Eldin, Lord Daer, Sir James Hall, and John Playfair, as well as Adam Ferguson, Robert Adam, Dugald Stewart, Henry Mackenzie, and Samuel Rogers. Hutton often dined on Sundays at Smith's and fortnightly at the home of James Burnett, Lord Monboddo, where Black, Thomas Hope, and William Smellie (first editor of the *Encyclopaedia Britannica*) were also regulars. Writing to Hope in 1779, Monboddo complained that Hutton, unlike himself, had "no historical faith at all," and denied the validity of all ancient traditions. Hutton also thought that what Monboddo scornfully called the artificial life, which we all lead, was both natural and best for man. Even so, he may have shared some of Monboddo's famous evolutionary speculations.[6]

On moving to Edinburgh by December 1767, Hutton almost immediately joined its intellectually far-reaching Philosophical Society, which had been organized around 1739 by his former mentor, Colin Maclaurin. Among the many topics presented to the society were medicine, botany, agriculture, technology, meteorology, chemistry, and astronomy. Hutton himself read several papers before the Society, but we know only two of them specifically (on artillery and botany). Several of the others almost certainly reappeared to some extent in Hutton's subsequent publications on natural philosophy, epistemology, language, agriculture, and geology. He also took part in the Society's attempt to observe the transit of Venus in 1769.[7]

The last major group with which Hutton associated was the Royal Society of Edinburgh. Superseding the Philosophical Society, it had been founded (primarily by William Robertson) in 1783 with Hutton among the officers; he is mentioned several times in its *Transactions*. Besides his "Theory of the Earth," Hutton's other contributions included his "Theory of Rain," a reply to Deluc's criticism of that theory, and the one

[6]The 1779 Monboddo-Hope letter comes from Knight 1900, 107; see also p. 18.

[7]Hutton's humorous role in the transit of Venus fiasco is recalled in Thompson 1927, 220; see also Ramsay 1918, 77.

Philosophical Society paper we have, on certain botanical phenomena observed on the slopes of Arthur's Seat, Edinburgh's extinct volcano. He also devoted considerable attention to the Forth and Clyde canal project and endeavored to secure more favorable taxation rates for Scottish coal by distinguishing it from English culm.

Hutton's interest in geology (a scarcely organized science not then so called) may have begun so early as 1748, when he was studying in Paris. Being interested primarily in chemistry, he might well have attended G. F. Rouelle's lectures on the science; if so, instruction on mineralogy and geology would have been included. Broadly speaking, the major geological issue in France at this time was whether the science should accommodate itself to the Creation narrative and Noachian flood of Genesis or go its own way. Fossil shells had by now been recognized as the remains of real animals, but some bold thinkers (de Maillet 1748; Buffon 1749) were suggesting that they had not resulted from the Flood. Such views, however, were widely regarded as atheistic. An Italian book published at this time (Generelli 1749) contained views surprisingly like Hutton's own later ones.[8]

According to Playfair, Hutton became interested in geology around 1753, while he was studying agriculture in East Anglia. At Yarmouth, the river Yare meets an often turbulent North Sea. Here, surely, Hutton first observed a river in flood carrying away "part of our land, to be sunk at the bottom of the sea" and a storm making its hostile attack on the coast; both helped to convince him that "in the natural operations of the world, the land is perishing continually" (Hutton, *Theory of the Earth,* I:185–186). Most of the rock exposed along that coast, however, belonged to the Crag, a loosely consolidated, geologically recent deposit full of fossil shells that had obviously once been under the sea. These formations are overlain by others (also fossiliferous) which then grade into the rich soil that farmers tilled. An awareness of the possibility of soil exhaustion was central to the Norfolk system of agriculture, which among much else stressed the necessity of crop rotation and good fertilizer. Because Hutton believed all of nature to have been designed for human habitation by an Infinite Wisdom, he could not accept the existence of purely counterproductive forces like the erosional ones so evidently at work around him. Realizing that rich soil was as necessary to agriculture as rich blood to human life, Hutton almost necessarily established an analogy between

[8]Hutton criticized Buffon's hypothesis in "Theory of the Earth," 1788, 264; and *Theory of the Earth,* 1795, I: 125–126. Though Hutton surely knew of Buffon earlier, William Smellie's translation of the latter's theory was published only in 1785 (as *Natural History,* vol. 1; London). Adequate surveys of eighteenth-century French and Italian geological thought, though much needed, have yet to be written. The comparison of Hutton's thinking with Generelli's originated, and ended, with Lyell.

them and began to seek the revivifying principle his conception of a benevolent earth required.

The earliest version of Hutton's theory of the earth available to us (though only by report, and of uncertain date) accords with what must have been his thinking at this time. Among Hutton's extensive papers (now lost), Playfair found "a few sketches of an essay on the natural history of the earth, evidently written at a very early period and intended, it would seem, for parts of an extensive work." Within these sketches, Hutton first observed that "a vast proportion of the present rocks is composed of materials afforded by the destruction of bodies—animal, vegetable, and mineral—of more ancient formation." He further saw that "all the present rocks are without exception going to decay, and their materials descending into the ocean." Naturally connecting these two complementary processes in his mind, Hutton compared "beds of grit or sandstone" with "banks of unconsolidated sand now formed on our shores," showing that "these bodies differ from one another in nothing but their compactness and induration." It followed, then, that "as the present continents are composed from the waste of more ancient land, so, from the destruction of them, future continents may be destined to arise." This "succession of continents," Hutton concluded boldly, was not a single historical event but an unbounded process, "the consequence of laws perpetually acting." The lost manuscript Playfair thus paraphrased (1805b, 55–57) goes on to further conjectures, but these clearly represent a later stage in Hutton's thinking.

When his presence was not required in Norfolk, Hutton "made many journeys on foot into different parts of England; and though the main object of these was to obtain information in agriculture, yet it was in the course of them that to amuse himself on the road, he first began to study mineralogy or geology" (p. 44). Playfair also paraphrased a letter of Hutton's to Sir John Hall attesting that he had "become very fond of studying the surface of the earth, and was looking with anxious curiosity into every pit, or ditch, or bed of a river, that fell in his way." Here and elsewhere, we should be careful not to take reflections of Hutton's casual banter too literally. Using a methodology he would repeat throughout his life, Hutton had formulated some far-reaching geological suppositions from his concentrated attention to a small geographical area. Because he believed all natural processes to be constant, ubiquitous, and virtually eternal, Hutton set out to discover corroborative evidence in other locales. Among his several foot journeys toward that end, he seems to have undertaken a major one that led him south along the coast of Suffolk (more Crag deposits), probably to London, south again to the Sussex coast (prominent chalk cliffs), then west to the Isle of Wight (spectacular erosion and contortions) and along the Dorset coast for some distance. He stopped short of Cornwall, which was almost the only part of English geology Hutton would never see.[9]

"Through the middle of the Isle of Wight," Hutton later wrote, "there runs a ridge of hills of indurated [hardened] chalk. This ridge runs from the Isle of Wight directly west into Dorsetshire, and goes by Corfe Castle towards Dorchester, perhaps beyond that place. The sea has broke through this ridge at the west end of the Isle of Wight, where columns of the indurated chalk remain, called the Needles; the same appearance being found upon the opposite shore in Dorsetshire" (*Theory*, I: 102). Dorchester, then, was probably the western terminus of Hutton's geological investigations; of everything he saw, the Needles apparently impressed him most.

Hutton's present geological speculations now centered on the nature of chalk and how this obviously marine deposit had been consolidated into solid rock. Fossiliferous chalk was used as fertilizer by farmers in southern England, so this geological question related directly to the broader one of how the soil's vitality was restored. During his primarily agricultural tour through Holland, Brabant, Flanders, and Picardy (where the sophisticated horticulture delighted him) in 1754, Hutton also analyzed local geology. Holland he found to be a fascinating example of how rivers can create, rather than destroy, land; nevertheless, he believed that the entire country would eventually be washed away.

In other places Hutton saw more of the chalk and, as in England, did not fail to notice that parts of it frequently contained large nodules of flint. He would later invite readers of his completed theory "to examine the chalk countries of France and England, in which the flint is found variously formed." In particular, Hutton recommended "an examination of the insulated masses of stone found in the sandhills by the city of Brussels, a stone which is formed by an injection of flint among sand" (*Theory*, I: 59). Hutton thought the presence of flints within both chalk and sandstone especially significant because it surely revealed how the consolidation of these rocks had taken place—if only he could discover it.[10]

The question was basically a chemical one, and, though Hutton continued to be interested in chemistry, no immediate answer appeared. In June 1755, however, after Hutton had returned from Norfolk to Slighhouses, his eventual best friend, Dr. Joseph Black (1728–99), presented a paper on magnesia (1756) to the Philosophical Society of Edinburgh.

[9]Playfair aside, reflections of Hutton's journey appear also in his *Theory* (e.g., I: 102, 213; II: 160–167, 251, 409) and letters (Eyles and Eyles 1951).

[10]See *Theory*, I: 59, and 51–62 more generally; II: 230–235. "Since flints are glassy nodules, any eighteenth-century mineralogist would have agreed that their glassy texture was evidence that they had cooled from a melt" (Laudan 1987, 119).

Within it, Black alluded to the fact that limestone (or chalk) could be made caustic by exposure to fire; since most limestones found naturally within the earth were *not* caustic, however, it followed that they had not been so exposed. Black then proved that the causticity of lime depended on the expulsion of carbonic gas (carbon dioxide). From this Hutton seemingly inferred that "strong compression might prevent the causticity of lime, by confining the carbonic gas even when great heat was applied, and that, as has been supposed of coal, the whole may have been melted in the interior of the earth, so as on cooling to acquire that crystallized or sparry structure which the carbonate of lime so frequently possesses." If this hypothetical reconstruction of his thought (by Playfair, 1805b, 59) is correct, Hutton was already speculating about both heat and compression during the latter 1750s.[11]

The first draft of Hutton's theory, that summarized by Playfair, was surely in existence by this time. Hutton's realization that loose geological materials can be converted into stone and elevated into land had led him to investigate the mechanism through which consolidation had been accomplished. After what Playfair called a long and minute examination, Hutton concluded that the fluidity seemingly required to achieve such consolidation could not have been produced by any solvent. No solvent known, he determined, was capable of dissolving all mineral substances—or even those often found within the same specimen; moreover, in rocks compounded from fragments (e.g., conglomerate or breccia), the completeness of the consolidation left no room for any solvent ever to have occupied. Thus, the only possible agent must have been one that could act on all substances, occupy no space within them, and yet work throughout them, however compact or hard. Heat alone had these properties and must therefore be the cause of mineral consolidation. When and where Hutton arrived at this conclusion are unknown, but Playfair himself suggested some time not long after 1760, while the theorist was still living on his farm.

Though the years Hutton spent at Slighhouses (1754-late 1767) have often been represented as a fallow period in his life, they actually included intensive fieldwork throughout much of Scotland. As in Norfolk, Hutton walked or rode out whenever he could, seeking further evidence to support his theory. "I have been examining the south alpine country of Scotland, occasionally, for more than forty years back," he would write no later than 1795 (*Theory*, I: 335). This deliberate research culminated in 1764, when he toured extensively throughout northeastern Scotland with his friend George Clerk-Maxwell. From Edinburgh they proceeded through Crieff, Dalwhinnie, Fort Augustus, Inverness, and East Ross

[11] See also Playfair 1802, 523–524.

into Caithness (at the very north of Scotland), then returned along the coast through Aberdeen. "In this journey," Playfair remarked prosaically, "Dr. Hutton's chief object was mineralogy, or rather geology, which he was now studying with great attention" (1805b, 45). Though we know little else about this trip, it seems fully comparable to that which took him from East Anglia to Dorsetshire and must have impressed on him both highly significant differences (the kinds of rock he saw) and equally significant similarities (geological processes).[12]

For Hutton, the most meaningful rocks now were no longer sandstone and limestone, nor, surprisingly, did they include granite. As he would later insist,

> I had examined Scotland from one end to the other before I saw one stone of granite in its native place. I had, moreover, examined almost all England and Wales (excepting Devonshire and Cornwall). . . . I had travelled every road from the borders of Northumberland and Westmoreland to Edinburgh. From Edinburgh, I had travelled to Portpatrick, and from that along the coast of Galloway and Ayrshire to Inverary in Argyleshire, and I had examined every spot between the Grampians and the Tweedale mountains from sea to sea. . . . I had also travelled from Edinburgh by Crieff, Rannoch, Dalwhinnie, Fort Augustus, Inverness, and through East Ross and Caithness to the Pentland Frith or Orkney Islands. (*Theory*, I: 213–214)

Only on his return by the east coast of Scotland, near Aberdeen, did he find any granite at all. It would interest him a great deal later on.

About six years after he had returned from his extended geological tour to Caithness, Hutton characterized his geological limitations and discoveries. Writing to John Strange early in 1770, he continued to see geology as primarily chemical: "My attention [he said] has been chiefly upon the various substances that enter into the composition of the mineral kingdom in general; and being neither botanist nor zoologist in particular, I never considered the different kinds of figured bodies [fossils] found in strata further than to distinguish betwixt animal and vegetable, sea and land objects, the mineralization of those objects being more the subject of my pursuit than the arrangement of them into their classes." Among some further observations, of which he then informed Strange, "The strata in this country are much more irregular and mixed with other masses, in which no form of strata is to be observed, than what I have observed in England, from whence I except Cornwall and Wales, in neither of which I have been." Scotland, so far as he knew,

[12]*Theory*, I: 213–214, 339, 578; II: 283–284. George Clerk-Maxwell was Clerk of Eldin's brother.

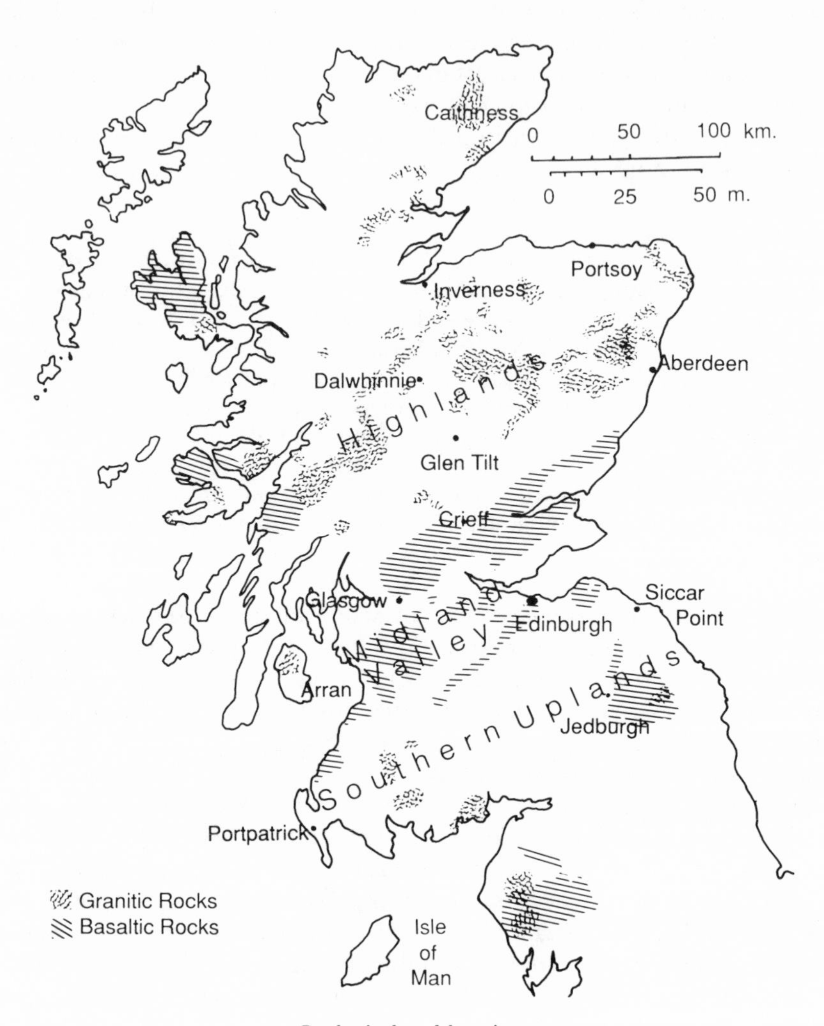

2. Geological and location map.

contained no chalk. There might be, he conceded, "in the west highlands, where I have not been, a limestone country where a much greater thickness of limestone may perhaps be found as in England, but that is not the case with the rest of Scotland." Hutton then divided the whole of Scotland into three distinct petrological regions (of freestone, schistus, and granite), citing specific localities for each.[13]

The rock of most importance to Hutton at this time was basalt (for us, a common form of lava), the definition and origin of which were much disputed. According to his primary source, Alex Frederic Cronstedt's *Essay Towards a System of Mineralogy* (1758; English translations 1770, 1772), Swedish trap, German schwartzstein, and Scottish whinstone were all basalt.[14] As quoted by Hutton, Cronstedt also pointed out that basalt was usually found as veins in rocks of another kind, "running commonly in a serpentine manner, contrary or across to the direction of the rock itself" (*Theory*, I: 151). Hutton himself found and described "a fine example of this kind" (152) in 1764, while on the road to Crieff. By this time he had also thoroughly investigated the basaltic phenomena closer to home. "On the south side of Edinburgh," he observed, "I have seen, in little more than the space of a mile from east to west, nine or ten masses of whinstone interjected among the strata" (153–154). This was primarily at Salisbury Crags, adjacent to Arthur's Seat, where part of the relevant exposure is now termed "Hutton's Section." The first significant geologist to assert the volcanic origin of basalt was Nicolas Desmarest in 1765, though his paper then remained unpublished for nine years. By 1774, however, several other geologists in France had also expressed opinions on the origin of basalt. (Those who did not accept the volcanic theory held, on the contrary, that basalt had been precipitated from water, like shale.) There would be later connections between Hutton and Desmarest, but Hutton's initial exposure to the basalt controversy probably took place through English sources.

"When first I conceived my theory," Hutton would later write, "naturalists were far from suspecting that basaltic rocks were of volcanic origin; I could not then have employed an argument from these rocks, as I may do now [around 1794], for proving that the fires which we see almost daily issuing with such force from volcanoes are a continuation of

[13]Hutton to Strange; Eyles and Eyles 1951, 323; see also Playfair 1802, 264–265 and *Theory*, I: 525. In Hutton's terminology, "freestone" meant any rock that split conveniently into layers, usually sandstone or limestone; "schistus" included shales, slates, greywackes, and schists; his "granite" did not necessarily exclude such similar rocks as syenite and gneiss. "Scotland is broadly divisible into what we now call Highlands, Midland Valley, and Southern Uplands. The rocks of the first and last were grouped by Hutton as Alpine schistus and granite; while those of the Midland Valley were described as later sandstones, coals, limestones, etc., accompanied by hills of whinstone exposed by erosion" (Bailey, 1967, 133).

[14]See Uno Boklund on Cronstedt in Gillispie 1970–80.

3. "Hutton's Section," Salisbury Crags, Edinburgh. Two unidentified figures (the one on the left is probably Hutton) investigate a site perhaps more important than any other to the development of Hutton's theory. Drawing by John Clerk of Eldin, early 1780s. (Sir John Clerk of Penicuik)

that active cause which has so evidently been exerted in all times, and in all places, so far as have been examined of this earth" (I: 246–247). Nevertheless, he would always distinguish between basalt and lava. By the 1770s, in any case, Hutton had become fully involved with the basalt controversy. During an important excursion to Derbyshire in 1774, particularly, he probably discovered another form of basaltic intrusion, in which strata of the parent rock had been separated rather than broken. Together with his friend James Watt (who would also furnish him a geological map of Cornwall), Hutton visited some halite mines at Northwich, in Cheshire, and proved to his own satisfaction that the rock salt they contained had been fused by heat rather than deposited by water. After returning to Birmingham, where he met several members of the Lunar Society, almost certainly including Erasmus Darwin, James Keir, and John Whitehurst, Hutton also visited Wales for some brief but important fieldwork.[15]

Hutton, Playfair tells us, could rarely be persuaded to read the geological theories of other authors, but he was an avid consumer of geological *facts*, which he sought out industriously by reading as many books of travel and description as he could find. While it is tempting to associate him with a great many current titles in which observers of diverse qualifications commented freely on the origin of basalt, the more important studies are surely those to which he specifically alluded or with which he can be at least plausibly associated. Among them we may include Sir William Hamilton, *Observations on Mount Vesuvius* (1772); Thomas Pennant, *A Tour in Scotland* (1774); John Strange, "Account of Two Giant's Causeways in the Venetian State" (1775); Rudolph Eric Raspe, *An Account of Some German Volcanos and Their Productions*, "with a new hypothesis of the prismatical basaltes, established upon facts" (1776, supplementing Hamilton); Barthelemy Faujas de Saint-Fond, *Recherches sur les volcans eteints du Vivarais et du Velay* (Researches on the extinct volcanoes of Vivarais and Velay, 1778); and, particularly, John Whitehurst, *An Inquiry into the Original State and Formation of the Earth* (1778). The general thrust of them all was not only to confirm the igneous origin of basalt but also to extend volcanic energies into times and places where they had not previously been observed. Whitehurst's book, moreover, the most substantial of all these publications, concentrated on Derbyshire, a region with which Hutton was recently familiar.

Since Whitehurst's *Inquiry* is one of the few geological theories that we *know* Hutton to have read (he cites it in *Theory*, I: 153), and apparently one of the more influential on him, a brief review may help us to see how much of Whitehurst's thinking Hutton opposed. In its way, Whitehurst's

[15]*Theory*, I: 77–79, 153, 263; II: 408–409; Playfair 1805b, 47–48. See also Schofield 1963.

book is the culmination of a long-lived but misguided endeavor to reason one's way into a "theory of the earth" (as the genre was termed) by combining aspects of Genesis with Newtonian physics, deductive logic, and some degree of field observation. Equally important, from Hutton's point of view, was the fact that Whitehurst utilized his science to affirm extremely distasteful theological and aesthetic positions.

Unlike many theorists, Whitehurst did not attempt to account for the creation of the earth, but he clearly assumed that such a creation took place, and at a distinct moment in time. Reasoning backward from the presently oblate form of a not quite spherical earth, he saw the history of our planet as a divine contrivance to create a habitable world. Once shaped, therefore, the original chaotic mass of the earth, mostly liquid, developed a solid core, a universal ocean, and (somewhat later) emergent continents. To account for the broken, disordered appearance of the present earth, Whitehurst imagined vast subterranean fires of uncertain origin suddenly elevating the ocean bottom into mountains and throwing hitherto concentric layers of strata into heaps of ruins, of which the Alps and all other mountain chains are such impressive evidence. To effect his catastrophe, he naturally assumed, the earth's subterranean fires must have been more powerful in the past than at present. On reading this ingeniously wrought history of the earth, Hutton would have disagreed strongly with much of it, though Whitehurst did endorse uplift by subterranean fire, accepted the igneous origin of basalt, and appended to his theory a fine analysis of Derbyshire strata.

While Hutton clearly favored observations supporting the igneous origin of basalt, he demurred from identifying basalt with lava. Though both flowed from volcanoes, lava, he believed, had been erupted from the volcano's mouth or some other surface aperture. Basalt, on the other hand, was wholly subterraneous, though it could be exposed subsequently through the erosion of overlying sediments. Even on the surface, he held, it was possible to distinguish lava from basalt; the first contained only empty vesicles, whereas the second had vesicles filled with such attractive minerals as zeolite, agate, and spar. When Hutton first conceived of this (erroneous) distinction is unclear, but he was investigating zeolites chemically by 1772, only two years after the word "zeolite" had entered English (via Cronstedt's *Essay*). Visitors to his personal collection of mineral and rock specimens would then notice how prominent agates were among them. In any case, the distinction between lava and basalt, as he conceived it, was among Hutton's last major concepts to be established. It had not been wholly original to him, having been suggested quite directly by Deodat de Dolomieu in 1783 and published the next year by Faujas de Saint-Fond. After quoting the relevant passage, Hutton fully endorsed the distinction between "subterraneous lava, in which zeolite and calcareous spar may be found, and that which has flowed from a volcano, in which neither of these are ever observed" (*Theory*, I: 155–156n). He argued this point with Saint-Fond himself in 1784, when the prominent French geologist toured Scotland.

Visiting the Western Isles, where Hutton had never been, Saint-Fond readily ascribed volcanic origins to columnar basalts found on Staffa ("Fingal's Cave") and elsewhere. His travels then took him to the capital, with its many luminaries. Among them, he thought,

> Doctor James Hutton is, perhaps, the only person in Edinburgh who has brought together in a cabinet some minerals and a large series of agates chiefly found in Scotland. But I observed that he had not been sufficiently careful to collect the different matrices in which they are enclosed, and which serve as a complement to the natural history of these stones. I therefore had much more pleasure in conversing with this modest philosopher than in examining his collection, which presented me with nothing new, since I had lately seen and studied in place the greater part of the objects in this collection. Doctor Hutton was at this time engaged in the calm of his study, writing a work on the theory of the earth. (1907, 2: 234–235)

From Hutton's home, on St. John's Hill, they probably walked together to Arthur's Seat, which Saint-Fond—on the basis of considerable geological experience in central France—immediately (and, for Hutton, unacceptably) identified as an extinct volcano. Its supposed basalts would then have been lavas instead. Not long afterward, Hutton was visited by another distinguished geologist, Sir William Hamilton, whose many vivid descriptions of Vesuvius and Etna (among other volcanic phenomena) had won him an appreciative audience, both lay and scientific, throughout Europe. A Deist like himself, moreover, Hamilton supported Hutton in many of his basic attitudes toward God, time, and nature. Thus, in July or August 1784 Hutton eagerly accompanied his renowned guest to Salisbury Crags. "Sir William Hamilton informed me," he then recorded with evident satisfaction, "when I showed him those mineral veins and spars in our whinstone, that he had never observed the like in lavas" (*Theory*, I: 160). Basalt and lava therefore appeared to be recognizably distinct. If so, Arthur's Seat was *not* an extinct volcano.[16]

Except for some further reading (of which the first volume of Horace Bénédict de Saussure's *Voyages dans les Alpes*, 1779, was most important), we know nothing more about Hutton's geological activities until 7 Febru-

[16]The time of Hamilton's visit to Hutton is established by Saint-Fond (1907, 1:241) and by surviving correspondence. Arthur's Seat is in fact an extinct volcano.

4. Arthur's Seat and Salisbury Crags, Edinburgh. In this remarkably analyzed landscape, thirteen separate layers of primarily volcanic rock are delineated, above and below ground. Drawing by John Clerk of Eldin, probably ca. 1785. (Sir John Clerk of Penicuik)

ary 1785, when, according to the minutes of the Royal Society of Edinburgh, "Dr. Hutton engaged to read at the Physical Branch of the Royal Society in March and April a paper containing an examination of the System of the habitable Earth with regard to its Duration and Stability." At the next meeting, on 7 March, with Dr. Cullen presiding and "Dr. Hutton not being able to attend, by reason of indisposition, the first part of his Discourse containing an Examination of the System of the habitable Earth with regard to its Duration and Stability was read by Dr. Black." In the following session, on 4 April, with Dr. Home in the chair, "Dr. Hutton read an abstract of that part of his discourse which was read at last meeting and then proceeded to read the second part of his Discourse on the System of the Habitable Earth with regard to its Duration and Stability." There were then no further meetings for three months, but on 4 July, Dr. Monroe presiding, "an abstract was read of Dr. Hutton's paper concerning the System of the Earth, its Duration and Stability." Although a second paper (on botany) was also read, by John Walker, the major business of the meeting must have been the discussion of Hutton's theory, which was the whole point of presenting an abstract, as required by the Society's procedures.

According to Playfair, Hutton had not previously revealed his geological ideas to anyone except Black, and Clerk of Eldin. Perhaps the names of Faujas de Saint-Fond and Sir William Hamilton should be added to this brief list, with John Strange, James Watt, and Erasmus Darwin as further possibilities. Even so, it is quite possible that only Black and Clerk knew Hutton's theory as a whole. Though chemistry and mineralogy were fairly common interests among the membership of the Society, geology itself was not. John Walker, for example, professor of natural history at the University of Edinburgh from 1779, openly disparaged geological theorizing of any kind. Like William Cullen, he was primarily a mineralogist. Walker and Hutton had known each other since around 1770, sharing interests in chemistry, agriculture, and natural history. As of 1785, Walker was first secretary of the Society's physical section, before which Hutton presented his theory. Another active participant in the physical section was John Playfair, a former student of Walker's, who became professor of mathematics at Edinburgh that year but was as yet little interested in geology. When Hutton appeared, therefore, it was to confront an audience in which certain members were at best indifferent, if not outrightly hostile, to his enterprise as such. No wonder Hutton was overcome with illness before his presentation.[17]

[17]Hutton was a familiar figure within the Royal Society and a close friend of such influential members as Black, Smith, Clerk of Eldin, and Robertson. In later years, moreover, he comfortably presented highly controversial papers on primarily nongeological topics. But there is good reason to believe that geological theorizing was unusually suspect in 1785. For John Walker, see H. W. Scott's edition (1966) of his lectures. This influential teacher's students included John Playfair, Sir James Hall, Robert Jameson, and John Macculloch, all of whom shared his antitheoretical bias for a time.

As revealed in 1785, Hutton's theory was less impressive than it might otherwise have been. At a time when original thought was expected to be elegantly phrased, when thinkers were evaluated also as writers, Hutton had written badly. Though a strict logician, he was almost entirely innocent of rhetorical accomplishments. More specifically, Hutton did not know how to construct expository paragraphs. Instead, he wrote crabbedly, sentence by sentence, making the development of his thought rather hard to follow. Significantly hampered as well by the lack of a developed geological vocabulary (a cultural accomplishment not yet in existence), he was only moderately successful at explaining what he saw. In too many cases, moreover, Hutton slighted empirical evidence to favor philosophical assertion. Utilizing habitually dichotomous logic, he contrasted past with present, the human record with natural history, land with sea, natural with supernatural, consolidation with dissolution, solution with fusion, water with heat, purpose with chance, good with evil, design with chaos, ends with means, and the general with the particular. Despite the field evidence from which much of it derived, therefore, Hutton's theory appeared to its first hearers as very much a philosophical or even theological exercise.

However rudimentary current geological knowledge may have been among the intelligentsia of Edinburgh, several members of Hutton's audience were already fine writers and accomplished philosophers. Though theological animosities remained generally under control, Deists like Hutton were probably outnumbered within the Society by well-placed Presbyterians (even if the latters' fervor might sometimes be in doubt). Walker and Playfair, for example, were both ministers of the kirk. They and other auditors had learned to suspect certain geological theories, and perhaps the genre as a whole, of outright infidelity. Ominously, the one Scottish philosopher who had displayed real interest in geological theorizing was the recently deceased (1776) David Hume; his posthumous *Dialogues Concerning Natural Religion* (1779) advocated uniform natural processes disturbingly like Hutton's. But, unlike Hutton, Hume also disparaged any attempt to find God's attributes in nature. If his artful use of dialogue in any way disguised Hume's real beliefs, the same could not be said of a second Scottish theorist, George Hoggart Toulmin, who had published much-disliked opinions in *The Antiquity and Duration of the World* (1780), *The Antiquity of the World* (1783), and most recently *The Eternity of the World* (1785), all of which seemed atheistic to the British and were stridently denounced as such. Despite their theological differences, the at least superficial kinship among Hume's, Toulmin's, and Hutton's ideas was definitely not overlooked.[18]

[18]For Hutton and Toulmin, see Appendix 2.

When Hutton's theory was first publicly presented, then, few members of his audience were prepared to argue the geological evidence with him. Those among them who were Deists probably realized that Hutton's theory accorded nicely with their own beliefs. Throughout the eighteenth century, largely in France, Deists had steadily attempted to undermine a literal adherence to Genesis by stressing, as Hutton did, the immense age of the earth revealed by geology, the inadequacy of the Noachian flood to explain fossils, the discrepancy between supposed "days" of Creation and geological periods, and the sufficiency of natural causes as opposed to special creations and other miracles. Like many Deists, Hutton did not believe that life had been created serially. But he did not believe in extinction either. At no time would he propose theories regarding the origin of life, though he would come eventually to suggest—in his unpublished "Elements of Agriculture," written ca. 1796—that species regularly improved themselves through something very like natural selection. On the other hand, those in Hutton's audience who were orthodox Christians would have taken his closing emphasis on the immensity of geological time, both in his "Theory" and in his "Abstract," for what it was—an implicit rejection of the Creation narrative in Genesis and an assertion of total reliance on what Hume had called natural religion. If Hutton could refute Hume and reestablish the validity of confirming divine goodness through scientific evidence, all would be well. But those who compared, as many perhaps did, might well have thought Hutton's theory (despite its author's intentions to the contrary) an attempt to substantiate Hume and Toulmin rather than refute them.

Hutton's "Memorial," 1785

The result of Hutton's first public presentation of his theory was that he became somewhat concerned about its religious implications. When it had been accepted for publication in the not yet emergent *Transactions* of the new society, Hutton thought to allay the theological suspicions of his opponents by supplementing his exposition with a preface, written July 1785:

> Memorial Justifying the Present Theory of the Earth from the Suspicion of Impiety
>
> It belongs to religion to teach that God made all things with creative power, that perfect wisdom had then presided in the election of ends and means, and that nothing is done without the most benevolent intention. But it belongs not to religion to give a history of nature or to inform

mankind of those things which actually are; it belongs not to religion to teach that natural order of events which man, in his science, may be able to unfold and, in the wise system of intellect, find means to ascertain. The object of revelation and that of natural philosophy being thus perfectly different, it must be absurd to suppose that these can truly interfere, as this may only happen in supposing them not strictly adhering to their respective subjects and one or other of them as not being just.

Let us now suppose that this interference, between that which is held to be religion in a country and that which is the result of philosophy, should happen; and that the question is, which of those two contradictory doctrines should be of the highest authority, or command belief? The one is supposed to be divine revelation, the other to be the produce of natural philosophy. Such a question will not appear strange to a person who is truly pious, who acknowledges that all his faculties are given him from above, that the laws of nature are true and steady as their author, and that man, made after the image of God, has been ordained to read the wisdom of his author in His works.

It may perhaps be alleged that as from the fallibility of human reasoning a physical disquisition may be erroneous, revelation (proceeding from a source which cannot err) should not only be reverenced as sacred but must be received without dispute and stand unaffected by every authority which is precarious. This is certainly a proposition that may be admitted, but it only goes to express an improper comparison—between a thing supposed perfect, on the one hand, and one acknowledged as subject to imperfection, on the other. Now it is not meant either to impeach the authenticity of revelation or to defend the fallibility of philosophic reasoning; these two things, as sources of human information, can only be compared properly when each, in its kind, is considered as perfect.

It must be observed that here is no question how far man, reading in the truths of nature the decrees of God, may reason falsely and conclude in error; we are supposing man to reason justly from his principles, and to take for principle nothing but what is supported by the truths of nature. In like manner, here is no question with regard to spurious revelations of the will of God; such things may be, and also be believed; but we are now supposing the revelation to be genuine, or the information true: and now it may be asked, with regard to those two different authorities, are they equal or unequal?

It must be evident that, if there be in the Author of Nature truth, and in the Author of Men benevolence, the word of God, whether revealed by the common faculties of man or given to human understanding in a preternatural manner, must be always one. Consequently, it were impious to suppose that either one or other of those two different means of information had superior authority or that the proper result of each had not an equal right to be believed.

Having thus shown upon what approach natural philosophy, if just, may be considered as an information of equal authority with that of revelation, let us now examine how far the doctrine in the present theory of the earth

> may be found consistent, or not, with the Mosaic history considered as a divine revelation.
>
> The Christian religion is founded upon what is contained in the ancient Jewish writings, and in those writings we find something which seems to have the appearance of natural history. But that this is not truly so, or at least that such a part of natural history has no concern whatever with the present theory of the earth, will appear from the following considerations.
>
> The Mosaic history of the creation, in giving a most succinct account of the order in which things were made, contains no chronological description of the beginning of things or such as might apply to our measure of time, which is by days and years. This will appear by considering that the sun, by which we measure our time, was not found until the fourth distinguished period of creation and that it would be unreasonable, or no less than absurd, to suppose the term "day," by which each of those periods is expressed in this Jewish history, means anything besides an indefinite period, or means any more than to signify that God made all things in a certain order.
>
> It is in the last of those six distinguished periods of the creation that man, according to this history, was made. Now, from this period the Jewish history of mankind contains a certain chronological register which certainly respects the habitable earth of this globe. But the present physical dissertation, in which are traced ancient operations of this earth, and by which operations the beginning of things is referred back to an indefinite period of time—this dissertation, I say, is not in any way concerned with the period at which man was made; for, as a physical investigation must proceed upon matter of fact alone, it is not pretended in this dissertation that any documents are found in natural history for the existence of man before that period at which commences the chronology of the Mosaic history; and, with regard to the mineral operations of the globe previously to this period, surely the Mosaic relation cannot be considered as a natural history. Consequently, although this theory of the earth gives a most distant view with regard to the operations of nature, it does not in any respect interfere with the chronology of the Old Testament.

Hutton began by arguing that religion and science, once they are properly defined, can never be in conflict because their legitimate spheres of authority are distinct. Revelation teaches us the reality of God's power, wisdom, and benevolence but should not concern itself with details investigable by human intellect. But, he then asked, if there *is* such a conflict, which authority ought we to prefer? (The proper reply would be, whichever is appropriate to that sphere.) Authentic relevation is perfect, whereas human philosophizing is admittedly subject to error; on the other hand, certain "revelations" may be spurious. But, if we suppose that the revelations are authentic and that the human mind has reasoned justly from true principles of nature, then it remains only to show that Scripture and the Huttonian theory of the earth do not conflict.

Though Hutton affirmed human reason and the empirical investigation of nature, he severely limited the role of Genesis to a general celebration of divine creative power; despite appearances, it was *not* a history of nature. He straightforwardly asserted, and surely believed, that science is a more reliable source of truth ("those things which actually are") than is religion—not an altogether successful strategy for quelling "the suspicion of impiety"! Furthermore, he wrongly represented his own theory as a historical statement applicable to a period antedating Adam when it was actually an ahistorical description of natural processes still going on. Despite his clumsy repetitions of an already hackneyed argument about the biblical meaning of "day," Hutton could not escape potential embarrassment concerning the creation of man (for which he apparently accepted the scripturally derived figure of about 4000 B.C.), but this was only one aspect of an even touchier controversy regarding the earth's age that he failed utterly to dispel. If Hutton had published this all too revealing apology, he would surely have increased rather than diminished religious opposition to his theory.

Hutton's good sense probably told him as much. Clearly not at ease with his intended preface, he sent it for comment to a judicious friend—William Robertson, principal of Edinburgh University, whose beautifully written historical studies (which made him famous) stressed general ideas in a way no doubt agreeable to Hutton. Robertson also opposed religious bigotry and distrusted religious enthusiasm while advocating, like Hutton, the same optimism regarding creation's essential rightness. In addition, Robertson's favorite daughter, Mary, was the wife of Patrick Brydone, another unorthodox Deist whose travel book on Sicily (1773) included a geological challenge to the Mosaic chronology that had given great offense.[19]

According to Alexander Carlyle, Robertson was fond of "skimming his friends' talk and giving it back to them in polished paraphrase." He ran true to form in this instance, presenting Hutton a more urbane version of his remarks designed to save the geologist from both stylistic and religious criticism:

> It is not the end of revelation to instruct mankind in speculative science, to communicate to them a history of nature, or to explain the true system of the universe. Intent upon inculcating the religious doctrines which we

[19]Robertson (1721–93) was principal of Edinburgh University from 1762 to 1792 and moderator of the Presbyterian General Assembly from 1763 to 1780. Fellow historians Hume and Gibbon were among his friends. A fine prose stylist himself, he was active in societies established to promote the reading and speaking of English in Scotland (McElroy 1969, 55–67) and became the effective founder of the Royal Society of Edinburgh (1969, 78–80).

ought to believe, and the moral virtues which we are required to practice, it rests satisfied with describing the phenomena of nature not according to philosophic truth but as they present themselves to our view.

The following attempt towards a theory of the earth is not in any respect repugnant to the Mosaic account of the creation. The sacred historian, in his sublime account of the formation of the Heaven and the Earth, has not given a chronological description of the order in which things were made—at least such a description as will apply to our measure of time, which is by days and years. This appears directly from considering that the sun, by which we measure time, was not formed until the fourth distinguished period of the creation, and, of consequence, the term "day," by which each of these periods is represented in sacred history, must signify an indefinite period and is employed with a view of conveying to us the idea that things were called into being successively, and in perfect order and arrangement.

It was in the last of these six distinguished periods that man was made. From the period of his formation the sacred writings contain a chronological history of mankind, and of the events which have happened on the habitable globe. But in the following physical dissertation, the sole object of my researches has been to trace ancient operations of this earth through an indefinite period of time and to discover their effect in preparing it to be such an habitation for man and other animals as Infinite Wisdom and Goodness destined it should be.

Most of the important changes come at the beginning, and the stylistic superiority of Robertson's considerably briefer version is evident. Having also diminished Hutton's opposition to Scriptural geology, however, Robertson probably realized that his friend was unlikely to accept even such modest dissembling as that. In returning the new version, therefore, he wrote prudently to Hutton as follows:

Dear Doctor

I send you a sketch of a short advertisement, which may be prefixed to your dissertation. The ideas are mostly taken from your own papers, only the style is rendered a little more theological. After all, it requires some debate perhaps whether the dissertation may not be printed without anything prefixed. I purposed to have called upon you today in order to have discussed this point but have so many things to do previous to my going out of town on Monday that it has not been in my power. I shall return in three weeks; if you should begin to print before that time, consult our friend Mr. Smith, and on following his advice you will be safe. I return your papers and am ever

Yours most faithfully
College July 22 William Robertson

Robertson's tactful candor shows that the geologist had chosen his critic wisely. Whether or not he actually consulted Adam Smith, whose works he was later to coedit, Hutton suppressed the preface.[20]

"Abstract Concerning the System of the Earth," 1785

The manuscript from which Black and Hutton read the latter's exposition of his own theory no longer exists; nor have we any contemporary report of the event. Fortunately, we do possess a few copies of Hutton's thirty-page "Abstract of a Dissertation Read in the Royal Society of Edinburgh, upon the Seventh of March and Fourth of April MDCCLXXV, Concerning the System of the Earth, Its Duration, and Stability" (1785) in which he summarized his longer paper for discussion on 4 July. Though Hume, Toulmin, or even the equally notorious Holbach may somehow have been involved, Hutton's title strongly suggests that he was replying more to John Whitehurst than to any of them. Copies of his "Abstract" circulated within the Royal Society of Edinburgh, among Hutton's friends, and as far away as France and Russia.

"The purpose of this dissertation," Hutton began, "is to form some estimate with regard to the time the globe of this earth has existed as a world maintaining plants and animals, to reason with regard to the changes which the earth has undergone, and to see how far an end or termination to this system of things may be perceived, from the consideration of that which has already come to pass" (1785, 3–4). Human records (i.e., Genesis) being inadequate for this purpose, Hutton proposed to examine the earth itself and to reason from principles of natural philosophy toward "some knowledge of order and system in the economy of this globe" and "a rational opinion with regard to the course of nature, or to events which are in time to happen" (4).

Because solid parts of the present land contain shells and other marine exuviae, it followed that the land is not original; before it was made, there must have been an earlier world composed of sea and land, with tides, currents, and other operations just like ours. While the present land was preparing at the bottom of the sea, moreover, the former land (or, at least, the sea) supported animate life, as now. He reasoned that changing sea bottom into solid land would require two important pro-

[20]For the whole episode, see Dean 1975, which includes unedited texts of the manuscripts involved. Alexander Carlyle on Robertson is from the latter's entry in the *Dictionary of National Biography*. To this striking evidence of Hutton's values one may add a line from his letter of ca. 12 March 1773 to Andrew Stuart of London (National Library of Scotland): "Is there any thing that truly may be called sacred amongst men but truth?" See also P. Jones 1984, 202.

cesses: the *consolidation* of loose bottom materials, and the *elevation* of the consolidated masses above the level of the ocean. For him consolidation must have been achieved through solution and subsequent concretion in water or through fusion and subsequent congelation by heat. He then eliminated the aqueous possibility, necessarily affirming heat and fusion. Since heat had now been established as a significant agency, it was surely responsible for elevation as well. In support of this hypothesis, he cited numerous geological phenomena, including mineral veins, present-day and ancient volcanoes, and the "subterraneous or unerupted lava" called basalt. Finally, Hutton confirmed his theory by "finding a perfect similarity in the solid land through all the earth" (21). In some further remarks, he stressed the immense amount of time his theory required; according to it, both the past and the future extended well beyond human comprehension. "With respect to human observation," therefore, "this world has neither a beginning nor an end" (28). In a final section (not present in the fuller, published version of Hutton's paper) Hutton endeavored to support his theory through moral arguments based upon the consideration of a Final Cause.[21]

"Theory of the Earth," 1788

Subdivided into four parts, Hutton's "Theory of the Earth; or an Investigation of the Laws Observable in the Composition, Dissolution, and Restoration of Land upon the Globe" was published as an author's separate sometime after January 1786 (at which time its text was still open for additions). A small number of copies dated 1786 were apparently run off; of them, only one is presently known. Others, also rare, are dated 1788, the year in which the first volume of the Society's *Transactions* finally appeared. An author would normally have received some advance copies, called "separates," to distribute as he saw fit.

"Theory" then appeared in the first volume of the *Transactions of the Royal Society of Edinburgh* (1788). Pride of place within that volume (edited by Henry Mackenzie) was given to a history of the young society; papers by Walker, Hutton himself (on rain), and Playfair, among others (the order being strictly chronological), preceded Hutton's on geology, which occupied pages 209–304 and two accompanying plates (a septarian nodule and graphic granite). Since first presented in 1785, Hutton's essay had acquired through its new designation more grandiose

[21]Hutton's "Abstract," together with important details concerning it, was first announced by V. A. Eyles in 1950. It has since been reprinted in White and Eyles 1970, Albritton 1975, and Craig 1978. I discuss its authorship in Appendix 4. For abstracts generally, see *Transactions of the Royal Society of Edinburgh*, 1:13. (The *Transactions* are hereafter cited as *TRSE*.)

stature as a comprehensive geological theory founded on natural laws. This change in status was probably not of Hutton's doing, though he surely consented to it. The phrase "theory of the earth," now in Hutton's title, recalled an outmoded genre of biblically tinged speculations that Hutton disliked as much as anyone did; the same phrase occurs nowhere within his text (but twice within his discarded preface) and, together with the rest of his paper's elegant label, may have originated with Robertson, who was a founding member of the Royal Society of Edinburgh and among Hutton's supporters within it. Once beyond his title, Hutton never repeated the phrase "theory of the earth." His preferred alternative, used throughout and no fewer than five times on the final page, is "system," which accords much better with his original title and the orderliness in nature that he wished to emphasize. The new word "geology" is entirely absent.[22]

We may summarize Hutton's famous essay as follows:

Part I: Prospect of the Subject to Be Treated of. As our perceptions of it affirm, the earth is a unique creation, specifically designed by its infinitely wise Creator to serve as a habitable world for life in all its forms. This world consists of four essentially spherical parts—core, water, crust, and air—each of which contributes to the maintenance of life. The world as a whole is subject to such basic powers as centripetal and centrifugal forces (which, being kept in balance, maintain our globe at a proper distance from the sun), light, heat, cold, and condensation, all of which foster motion and activity.

A solid body of land could not have answered the purpose of a habitable world because soil is necessary to the growth of plants, and soil is nothing more than materials collected from the destruction of land. Necessarily, then, our seemingly solid continents must be subject to erosion by water and air. Paradoxically, the *removal* of fertile soil from the continents by natural agents is actually part of a great cycle by which the continuing fertility of the earth is assured. That this process is extremely slow, involving lengths of time beyond human experience and comprehension, should not disguise the benevolence underlying it. Fossil shells and other such relics prove that the earth has a history far antedating all human records.

The solid parts of the globe are, in general, composed of sand, gravel, shale, and usually fossiliferous limestone, all of which are productions of water, wind, and tides. (There is also a granitic part, to be considered

[22]It may be that the title of Hutton's paper was changed to avoid any kind of implied associations with Holbach's atheistic *Système de la nature*. For theories of the earth as a genre, see Porter 1979; for the word "geology," see Dean 1979.

Like those throughout this book, the following summary of Hutton 1788 attempts only to establish his major assertions and arguments, without regard for their acceptability today.

later.) Though originally loose, they are now firmly compacted rock. Evidently, then, some sort of consolidating power has been at work.

Part II: An Investigation of the Natural Operations Employed in Consolidating the Strata of the Globe. Logic suggests only two possible means of consolidation: strata once at the bottom of the sea have been lithified either by aqueous solution and crystallization or by heat and fusion. But water can dissolve only some of the materials involved and can deposit them only in simple ways. If, however, heat and fusion are substituted for water and solutions, all these difficulties soon disappear. Thus, heat is competent to consolidate strata whereas water alone is not. Consider, for example, such indissoluble silicates as feldspar, flint, and fossil wood. Alternatively, what of sulfurous minerals, metals, and coal? (Hutton discussed a number of specimens from his own collection.) Three further cases especially worthy of attention are rock salt (clearly a product of fusion), ironstone septaria, and agates. A geological case in point is Calton Hill, Edinburgh, a basaltic outcrop evidently brought into being by fusion, its basalts being filled with nodules.

Let us also consider strata consolidated without the introduction of foreign matter, that is, merely by the softening or fusion of their own materials. Such are the sandstone and limestone strata which constitute all but a small portion of the globe. Though Scotland includes a great deal of sandstone, however, there is surprisingly little limestone. Spanish marble and English chalk prove, nonetheless, that consolidation is not only general but universal.

Granite, unlike the examples previously considered, is not generally stratified, but a specimen from Portsoy (depicted in plate II) proves that it has crystallized out of a fluid state by means of fusion. Returning to a more general consideration of strata as such, Hutton cited the existence of perpendicular fissures and veins, and fragments of former strata, as further evidence of fusion.

Part III: Investigation of the Natural Operations Employed in the Production of Land above the Surface of the Sea. That strata deposited at the bottom of the sea are consolidated by heat and fusion has now been established. How, then, are they elevated into continents? How else but by the same force? Strata at the bottom of the sea are necessarily horizontal, but those we see are in every possible position; they could not possibly have been so created. Other proofs derive from mineral veins. Earthquakes and volcanoes, products of subterranean heat, prove that the same force is sufficient and universal. Though a normal part of nature, volcanoes are less important manifestations of subterranean heat than are consolidation and elevation. Yet naturalists have discovered the remains of ancient volcanoes in many different countries. Lavas exist where no vestiges of volcanoes can now be found. Other melted matter, analogous to lava, was forced among strata still consolidating at the bottom of the sea; now on dry land, it forms the rock called basalt. A review of examples from Scotland and Derbyshire leads to a fuller explanation of how basalt differs from lava. Virtually identical in origin and composition, the two rocks are distinguished only in the manner of their production. Lava was erupted into the atmosphere while fluid; basalt congealed within the earth under immense compression (and may be called subterraneous lava). As for telling them apart, only basalt contains solid inclusions. "The flowing of basaltic streams among strata broke and displaced," Hutton affirms, "affords the most satisfactory evidence of those operations by which the body of our land had been elevated above the surface of the sea" (1788, 282); signs characteristic of *volcanic* activity, however, are entirely lacking. Though his examples are limited, moreover, Hutton's argument is global, for "The great masses of the earth are the same everywhere, and all the different species of earths, of rocks or stone, which have as yet appeared, are to be found in the little space of this our island" (283). To know the construction of Britain, therefore, is to know the construction of the world.

Part IV: System of Decay and Renovation Observed in the Earth. Philosophers in the past have imagined an earth more regular than ours, one subsequently spoiled by some natural disaster or exercise of divine wrath. Yet the present earth, in all its beauty and diversity, is ideally suited for habitation, its processes being stable, unchanging, and purposeful. Working in concord with this physical system is an animate one. Among known fossils, we find remains of every plant or animal genus (and probably species) now living on the earth, even some with which we are otherwise unacquainted. Together with gravel, sand, and clay (all derived in a former world by processes observable in ours), these relics affirm that the former world was in all respects essentially a duplicate of our own.

Hutton's closing remarks stress the immensity of time required for one world to succeed another. We cannot see the ocean bottom, to observe nature at work there. But that part of the total process we *can* see, the erosion of the land, takes place with incredible slowness, as comparisons of classical and modern writers on geography substantiate. "To sum up the argument," he concluded, "we are certain that all the coasts of the present continents are wasted by the sea and constantly wearing away upon the whole, but this operation is so extremely slow that we cannot find a measure of the quantity in order to form an estimate" (301). He conceded that the operations of nature may be less systematic than his theory proposes, but "It is not necessary that the present land should be worn away and wasted exactly in proportion as new land shall appear; or, conversely, that an equal proportion of new land should always be produced as the old is made to disappear" (302). Nor is it necessarily true that *all* the land will be worn away. Yet there is an essential balance, between lands being created and destroyed, which allows the earth to remain fully habitable and permits animals and plants to migrate from old land to new.

The continent destined to succeed ours has probably begun already to appear—in the middle of the Pacific Ocean. (Captain Cook had discovered the Hawaiian Islands in 1778). Obviously, the emerging continent these islands represent must have derived from the destruction of another not now in evidence; we may therefore envision a three-part cycle rather than two. In either case, "The system is still the same." Given the fact of successive replacements, however, we cannot hope to discover either the first cycle or the last. "The result, therefore, of our present enquiry is that we find no vestige of a beginning, no prospect of an end" (304).

With this memorable and usually misunderstood affirmation, Hutton concluded his "Theory of the Earth," which had clearly been revised and elaborated further after its initial presentation in 1785.

[2]

"Theory of the Earth": The Middle Phase

Though complementary, successive, and developmental rather than contradictory, there were several Huttonian theories of the earth. His first, begun in Norfolk, affirmed the relatively simple deduction that much of the land had once been at the bottom of the sea. His next added that land and sea alternated their positions in an endless exchange of substance. When Hutton then began to speculate on consolidation and uplift, with subterranean heat as the agency of both, he embodied these essential convictions in the ambitious but eventually abandoned essay on natural history that Playfair would later see. The theory continued to grow thereafter, but its essentials did not change.

Between 1768 and 1785 the most important of Hutton's new concerns was basalt, which occupied much of his attention after he moved to Edinburgh. Stimulated by the founding of the Royal Society of Edinburgh in 1783, Hutton embodied his current understanding of the earth in his "Abstract" of 1785. Yet even this précis was obsolete by the end of that year, for the revised version of his paper clearly went beyond it in some respects—by proposing a three-stage cycle, for instance. As we have seen, a specifically theological concluding section was also dropped.

Characteristically, Hutton had no sooner relinquished his "Theory of the Earth" to the press than he began work on further writings intended to augment it. That he already had a book in mind is highly probable but not quite certain. (By 1787, the issue was no longer in doubt.) Whatever his original intentions, Hutton spent the next few years taking geological excursions and promptly writing up their results as ten sequential but largely self-contained essays that have since achieved publication in various ways: one became a regularly issued paper, three appeared later in Hutton's two-volume *Theory* of 1795, and six others remained in manuscript as part of a never-completed third volume published only in 1899.[1]

"Of Granite," 1785 [*Theory*, I: 311–319]

In the first and most easily identifiable of his subsequently incorporated fragments—this one evidently intended to follow a reprinting of his "Theory"—Hutton refuted the erroneous belief of some naturalists in the existence of "primitive" mountains, meaning granitic ones supposed to have survived since the Creation, or nearly. (A series of German theorists, beginning with Lehmann [1756], had held this view.) If the existence of primitive mountains could be established, Hutton's theory would necessarily fail, for he held no rocks or landforms to be "original." The problem of supposedly primeval granite was therefore among Hutton's most serious challenges; after doing what he could to minimize it in his 1785 "Theory," Hutton was further moved (in part by passages in Saussure) to face the issue more directly.

Being a variable composite of several minerals, granite, he asserted, has no inherent claim to primacy; it is found in a wide variety of colors and textures. Were it actually original, nature would have acted without either order or wisdom, indulging in change without principle and variety without purpose. As it happens, however, at least one kind of granite (called gneiss by some) is *stratified*; it could not, therefore, have been an original deposit and is not necessarily older than the also stratified schistus that so often accompanies it. The prominence of granite among Alpine peaks is a function of its durability, not its age. Among the many varieties, there are basically two kinds of granite, one of them stratified and one of them not. Unstratified granite, porphyry, and basalt are all kinds of subterraneous lava; though distinct, they grade into each other and may be considered basically identical, with no one of them necessarily oldest. Thus, we can now explain the irregularity of unstratified granite masses and better understand the wavy structure of the stratified type, which though not made to flow was softened and bent by heat. "This," Hutton then added to his 1795 version in a footnote, "is what I had wrote upon the subject of granite before I had acquired such ample

[1]Though I am the first of Hutton's commentators to identify and consider these essays as such, textual evidence readily establishes their identity, approximate dates of composition, and sequence. Volume III of Hutton's *Theory* remained in manuscript until 1899. The essays within it manifest at least two stages of composition, apparently separated by as much as a decade. Chapters i through iii are missing; they may have discussed (or been reserved for) basaltic lowland geology.

testimony from my own observations upon that species of rock" (I: 319n). Those observations appeared in several further essays not published during his lifetime (*Theory*, vol. III) and in others that were.

"Observations Made in a Journey to the North Alpine Part of Scotland in the Year 1785" [*Theory*, III, chapter iv]

Hutton's 1785 observations on alpine Scotland begin with a subdivision of the country into regions. In the Highlands, he asserted, "is to be found everything requisite for establishing a natural history not only of this, but of every other alpine country" (*Theory*, III: 4). The Highlands and their strata are surprisingly regular, stretching across Britain from southwest to northeast. They consist generally of granite (unstratified) and "schistus" (stratified rocks of various kinds). There is also a kind of stratified granite (gneiss), which must be carefully distinguished from the unstratified kind. For some naturalists, granite and schistus were "primary" rocks—preceding all others in age and antedating all fossils as well. But having "just now found evidence to the contrary in a journey . . . to the Highlands" (9), Hutton immediately presented it, so that anyone wishing to verify his discoveries might do so. Knowing that much granite could be found adjacent to the Dee and Tay rivers, he had gone at harvest (September) 1785 with his friend John Clerk of Eldin to visit the Duke of Atholl at Blair. Hutton wished to examine the geology of his host's deer forest and to trace granite boulders lying along the river to their source. Having obtained the Duke's generous cooperation, he and Clerk proceeded to the long, narrow valley of Glen Tilt, where the river had exposed several sections of great interest.[2]

Once there, Hutton quickly discovered granitic veins within the schistus; thus, granite was, like basalt, a form of subterranean lava. He then found what appeared to be stratified granite underlying the schistus. After studying several examples in Glen Tilt and adjacent Glen Tarff—the Tarff is a tributary of the Tilt—Hutton confirmed that the granite, though below it, was not older than the schistus. On the contrary, the schistus was older and had been intruded by the granite. Veins of porphyry (a rock composed of the same materials as granite, Hutton tells us, but formed differently) similarly established the correctness of his theory. Therefore, he concluded, "our alpine country consists of indurated or erected strata of slate, gneiss, and limestone, broken and injected with granite and porphyry" (24). A comparison with sites south

[2]To this and other volume III essays, Geikie (1899) appended more recent geological analyses. For Glen Tilt, see also Craig 1978, 30–39. The Tilt is a principal tributary of the Tay.

5. Forest Lodge, Glen Tilt. This spectacularly situated shooting lodge served Hutton and Clerk as their headquarters during the Glen Tilt expedition. Despite some vertical exaggeration, the erosive powers of running water and of gravitational collapse are clearly indicated. Drawing by John Clerk of Eldin, 1785. (Sir John Clerk of Penicuik)

of the Grampians established the similarity of basaltic intrusions with those of granite and porphyry. Thus, whatever be the materials, Hutton concluded, "*Nature acts upon the same principle in her operations in consolidating bodies by means of heat and fusion, and by moving great masses of fluid matter in the bowels of the Earth*" (25–26).

Hutton next turned to a remarkable vision of Highland geology as a whole. "This alpine country of Scotland," he proposed, "may be considered as a mass of strata elevated in their place and situation, being now, instead of horizontal, almost vertical and inclined sometimes toward the northwest, sometimes again toward the southeast" (26). He then emphasized the long, slow, but efficacious work of erosion, especially by the Tay, with its downward abrading powers and evident terracing. Its work, he specified, was not the product of some brief catastrophe; it required an immense period of time. The purpose of all such erosion is "to diminish the heights of mountains, to form plains below, and to provide soil for the growth of plants" (30).

"Observations Made in a Journey to the South Alpine Parts of Scotland in the Year 1786" [*Theory*, III, chapter v]

"Having last year got satisfactory proof of the Theory, from the examination of the north alpine region of Scotland," Hutton began his 1786 essay "and having a more particular knowledge of the south alpine region, as I had often traversed it at different places, I was anxious to find also something decisive with regard to granite in this region, where I had hitherto seen but little of that substance" (31). Accordingly, he and Clerk again set forth at harvest (7–29 September), hoping to visit the western island of Arran. But they had come too late in the season, so the two geological enthusiasts contented themselves with a tour along the coast from Glasgow through the shires of Ayr and Galloway.[3] For Hutton, an outstanding discovery along this route was the numerous basaltic dykes (his word) within the coal and sandstone strata of Ayrshire. Except for isolated boulders, he did not initially succeed in finding a significant exposure of granite. The castle at Ballantrae, he noted, stands atop a basaltic hill; "one would imagine from a superficial view of things here, as well as at Arthur's Seat by Edinburgh," he wrote unwarily, "that there had been volcanic eruptions, but from a more accurate view of things, this will appear to be the operations of subterraneous or unerupted lava,

[3]The notebook kept by Clerk on this expedition is preserved in the Scottish Record Office (GD 18/2120); see Craig 1978, 40–51.

6. Cairnsmore. Two figures, probably Clerk (left) and Hutton, investigate veining in granite. Drawing by John Clerk of Eldin, 1786. (Sir John Clerk of Penicuik)

for it contains much calcareous matter in a sparry state" (37). That the Rinns of Galloway, now a peninsula, had obviously once been an island revealed to Hutton how sea level had formerly been higher in the past than now. Granitic boulders, easily found in Galloway rivers, suggested a parent outcropping among the mountains. Inquiries at a local lead mine, however, led to nothing. Finally, the desired granite was found in abundance at Cairnsmore, "a mountain which seemed as if cut asunder in order to gratify our particular desire" (46).[4] The jointed granite there, Hutton ascertained, had (like basaltic columns) been formed by contraction, not stratification. Granitic veins intruding the schistus again proved that the schistus was older. Except for one further outcrop the next day, however, they saw no more granite until near Kirkcudbright, where fine porphyry dykes were observed as well. Some further examples turned up near Solway Firth.

As a result of these investigations, Hutton felt able to conclude "that, without seeing granite actually in a fluid state, we have every demonstration possible of this fact: that is to say, of granite having been forced to flow, in a state of fusion, among strata broken by a subterraneous force, and distorted in every manner and degree" (60). The best examples, however, had been found in Glen Tilt. "I have been the more anxious about this subject," Hutton confessed, "as I was long . . . uncertain if granite should be considered as a stratification of matter collected at the bottom of the sea, and afterwards consolidated by fusion in its place, or if it should be considered as a mass of subterraneous lava, which had been made to flow in the manner of our whinstones or basalts" (61–62). Concerned too about claims that granite was part of the originally created earth (or its core), Hutton reiterated that stratified granite *did* exist. He then compared aspects of Highland and Upland geology. The unusual regularity of the Highlands, he concluded, could be accounted for only by supposing "an internal power in the globe, a power of producing land where sea had been before, and of consolidating masses which had been formed of loose and incoherent bodies" (81). The essay then ends with affirmations of design and his theory as a whole.

"A Comparison of M. De Saussure's Observations in the Alps with Those Made upon the Granite Mountains of Scotland," 1786–87 [*Theory*, III, chapter vi]

Hutton remained particularly anxious about granite, and not only to refute the notion of primitive mountains. He had long desired to see the

[4][First week in July, 1792] "We dined with Dr. Black. Dr. Hutton, who was there, showed me some specimens of granite invading schistus; they were taken from a very large quarry at Cairnmuir, Galloway. In another part of the country (Lord Bute's estate), he had found limestone, schistus, and granite in strata" (Lubbock 1933, p. 236).

second volume (1786) of Saussure's *Voyages dans les Alpes*, which his previous remarks were written without. But when it had arrived, he was disappointed to find that junctions of schistus and granite were mentioned only rarely. Though the Swiss geologist's theory certainly differed from his (Saussure thought granite an aqueous precipitate), Hutton esteemed the author of *Voyages* and was sure he would never allow theory to distort observation. By comparing his own observations in Scotland with Saussure's in the Alps, therefore, Hutton believed it possible to establish universal truths. Thus, after quoting Saussure's description of Chamonix at length, he proposed that some granite taken by the Swiss naturalist to be stratified was in fact intrusive. Further quotations and comments then established that granite in the Alps forms structures very like those in Scotland. But Hutton almost rejected his previously conceded stratification in granite of any kind; the supposed layering observed by some in *massive* granite, particularly, was actually nothing more than contractive fracturing, such as one sees in basalt.

"Theory Confirmed by Observations Made upon the Pyrenean Mountains," 1786–87 [*Theory*, III, chapter vii]

For Hutton, greatly distorted limestone strata in the Pyrenees, as described by the Abbé Palassou in his anonymous *Essai sur la minéralogie des Monts-Pyrénées* (Essay on the mineralogy of the Pyrenees; Paris, 1781), establish the existence of a ubiquitous subterranean power responsible for mountain building everywhere.

"An Illustration of the Theory from the Natural History of Calabria," 1786–87 [*Theory*, III, chapter viii]

Deodat de Dolomieu, in his *Mémoire sur les tremblemens de terre de la Calabre, pendant l'année 1783* (Memoir concerning the Calabrian earthquake of 1783, 1784), described granitic mountains and massive accumulations of marine debris in Calabria and Sicily, attributing the presence of shells to a gradually retreating sea. Hutton suggested in reply that either earth movements or the subaerial erosion of fossiliferous mountain strata may have been responsible.

"An Examination of the Mineral History of the Island of Arran," 1787 [*Theory*, III, chapter ix]

The geological excursion to Arran that Hutton had postponed in 1786 took place the following August with John Clerk of Eldin's son as his companion.[5] Hutton's original goal had been to locate additional examples of granite and contiguous strata, but the island as a whole proved so interesting that he began to consider its natural history more generally, being also resolved to further discredit the concept of primitive mountains. As part of his detailed survey of the island, Hutton located some fine junctions of granite with schistus. The granite was posterior to the schistus and must therefore have been in a state of fusion by means of subterraneous heat; once again, supposed stratification in granite proved to be only contractive fracturing instead. He found further substantiation for his theory in the numerous dykes of basalt and felsite so prominent on Arran.

"Observations on Granite," 1790

Hutton returned to his Arran essay three years later to compare his results with those of Abraham Mills, who published "Some Account of the Strata and Volcanic Appearances in the North of Ireland and Western Islands of Scotland" (1790) in the Royal Society's *Philosophical Transactions* after reading his paper in London on 21 January 1790. Earlier that same month, on the fourth, Hutton had already brought his views on granite before the public in Edinburgh, in "Observations on Granite." They had probably been drafted about three years beforehand. Since reading his paper on the theory of the earth in 1785, Hutton told his audience, he had examined many parts of Scotland, particularly to investigate the nature of granite. Having succeeded beyond all reasonable expectations, he was now communicating the results of his observations to the Society. In his previously given theory, Hutton recalled, he had maintained that the earth's strata had been consolidated by subterraneous heat; in many places these consolidated strata had been broken and invaded by fluid masses resembling, but distinct from, lava. He thought granite, too, consolidated by heat, but had been uncertain about its nature: whether granite was originally stratified, then consolidated by fusion; or a subterraneous body made to break and invade strata, like basalt and porphyry. (Hutton was thus echoing his unpublished 1786 report. He did not now doubt that stratified granite existed, for Saussure had certainly described it and Hutton himself owned specimens.) To resolve his dilemma regarding unstratified granite, Hutton had set out to find examples of it in association with stratified rocks; he could

[5]For the Arran excursion, see Craig 1978, 51–54; Tyrrell 1950; and Tomkeieff 1962. "Mr. John Clerk, the son of his friend Mr. Clerk of Eldin, accompanied him on this excursion, and made several drawings, which, together with a description of the island drawn up afterwards by Dr. Hutton, still remain in manuscript" (Playfair, 1805b, 70).

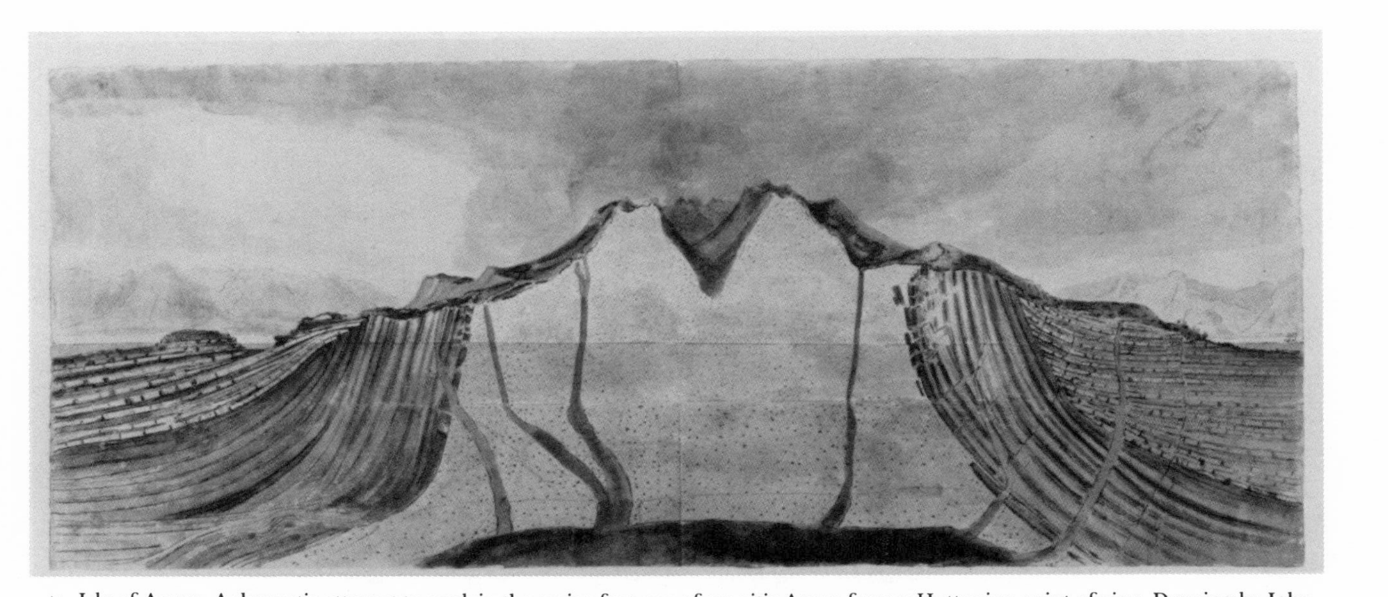

7. Isle of Arran. A dramatic attempt to explain the major features of granitic Arran from a Huttonian point of view. Drawing by John Clerk of Eldin, Junior, 1787. (Sir John Clerk of Penicuik)

then determine whether the granite had formed before or after them. Hutton next recounted his successful expeditions to Glen Tilt in 1785, Galloway in 1786, and Arran in 1787. "Granite, which has been hitherto considered by naturalists as being the original or primitive part of the earth," he concluded, "is now found to be posterior to the alpine schistus—which schistus, being stratified, is not itself original" (81). He would, at some future time, present this evidence in more detail, describe the natural history of Arran, and discuss "the successions of strata, or a certain order of geological periods, which may be ascertained by the natural history of our minerals" (81). Unfortunately, this intriguing promise was never fulfilled.

"Concerning That Which May Be Termed the Primary Part of the Present Earth," 1787 [*Theory*, I, chapter v]

Having satisfied himself with regard to granite, Hutton moved on to the broader question of the primitive rocks as a class. In a quotation-filled essay probably written in 1787, he affirmed only one kind of major land-building rocks. All of them had the same origin, being collected at the bottom of the sea and afterwards elevated into land (along with various melted substances) by the operation of mineral causes. But, as he stressed, vastly different periods of time and a wide variety of subsequent operations might have been involved.

In a significant addition to his previous thinking, Hutton agreed that the destruction of continents might sometimes be sudden. Just how the elevated continents were supported from beneath remained unclear, but he supposed that they might literally be propped up, and that the props might on occasion fail. Yet after accumulating new layers of superincumbent strata the same fallen continents might then be elevated once again. If so, all marine fossils within them would be obliterated and their geological structure would then have become extremely distorted and complex. The mountain schistus strata of Scotland, previously regarded as primitive, might well have been formed in this way:

> If, in examining our land, we shall find a mass of matter which had been evidently formed originally in the ordinary manner of stratification, but which is now extremely distorted in its structure and displaced in its position—which is also extremely consolidated in its mass and variously changed in its composition—which therefore has the marks of its original or marine composition extremely obliterated, and many subsequent veins of melted mineral matter injected—we should then have reason to suppose that there were masses of matter which, though not different in their

origin from those that are gradually deposited at the bottom of the ocean, have been more acted upon by subterranean heat and the expanding power. (I, v, 375–376)

Later, in chapter ii (1793–94), Hutton would allude to the same idea (i.e., of metamorphic rocks), accepting both "mechanical comminution" and "chemical operations" (220) as causes.

Hutton then endeavored to substantiate this startling proposal by quoting a lengthy passage from the *Lettres physique et morales sur l'histoire de la terre* (Letters physical and moral regarding the history of the earth, 1779) by Jean André Deluc (1727–1817), who had therein recanted his previous belief in the marine origin of Alpine strata. But Deluc's recantation was itself an error. "Reasoning from his principles," Hutton declared, "this author could not see the truth, because he had not been persuaded that aquiform strata could have been so changed by the chemical power of fusion, and the mechanical force of bending, while in a certain state of softness" (380). Had Deluc paid similar attention to highly distorted Alpine limestones, he would have seen the truth about schistus: "*If one species of strata may be thus changed in its texture and its shape, may not another be equally so? Therefore, may not the origin of both be similar?*" (380–381).

In his book Deluc had distinguished two types of mountains: aquiform, shaped by water; and primordial, of unknown origin. Hutton chastised him for not attempting to explain the latter, then turned to Saussure, who had theorized that primordial mountains had been formed by crystallization. (Hutton, again, believed in only one cause.) A lengthy quotation from Saussure emphasized the origin of mineral veins. "That mineral veins have been filled with matter in a fluid state," Hutton surmised, "is acknowledged by everybody who has either looked at a mineral vein in the earth or in a cabinet specimen; mineralists and geologists, in general, suppose this to have been done by means of solutions and concretions, a supposition by no means warranted by appearances—which, on the contrary, in general demonstrate that the materials of those veins had been introduced in the fluid state of fusion" (394). Saussure's theory, however, failed to identify the power by which matter is forced from the bottom of the sea to the tops of mountains and ignored the problems of softening and consolidating those masses.

While again quoting Saussure and others at length, Hutton responded to a variety of issues in prodigious footnotes. Saussure, for example, endeavored to explain the well-known transported blocks of Alpine granite to limestone plains by postulating a grand debacle (or flood) that Hutton thought incredible; Deluc tried to account for the same boulders through some equally unlikely explosions. In the Harz mountains of

Germany, Saussure had found evidence of volcanic activity but still believed mineral veins to be filled by the action of water. "There is not a mineral vein (so far at least as I have seen)," Hutton retorted, "in which the appearances may be explained by anything else besides the operation of fire or fusion" (405n). When Saussure next affirmed that wrinkled limestone strata atop a summit had been formed as they now appear, Hutton went after him again. He then turned directly to the supposed distinction between primordial and secondary strata but could not resolve the issue convincingly. Yet, "Since writing this chapter," he informs us unexpectedly, "I am enabled to speak more decisively upon that point, having acquired more light upon the subject, as will appear in the next chapter" (420).

"The Theory of Interchanging Sea and Land Illustrated by an Investigation of the Primary and Secondary Strata," 1787–88 [*Theory*, I, chapter vi]

Section I: A Distinct View of the Primary and Secondary Strata. Having surveyed the observations of others in his previous essay, Hutton was now able to include relevant ones of his own made in the fall of 1787. "From Portpatrick, on the west coast, to St. Abb's Head, on the east," he revealed impressively, "there is a tract of schistus mountains, in which the strata are generally much inclined or approaching to the vertical situation, and it is in these inclined strata that geologists allege that there is not to be found any vestige of organized body. This opinion, however, I have now proved to be erroneous" (*Theory*, I: 421–422).

The Southern Upland mountains in question were, for Hutton, composed of solidified gravel, sand, clay, mica, and other materials clearly derived from earlier rock; adjacent lowlands were similarly composed, but their more horizontal strata contained evident fossils. Though both mountains and lowlands had been consolidated by heat and fusion, the mountains were much more solid. Hutton wanted to find out why. His first hypothesis, necessarily, was that the consolidating power of heat had not been equally applied.

Evidently, the consolidating power had acted more than once. How was it, then, that the mountain schistus of Scotland "had been twice subjected to the mineral operations, in having been first consolidated and erected into the place of land, and afterwards sunk below the bottom of the sea, in order a second time to undergo the process of subterraneous heat, and again be elevated into the place where they now are found"? (427). He knew that the question would not be easy to resolve.

Despite years of careful inquiry among the schistus mountains, Hutton had never been able to explain their formation convincingly and therefore continued to suspend his judgment. He had long searched also for a junction between the secondary or low-country strata and the mountain schistus; this he finally found at the north end of the island of Arran, at the mouth of Loch Ranza. "It was but a very small part that I could see," Hutton recalled, "but what appeared was most distinct. Here the schistus and the sandstone strata both rise inclined at an angle of about 45°, but these primary and secondary strata were inclined in almost opposite directions" (429). Though the schistus was evidently older, neither stratum could possibly have been formed in its present position. Hutton's discovery of this now famous unconformity has often been lauded by historians, but his own words (quoted below) survive to remind us that further evidence was required.

After returning from Arran in 1787 and writing his essay on it, Hutton visited a friend at Jedburgh, in southeast Scotland. The region's relatively horizontal strata, sandstone and marl run through with veins of basalt, were beautifully exposed in steep banks along the river Jed. While walking along these banks, Hutton was surprised to find some distinctly *vertical* strata, with the more usual horizontal layers atop them. He rejoiced at having had the good fortune to stumble on "an object so interesting to the natural history of the earth, and which [he] had been long looking for in vain" (432). It was of course necessary to determine which group of strata was the older. Hutton was thereafter led to conclude that the sea bottom surface of the vertical strata had been subjected to very different conditions at two distinct periods: first, when detritus from the distorting of the strata had been carried away; and second, when the horizontal strata had subsequently been deposited atop the result. It would then be fair to suppose, Hutton concluded brilliantly, "that the disordered strata had been raised more or less above the surface of the ocean; that, by the effects of either rivers, winds, or tides, the surface of the vertical strata had been washed bare; and that this surface had been afterwards sunk below the influence of those destructive operations, and thus placed in a situation proper for the opposite effect, the accumulation of matter prepared and put in motion by the destroying causes" (434–435). This analysis and some related points were then confirmed by eighteen more pages of observations, quotations, and analysis, much of it added later.

Section II: The Theory Confirmed from Observations Made on Purpose to Elucidate the Subject. Intent on further investigation, Hutton wondered where he could find a well-exposed example of the junction between low-country sandstone and alpine schistus. In response to a definite suggestion, he and Playfair (in their first known fieldtrip together) vis-

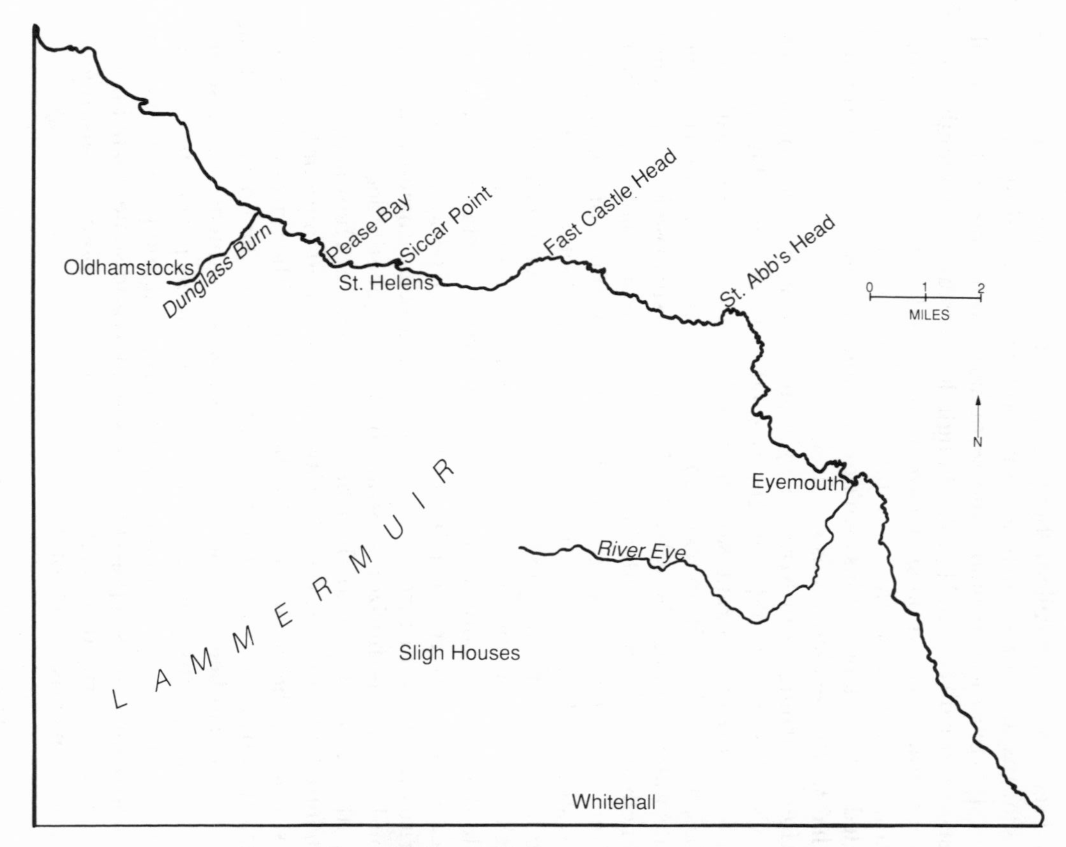

8. Siccar Point excursion map.

ited Sir James Hall at Dunglass around the beginning of June 1788, "the most proper time for a mineral expedition both upon the hills and along the sea shore" (*Theory*, I: 454). The boundary between the vertical and horizontal strata occurred somewhere in the vicinity of Dunglass Burn, but actual junctions between the two were seldom well exposed. Hutton set out to find some good examples. The first such junction he discovered, along the banks of converging rivulets, delineated the boundary more precisely but established also that its western portion was sure to be obscured by deposits of gravel. Eastward, along the seacoast, on the other hand, strata were fully exposed in cliffs as much as two hundred feet high.

From Dunglass Burn (explored later), Hutton, Playfair, and Hall embarked in a small boat, following the originally horizontal sandstone as it rose toward the schistus. At St. Helens they found a junction between the two, much like the one seen earlier along the burns. Then, at Siccar Point, they came upon "a beautiful picture of this junction washed bare by the sea." As Hutton observed, "The sandstone strata are partly washed away and partly remaining upon the ends of the vertical schistus; and, in many places, points of the schistus strata are seen standing up through among the sandstone, the greatest part of which is worn away. Behind this again we have a natural section of those sandstone strata, containing fragments of the schistus" (458). Hutton, Playfair, and Hall continued to the southeast beyond Fast Castle but found little other than schistus, until arriving at St. Abb's Head, which proved to be basaltic.

Having now garnered a series of examples, Hutton began to analyze their significance. He had been pleased to find the schistus associated with strata of recent sand, for schistus was consolidated sand. The highly warped strata, moreover (of which Sir James Hall contributed a sketch), seemingly established once more that the sand had been consolidated into schistus through heat and fusion. Hutton and Playfair then sought further data; walking upwards along Dunglass Burn, they looked for additional junctions of horizontal and vertical strata and the source, if possible, of the streambed's many basaltic boulders. Despite finding the latter (a large basalt dyke intersecting the burn) at Oldhamstocks, they were unable to discover a visible junction between the two types of strata. A later expedition to Whitehall was rained out.

In both parts of this essay Hutton was also concerned with puddingstone (or conglomerate), which he took to be the consolidation of flinty debris. He had seen and collected specimens on Arran; along the south side of Loch Ness, whole mountains were made of it. Surely they helped to prove that nothing visible on the face of the earth was "primitive" or original. The presence of puddingstone near Dunglass, moreover, permitted Hutton to confirm his more general interpretation of the landscape. "It is plain," he wrote, "that the schisti had been indurated, elevated, broken, and worn by attrition in water before the secondary strata, which form the most fertile parts of our earth, had existed. It is also certain that the tops of our schistus mountains had been in the bottom of the sea at the time when our secondary strata had begun to be formed; for the puddingstone on the top of our Lammermuir mountains, as well as the secondary strata upon the vertical schisti of the Alps and German mountains, affords the most irrefragable evidence of that fact" (470–471). Consolidation, uplift, and new land must have followed, thus confirming Hutton's theory in every respect.

With this seemingly definitive affirmation of his views, Hutton, now elderly and often frail, effectively ended his geological explorations. Later that summer (1788), it is true, he and Clerk of Eldin accompanied the Duke of Atholl to the Isle of Man, but nothing Hutton saw there modified his theorizing. On the way back, he also visited the Lake District. After traveling through the schistus mountains of Cumberland and Westmoreland, Hutton explored a limestone quarry near Windermere; on finding a piece with crinoid impressions in it, he swiftly dismissed the supposedly primitive mountains around him as one more illustration of his theory's truth. A subsequent visit to the same place by Playfair in 1791 garnered a specimen of schistus that was unquestionably fossiliferous. Thus, it seemed, the last remaining objection had been decisively refuted.

Hutton and His Public, 1787–95

Flattering as these successes may have been to Hutton and his immediate circle, we must remember that none of the findings reviewed thus far in this chapter were public as of 1788. Indeed, it was only then that the "Theory" itself became available to all but a few. On their official appearance in the *Transactions* of the Royal Society of Edinburgh, however, Hutton's ideas achieved a broader and farther-reaching dissemination than many historians have acknowledged.[6]

It is still too commonly assumed that Hutton's geological theory remained relatively unknown prior to the publication of John Playfair's *Illustrations of the Huttonian Theory* in 1802. Playfair himself claimed that his book was necessary because the obscurity of Hutton's writings "has been often complained of, and thence, no doubt, it has risen that so little attention has been paid to the ingenious and original speculations which they contain" (1802, iii). Critics in his own time and since have agreed

[6]The remainder of this chapter is a slightly revised version of Dean 1973.

that Hutton's style is at best undistinguished—it certainly lacks the wonderful lucidity of Playfair's—but we have been quite wrong in supposing that Hutton's ideas were available to the public only in their author's muddy exposition of them and were therefore substantially ignored. Because the remainder of this chapter demonstrates that Hutton's theory was actually well known before 1802, we must look elsewhere for some of the additional reasons that brought Playfair's *Illustrations* into being.

There are at least three references to a printed version of Hutton's theory in 1787, the year before it appeared in the *Transactions* of the Royal Society of Edinburgh. Adam Ferguson notified Saussure that he was forwarding "papers" by Hutton, who admired Saussure's work: "He has long worshipped the same divinities as you, and embraced every specimen of stone and earth with the most pious attention. His ideas are magnificent and, what is more precious and more difficult in science, formed with a scrupulous regard for reality." But these papers were not further identified, the specific date of this letter is unknown, and the original is unavailable. On 25 August 1787, however, Josiah Wedgwood wrote James Watt a letter informing him that Josiah Wedgwood, Junior, would soon send "two books to Mr. Watt's address, one for himself and the other for Mr. [Matthew] Boulton; they are 'Theory of the Earth,' which he has brought with him from Scotland from Dr. Hutton." These references are not to the brief resume of 1785 but rather to offprints of Hutton's forthcoming article; in Wedgwood's case, his correct citation of its title places the matter beyond doubt. Journal articles were often issued separately in advance of formal publication and, since this one was lengthy, it may well have seemed a book to the elder Wedgwood or his son. V. A. Eyles (1950, 382) located three copies of Hutton's ninety-six page separate but could not establish the date of its appearance. August 1787 must be correct, however, for it was then that Dr. Joseph Black wrote the Princess of Dashkow (director of the Imperial Academy of Sciences at St. Petersburg) to tell her of Dr. Hutton and a "Theory of the Earth" by him. Though ostensibly based on the published text, Black's summary also included more recent material derived from private conversations. In Hutton's view, he explained, there are two grand operations going on perpetually in nature: the extremely slow demolition of exposed continents by air, water and frost; and the constructive rebuilding of subterranean fire, which consolidates and hardens the materials of former lands through deep-seated processes impossible for us to imitate. Hutton, Black continued, believed granite to be one of those rocks that had been melted by subterranean fire at a great depth. Granitic mountains, therefore, were to him by no means primary but rather the product of explosive lateral protrustions raising, and often bursting through, strata. This concept agreed very well with phenomena record-

9. Hutton and his opponents, 1787. As Hutton geologizes the escarpment of Salisbury Crags, Edinburgh, the rock face transforms into silhouette profiles of three sternly disapproving critics. From John Kay, *A Series of Original Portraits* (1837), I.

ed by Saussure and likewise with what Hutton himself had seen in the granitic mountains of Scotland. In both cases, the so-called secondary strata were in fact older than the granite. After alluding to such grand and sublime aspects of the Huttonian theory as its "boundless pre-existence of time," imaginative use of subterranean fire, and dramatic applications of present-day forces, Black noted (and no one was better qualified to state) that "Dr. Hutton had formed this system or the principal parts of it more than twenty years ago" (124). The paper he had just published—not yet in the *Transactions* but as a separate—moreover, was "but a specimen; he is preparing a larger work" (125). Black's letter was written on 27 August, only two days later than Wedgwood's. Thus, the approximate date at which Hutton's separate appeared is no longer a mystery. We also know that by this time its author was definitely working on a book.[7]

We do not know as reliably just when the first volume of the *Transactions* of the Royal Society of Edinburgh, containing the official version of Hutton's paper (textually identical to the separate, excepting page numbers), appeared. But Hutton's paper of 1788 was reviewed by at least three periodicals, the most influential of which, *Monthly Review*, allowed it just over two pages.[8] Parts I and II of the theory were there sketched briefly, with more emphasis than Hutton's on his statement that man had a recent origin. The reviewer doubted that fusion consolidates all rocks, though "subterraneous fires are . . . the most probable cause of the irregularities in the surface of our present earth" (37). But it was shocking to find Hutton, in the fourth section of his memoir, supposing "a regular succession of earths from all eternity! and that the succession will be repeated for ever!!" (37–38). The reviewer then quoted, as so many others would, Hutton's already infamous conclusion about finding "no vestige of a beginning, no prospect of an end."

Analytical Review dismissed Hutton's article in a paragraph, the substance of which is simply a general complaint (common at this time) against geological theorists as such. The productions of these dreamers, it claimed, "usually contain a selection of interesting facts, but their general systems wanting proof can be considered only as philosophical romances. The present theory may pass among the rest." Its particular features are not considered.

[7]The letter from Ferguson to Saussure is in Freshfield and Mantagnier 1920 (466a); that from Wedgwood to Watt is in the Boulton and Watt collection, Birmingham Public Libraries. Black's letter is in Ramsay 1918 (117–125) and Craig 1978 (3–5). Note the close association of Black's letter with Hutton's "Observations on Granite" of 1790. In August 1787 Hutton had just returned from Arran—and had yet to discover Siccar Point. The Princess's son graduated from Edinburgh University in 1779. Black was elected an honorary member of the Imperial Academy of Sciences of St. Petersburg in 1783 (Craig 1978, 3; Eyles and Eyles 1951, 338).

[8]G. L. Davies (1969) located the three periodical reviews of Hutton's "Theory": *Monthly Review* 79 (1788): 36–38; *Analytical Review* 1 (1788): 424–425; and *Critical Review* 66 (1788): 115–120.

Critical Review, like the two preceding journals, commented on each article in the *Transactions*, but in a more responsive fashion. As its four-page commentary observed, Hutton's "is an extensive treatise rather than a paper for a philosophical society, and if it carries us somewhat farther than we usually proceed in the examination of such memoirs, it must be attributed to the importance of the subject, and the merit of the author; for we can allow him great merit, though we differ from him in many respects" (115). It seems difficult, in the face of such statements, to argue that Hutton was universally ignored or that no reviewer endeavored to give him a fair hearing.

According to *Critical Review*, "There are two great hinges on which Dr. Hutton's system turns; the one is, that all calcareous matter is of animal production; and secondly, that the cementation of calcareous and other earths is from fire, which is also the agent that has raised them from the bottom of the sea. These principles are probably true only to a certain extent," as it then argued (116). Some calcareous rocks have simply been deposited from water, and fire cannot consolidate substances that are not fusible. The formation of flint (one of Hutton's weakest points) was discussed at length. But this periodical, unlike others, fully appreciated the imaginative grandeur of Huttonian decay and renovation: "The mind," it said "cannot comprehend so vast a system" (120). Though his theory failed to be specifically affirmed, Hutton was not condemned on religious grounds nor charged with believing in an eternal earth, making this one of the fairest and most perceptive criticisms he was to get.

Hutton's paper of 1788 could scarcely have been ignored after the publication of such reviews. In fact, comments continued to appear. At the end of his original preface, for example, John Williams (*The Natural History of the Mineral Kingdom*, 1789) added that he had just "perused a new theory of the earth by James Hutton" and spent the next forty pages refuting it, this being the first substantial discussion of Hutton's theory in book form. Hutton was, to be sure, "a naturalist of eminent abilities, whose knowledge in several branches of mineralogy does honor to his country," but his theory can be reduced to four propositions:

> 1st, That all our rocks and strata have been formed by subsidence under the waters of a former ocean, from the decay of, and waste of a former earth, carried down to the sea by land-floods.
>
> 2d, That these submarine rocks and strata were heated to the degree of fusion by subterraneous fire, while immersed under the waters of the ocean, by which heat and fusion, the lax and porous sediment was consolidated, perfectly cemented, and all the pores and cavities filled up by the melted matter, while the whole mass was in a state of fusion.
>
> 3d, That the rocks and strata, so formed and consolidated under the waters of the ocean, were afterwards inflated and forced up from under

> water by the expansive power of the subterraneous fire, to the height of our habitable earth, and of the loftiest mountains upon the surface of the globe.
>
> 4th, That these operations of nature, viz. the decay and waste of the old land, the forming and consolidation of new land under the waters of the ocean, and the change of the strata now forming under water to future dry land, is a progressive work of nature, which always did, and always will go on in a perpetual succession, forming world after world. (1789, xxiv-xxv)

Williams then refuted each of these four propositions at length, the most strident objection being his habitual one that Hutton "warps and strains everything to support an unaccountable system, viz. the eternity of the world, which strange notion is the farthest of all from being defensible," because we can see God everywhere in His works. For Williams, "The wild and unnatural notion of the eternity of the world leads first to scepticism, and at last to downright infidelity and atheism" (lix), against which he fulminated, as one might in the perilous milieu of 1789, the year of the French Revolution. Of the eleven pages *Monthly Review* devoted to Williams, the first six concern his analysis of Hutton, suggesting that it may have been the most influential portion of his book.[9]

Part of that periodical's nervous interest in Hutton may have come also from its awareness of the kinship between this new Scottish theory and the reprehensibly atheistic one already promulgated in three different versions (1780–85) by Toulmin, whose possible influence on or derivation from Hutton has been debated by twentieth-century scholars. What *can* be ascertained is that Toulmin specifically mentioned the 1788 paper in a fourth, considerably revised redaction of his own theory, which, unfortunately for Hutton, was entitled *The Eternity of the Universe* (1789) and advocated precisely that. Though Hutton relied too much on the efficacy of subterranean fire, Toulmin commented, they agreed on the basically cyclical nature of geological processes (203–204).

Before Hutton's theory appeared, *Monthly Review* had already commended Deluc's opposition to the materialistic geological theories of Buffon and Mairan, thereby affirming that "allegiance between *Nature* and *Revelation*, which the wisest men of all ages have discerned and admired, and which the minute philosophers of the present times have made many impotent efforts to destroy."[10] Hutton would often be paired with Buffon by adversaries of both. Being of that persuasion, in 1790–91 *Monthly Review* published four original letters from Deluc to Hutton criticizing the latter's theories. Covering more than seventy

[9] *Monthly Review*, n.s. 6 (1791): 121–131.
[10] *Monthly Review* 64 (1781): 487.

pages, they constituted the lengthiest and perhaps most influential discussion of Huttonian theory before Playfair.

In his first letter, Deluc, agreeing that "our continents have been once the bottom of the sea, which is the grand phenomenon to be explained in a theory of the earth," reduced Hutton's to three fundamental hypotheses:

> 1st, That no solid stratum of the earth can have been formed but by its substance having been first melted and then cooled. 2d, That the same heat by which our strata have been melted has raised them above the level of the sea. 3d, That new continents, similar to ours, are now forming at the bottom of the ocean, of the materials of the present ones, as these have been formed of the materials of former continents; and that the continents now forming will rise also, when ours shall be entirely wasted. (Deluc 1790a, 207)

Deluc then examined and attacked each hypothesis, quoting frequently from Hutton in order to controvert him. Since Hutton's entire theory depended on whether or not the continents are in a "state of decay," Deluc in his second letter (1790b) isolated the five arguments Hutton had used to support this thesis and refuted each, to his satisfaction if not to ours. His third letter (1791a) argued that the effects Hutton attributed to subaerial erosion actually took place at the bottom of the sea, and the fourth (1791b) disputed the age of our continents. In closing, Deluc promised to oppose Hutton's dismissal of "accidents" and "preternatural causes" (i.e., miracles) in a later publication.[11]

Hutton received both publicity and a measure of support during the 1790s from Erasmus Darwin (later to be Charles Darwin's grandfather), who was not only an evolutionist himself but also the most popular English poet of that decade. His *Botanic Garden* (1789–91) is not merely a versification of Linnaean botany but a geological treatise in its own right and a compendium of contemporary science. In his extensive notes to the poem (probably written in 1788), Darwin mentioned Hutton's paper of that year no fewer than twelve times, taking repeated notice of its author's ideas on the origins of siliceous nodules, rock salt, marble, and granite. Regarding the latter he asked: "Has the granite stratum in very ancient times been produced like the present calcareous and siliceous masses, according to the ingenious theory of Dr. Hutton, who says new continents are now forming at the bottom of the sea to rise in their turn, and that thus the terraqueous globe has been, and will be, eternal?" (1806, 1: 357). This is as near as Darwin came to summarizing the whole

[11]"Playfair (1805b, 85–86n) uncertainly recalled that Hutton sent a reply, which *Monthly Review* would not print.

of Hutton's theory. Though the two men were friends and correspondents, they disagreed about geology, and of their separate theories Darwin preferred his own.

One of the many ways scientific thought reached the public during Hutton's lifetime was through summations, as we have seen, but these were not invariably accompanied by criticism. A condensed reprint of Hutton's 1788 article, for example, appeared in the initial volume of *Memoirs of Science and the Arts* (1793), a periodical that abridged the transactions of learned societies but failed to prosper, perhaps because so many already existing ones were doing much the same. The abridgment in this case took seven pages and consisted, insofar as possible, of Hutton's own words; there was no commentary.

Hutton's words were put to quite a different use by Richard Kirwan (1733–1812) in his important "Examination of the Supposed Igneous Origin of Stony Substances" (1793), a thirty-page paper read earlier that year before the Royal Irish Academy in Dublin. Hutton, "a gentleman advantageously known to the philosophic world through an important meteorological discovery, the General Law of the Formation of Rain, . . . endeavors to prove that previous to the present state of our globe [stones] were utterly deprived of solidity, and have since acquired it by fusion, and subsequent congelation on cooling." For Kirwan, there could be only two possible positions with regard to the formation of stones. One, the more common, held them to have been dissolved in and later precipitated from water; the other, advocated by Hutton and a few predecessors, traced their origin to heat. If Kirwan could refute Hutton, his own position would seemingly be established. Through direct quotation, paraphrase, and assertion, Kirwan attributed to Hutton a series of ideas that he regarded as false, among which were the following: (1) That "the terrestrial part of the globe was originally a solid compact mass, from the dissolution of which the less compact and looser earths . . . have arisen." Kirwan agreed that decaying rock augments soils but denied that all soils originate in that way (1793, 54). (2) That "soil is necessarily washed away by the continual circulation of water running from mountains to the sea." Kirwan denied that such erosion is constant or that all such water necessarily flows to the sea, much being lost to evaporation. Deposition, moreover, takes place along rivers on their way to the sea and at their mouths; much of whatever silt actually reaches the sea is then deposited along the coast, marine erosion being far less important than Hutton and others had supposed (55). (3) That "the solid parts of the globe are in general composed of sand, gravel, argillaceous and calcareous strata, or of various combinations of these with other substances." Below the surface, however, Kirwan replied, the basic rock is granite, not only in Scotland but throughout the world (56).

Kirwan also contested Hutton's assertions that "all masses of marble or limestone are composed of the calcareous matter of marine bodies" (56), and that "all the strata of the earth, not only those consisting of calcareous masses, but others superincumbent on these, have had their origin at the bottom of the sea" (58).

All these ideas were extracted from the first part of Hutton's essay. Kirwan's meticulous opposition to the second part and his corresponding advocacy of water as a consolidating agent or medium need not be detailed. He went on to dismiss Hutton's succession of worlds as "contrary to reason and the tenor of the Mosaic history" (63), because this succession must have existed from eternity; and since "succession without a beginning is generally allowed to involve a contradiction, therefore the system that forces us to adopt that conclusion must be false" (64). Finally, Kirwan attacked Hutton's agency of heat as a gratuitous and inconsistent assumption incompatible with genuinely scientific inquiry. After fairmindedly including several objections to the aqueous theory (still admittedly imperfect), Kirwan devoted his last few pages to the origin of granite.[12]

A rather favorable summation of Hutton's theory was published the same year. Richard Joseph Sulivan's *A View of Nature, in Letters to a Traveller among the Alps* (1794) reflected on the "atheistical philosophy, now exemplified in France" and sought to exonerate natural history from such unsavory associations. Sulivan's precis of Hutton (I: 91–92) was obviously borrowed from Williams's earlier one of 1789, but later sections of the book suggest that Sulivan read Hutton in the original and found much to agree with.

More important, the very influential *Encyclopaedia Britannica* (3d ed., 1797) summarized Hutton's paper of 1788 over twelve large double-column pages, the most detailed paraphrase it was ever to receive. This summary is perfectly objective, marred only by a note opposite Hutton's conclusion (which is fully and fairly quoted) that reads: "Eternity of the world the final result of this theory." The *Britannica* also summarized the theories of Thomas Burnet, John Woodward, William Whiston, Buffon, and John Whitehurst, but Hutton was given as much space as the rest of them combined. Each summary was followed by an evaluation. Whereas Burnet, Woodward, Whiston, and Buffon were quickly disposed of,

[12]Both Hutton and Playfair responded at length to this highly influential criticism; see Chapters 3 and 5 below. Kirwan, meanwhile, had advertised it himself in the second edition of his *Elements of Mineralogy*: "There is another system which attributes not only to basalts but to all stony substances an igneous origin; it is that of Lazzaro Moro, revived and wonderfully improved by Dr. Hutton of Edinburgh, well known by his excellent essay on the origin of rain. This may be called the *Plutonic* system [first known use of the word]. I have endeavored to show its fallacy in a paper lately read to the Royal Irish Academy, which will appear in the next volume of its *Transactions*" (1794, 1:455).

> the theory laid down by Dr. Hutton is of a different nature from the rest; and as it has been supposed directly to militate against revelation, merits a very particular consideration. The expression, however, with which he concludes his dissertation, that "we can find no vestige of a beginning, no prospect of an end," might be supposed to relate only to the deficiency of our understandings or mode of inquiry, had he throughout the whole course of his work given a single hint of any materials from which the world was originally formed. In this he differs most essentially from the other theorists whom we have mentioned; for all of them suppose a chaos to have been originally created, from whence all the variety of substances we see at present have been formed. But as the Doctor makes no mention of anything prior to a world nearly similar to what we see just now, we must necessarily conclude that its eternity is a part of his creed. (*Encyclopaedia Britannica*, 1797, 6: 254–255)

So began a five-page refutation, whose author is unknown.

For the *Britannica*, the world is not eternal and its creation could not have been brought about by natural or even knowable forces. The beginning of the world "was occasioned by a power which cannot possibly be investigated, because it lies without the bounds of Nature itself, and far beyond the reach of our faculties"; this objection "militates invincibly against all theories of the earth which seek to derive its original from natural causes" (255). These admonitions were directed against Hutton, who might very well have agreed with both of them.

Though the strata were originally formed by divine power, however, they "are certainly preserved, repaired, and changed by natural causes, of which aqueous solution is a principal, though not the only one" (258). The *Britannica* dismissed several of Hutton's examples to the contrary and summed up its case as follows:

> Thus we have seen that, contrary to our author's hypothesis, the world has undoubtedly had a beginning; that our dry land has not, for ages, been the bottom of the sea; that we may reasonably suppose the Deluge to have been the cause of all or most of the fossil appearances of shells, bones, &c. we meet with; that our author has erred in denying to aqueous solution the effects which experience has shown it capable of producing, and in ascribing to fusion effects which experience doth not warrant; and that his theory, far from having any foundation in chemistry, is directly contradicted by that science. (260)

In this and subsequent editions, the *Britannica* failed even to mention Hutton's fuller *Theory of the Earth, with Proofs and Illustrations* (1795), the first two hundred pages of which are a revised version of his 1788 paper, followed by a detailed response to Kirwan's attack of 1793 and further elaboration of earlier positions.

All these reactions emanated from the British Isles, but there were a handful of important foreign ones as well. V. A. Eyles, for example, discovered a detailed exposition of Hutton's 1788 paper in the *Magazin für das Neuste aus der Physik und Naturgeschichte* (Gotha, 1790), an abstract of it among Werner's papers at Freiberg, and a complete translation in the *Sammlunger zur Physik und Naturgeschichte* (Leipzig, 1792). In France a partial translation with commentary of Hutton's 1785 "Abstract" appeared in the *Journal de Physique* (July 1793) and was then subjected to further remarks by Nicolas Desmarest in volume 1 of the *Encyclopédie méthodique* (1794) and by Jean-Claude de Lamétherie in his *Théorie de la terre* (1795). Desmarest's fifty-page discussion of Hutton also included translated extracts from the latter's "Theory." Finally, Barthelemy Faujas de Saint-Fond, who had visited Hutton in 1784, published his belated travel account in 1797, noting therein that the "Theory" on which Hutton had then been at work was now published, though more "a memoir containing general views of the subject than a body of observations" (1907, 2: 235n). By this time too, Hutton's theory was known in America—and, as we have seen, in Russia.[13]

As one inspects these initial reactions to Hutton, a few general conclusions seem inescapable. The first is that, by and large, his critics tried to be objective. Ironically, only the misnamed *Analytical Review* failed to consider Hutton's specific arguments. Most of the others attempted genuine analysis, though their various summaries of his theory tended to be sounder than the discussions that derived from them. These summaries differed, just as those by modern scholars do, but none except that by *Analytical Review* resorted to deliberate caricature. Though Hutton's views were certainly disliked, I am not convinced that he was fundamentally misread.

Hutton's critics, furthermore, sensed almost immediately that this man and his theory were of unusual significance. Several went out of their way to praise him as a man of science, and Hutton was never subjected to the deplorable personal attacks that Joseph Priestley, Sir William Hamilton, Sir Joseph Banks, and later even Playfair were. A disproportionate amount of space was given Hutton's theory by the press. Far from being ignored, it became an issue almost on publication, and we may note that other geological works were sometimes valued more for what they had to say on Hutton than for any original contribution of their own.

[13]Eyles 1955; 1969; 1970; Rappaport 1964; for America, see Gerstner 1971 and Davies 1969, 186. According to Kirwan, Hutton's theory was "but little known, or at least noticed, on the Continent" (1793, 52). Leopold von Buch, moreover, subsequently informed Charles Lyell (K. Lyell 1881, 2: 48) that German scholars did not have access to Hutton's and Playfair's works during the Napoleonic wars.

One reason why discussions of Hutton tended to be lengthy is that rebuttals needed to be careful and systematic. Neither Hutton nor his theory could have been dismissed offhand, and most of his reviewers were quite aware that they were doing battle with a major adversary. We need not be surprised, then, by the number of specifically geological objections to Hutton, for much of his opposition was couched in purely scientific terms. The geology of Hutton's day is not ours. For example, those who believed as he did in the efficacy of present-day forces were a distinct minority, but he was in accord with many of his contemporaries in finding geological change more orderly than we do. Nowadays we tend to dismiss Hutton's geological opponents, forgetting that his logic is not always unassailable, that his evidence is often scant or hypothetical, and that in some places he is simply wrong. Contemporaries pointed out genuine weaknesses in his theory that later historians have also recognized. The opposition was more wrong than right overall, but its proponents gave Hutton an honest probing, and from the rough and tumble of such controversy our modern beliefs eventually emerged.

[3]

Theory of the Earth, 1795

Between 1788 and 1795, Hutton remained diversely busy. In 1789, for example, he accompanied old Adam Smith, Henry Mackenzie, and a young London visitor (the poet Samuel Rogers, then twenty-six) to a meeting of the Royal Society of Edinburgh. A total audience of seven there heard a long paper on debtors' laws; Smith, as usual, fell asleep. When he died on 17 July 1790, Black and Hutton jointly edited—and then destroyed—his papers. Hutton himself published three essays that year, one of them a reply to Deluc's criticisms of his theory of rain (1788), and presented his "Observations on Granite," which had been completed several years earlier. He may also have replied briefly to Deluc's four letters in *Monthly Review* (1790–91) attacking his geological theory. If so, that journal failed to publish his remarks.[1]

From 1791 onward, Hutton was subject to recurring periods of serious illness (probably from bladder stones) that often required him to remain bedridden. Nonetheless, he spent the last six years of his life writing prodigiously, and publishing as well. Hutton's first book was his *Dissertations on Different Subjects in Natural Philosophy* (1792), which reprinted his paper on rain and reply to Deluc's criticism of it while adding a third essay on the same topic, again refuting Deluc. A fourth dissertation considered vernal and autumnal winds. This essay was followed by others (probably written much earlier) on phlogistic and phosphoretic bodies; the manner in which specific gravity, hardness, and ductility are

[1]Thompson 1927, 332. For Smith and Hutton, see also *TRSE* 3 (1794): 131n, 137. On Hutton and Deluc, see pp. 51–52 above.

affected by phlogistic matter; and the influence of phlogistic matter on light and color. Hutton concluded this 740-page volume with an appendix on the weather of Bengal, to confirm his theory of rain. He feelingly dedicated the book to Black, publicly acknowledging "that obligation which men of science owe you for your philosophical discoveries, particularly for that of latent heat, the principle of fluidity—a law of nature most important in the constitution of this world and a physical cause which, like gravitation, although clearly evidenced by science, is far above the common apprehension of mankind." In part, Hutton was likewise alluding both to his own geological theory and to its checkered reception.[2]

On 7 April 1794 Hutton read to the Royal Society of Edinburgh the first part of a lengthy disquisition on physics; it was afterward continued at the Society's meetings in May, June, July, August, and December. This effort, apparently unaltered, then became his second book, *A Dissertation upon the Philosophy of Light, Heat, and Fire* (1794). "In my dissertation upon the subject of phlogiston [i.e., 1792]," he began, "I have shown the error of the new antiphlogistic theory, for there I have proved that the light and heat of fire does not proceed from the condensation in vital air in burning or from anything that might properly be termed *Calorique*, but that it necessarily requires another cause, which may be properly termed *phlogiston*" (xvii). His intent in this new work was to proceed a step further, establishing that "though heat be necessary in general to the burning of bodies or kindling of fire, it is not heat which is immediately produced in fire but . . . the solar substance lodged in those phlogistic bodies, which is then made to emerge in light and to excite that heat which appears on those occasions as the effect of fire" (xviii). He hoped to destroy the antiphlogistic theory and to reaffirm his own. "Philosophy necessarily rests upon this principle," Hutton added, "that nature is uniform or consistent; therefore, an apparent inconsistency in our experience is not to be rashly imputed to the order of nature, nor is science to be blamed for the discordant opinions of generalizing men. The errors or inadvertences of man, so far from being a proper ground for scepticism, must contribute for establishing the certainty of science, when these are properly corrected" (5–6). A good summary of Hutton's argument, which many think included his discovery of infrared light, appeared in the Edinburgh *Transactions*, IV (1798).[3]

[2]For Hutton's theory of rain, see Middleton 1965, 106–110; for the book as a whole, see *Monthly Review* 16 (1795): 246–254; Playfair 1805b, 62–67, 74–81, 92n; Gerstner 1968; and Heimann and McGuire 1971, esp. 281–293. Black 1786 includes theorizing on evaporation by Joseph Black and Hutton.

[3]Hutton to Cadell and Davies, 22 Jan 1795 (Edinburgh University Library); Playfair 1805b, 81; Hatch 1975 and Watanabe 1978 are modern commentaries on the *Dissertation Upon Light, Heat and Fire*.

Hutton's third and longest book, published in three large volumes that same year, is also his least penetrable. Formidably titled, *An Investigation of the Principles of Knowledge, and of the Progress of Reason, from Sense to Science and Philosophy* (1794) began with a wide-ranging discussion of human knowledge and its sources in sensation, perception, conception, passion, and action; we also learn the nature of ideas and reason. Man alone is granted science, by which Hutton meant conscious principles leading to wisdom and truth. These principles include time and space, unity and number, and cause and effect. Next, Hutton considered various kinds of proof, balancing evidence with doubt, truth with probability, analogy with testimony, the real with the imaginary, and physical principles with mathematical truths. This discussion led him to relate power and matter, especially as manifested in the system of nature: man has the wisdom to use natural power for his own well-being and pleasure. The progress of intellect, Hutton saw, tends continually toward the perfection of mind. Consideration of both nature and art by mind is the essence of education, which, if rightly pursued, leads intellect through the workings of nature to its End or Final Cause, and the possibility of a future life. Finally, man is drawn by his mindfulness of both God and immortality to accept morality, piety, and religion, but must not allow religion to be corrupted by those ignorant of science and philosophy. Hutton's conclusions, therefore, were theistic but not Christian.[4]

After editing Adam Smith's essays (1795) and publishing his enlarged theory of the earth (reviewed below), Hutton devoted the last two years of his life to what would have been another book, his thousand-page manuscript "Elements of Agriculture." It was not written in expectation of a large audience, he admitted, but rather for the author's pleasure in reviewing a subject that had been "in a manner the study of [his] life." Applying what he had previously established in his *Principles of Knowledge*, Hutton proposed that agriculture represented a certain controlling power through which the science of man cooperated with means employed by nature to further the divine end of a fertile, habitable globe. He began, therefore, with soils and their fertility. Additional sections discuss climate (heat accelerates the growth of plants), breeding, varieties of animals (which, like the human mind, tend toward perfection), and many aspects of farm management. He also opposed his late friend Adam Smith's advocacy of unlimited agricultural exploitation; that, Hutton saw, would lead only to exhaustion of the soil. With this never-completed book (in which Norfolk is often mentioned), Hutton's geological theorizing had, as he doubtless appreciated, come full circle.[5]

[4]Commentaries on the *Principles of Knowledge* include Playfair 1805b, 87; McCosh 1875; Ellenberger 1972a; Olson 1975; O'Rourke 1978; and P. Jones 1984. For Hutton's definition of science, see 1794: 19. A copy of Hutton's work with interesting annotations by S. T. Coleridge is in the British Library.

[5]There are informative commentaries on "Elements of Agriculture" by E. B. Bailey (1967) and

Theory of the Earth, with Proofs and Illustrations, 1795

Most of the important papers that Hutton presented before the Royal Society of Edinburgh eventually reappeared in his books. This was his intention also with "Theory of the Earth" as presented in 1785; no sooner had its publication been assured than he began to supersede it with a longer version (as Black noted in 1787). Presumably, that version would have reprinted the "Theory" and some additional essays already written by 1788, including the immortal finds in Glen Tilt, on Arran, and at Jedburgh and Siccar Point. After these essays were written, however, Hutton turned abruptly to other concerns, writing and publishing the books we have just reviewed.

Any feelings of complacency regarding his own theory Hutton may have fostered through his as-yet private researches probably suffered little from the largely adverse attention his "Theory" of 1788 received; in all likelihood, he saw only a fraction of the criticism directed against it. On 3 February 1793, however, Kirwan read to the Royal Irish Academy in Dublin his "Examination of the Supposed Igneous Origin of Stony Substances," a major onslaught probably unknown to Hutton before its publication that year or the next. As we have seen, Kirwan's essay attempted straightforwardly to demolish Hutton's geological theory, which was quoted at length and ruthlessly dissected. Arguing largely on chemical grounds, Kirwan opposed the igneous origin of granite and all other rocks (except volcanic ones). Shortly after Kirwan's paper was read, and well before it had been published, Hutton became severely ill and had to undergo a serious operation (without anesthesia); he was then 67. His slow convalescence required many months, during which Hutton remained in his room and was eventually able to correct proofs for his *Principles of Knowledge*. While he was still recuperating, Kirwan's "Examination" appeared, probably as an author's separate. "Before this period," Playfair remembered, "though Dr. Hutton had been often urged by his friends to publish his entire work on the theory of the earth, he had continually put off the publication, and there seemed to be some danger that it would not take place in his own lifetime. The very day, however, after Mr. Kirwan's paper was put into his hands, he began

---

Jean Jones (1985); the latter furnishes much subsidiary information as well. As Jones demonstrates, Hutton had some reputation as an improving agriculturalist, though his unfinished treatise was never published (despite an attempt in 1806 to do so, as "Dr. Hutton's Principles and Practice of Agriculture, with Life by Professor Playfair"; Besterman 1938, 34). For Hutton and natural selection, see also his *Principles of Knowledge*: "In conceiving an indefinite variety among the individuals of that species, we must be assured that, on the one hand, those which depart most from the best adapted constitution will be most liable to perish while, on the other hand, those organised bodies which most approach to the best constitution for the present circumstances will be best adapted to continue in preserving themselves and multiplying the individuals of their race" (1794, 2: 500). P. Jones (1984, 193) suggests influence from Erasmus Darwin.

the revisal of his manuscript and resolved immediately to send it to the press" (1805b: 86). The exact date has not been recovered.

We know little else about the writing and publication of Hutton's masterpiece, except that its author wrote his London publisher, the firm of Cadell and Davies, on 28 September 1795 to announce that all 1,187 pages of his two-volume *Theory of the Earth* were now printed and lacked only the six plates (four quarto and two folio), which were not yet run off. Two hundred copies of the work were to be sold in London, but at what price? On learning from a reply of 1 October that the two volumes of *Theory* would be sold at fourteen shillings in boards (and therefore at nine shillings eight pence wholesale), Hutton assented in a letter of 6 October. He also specified that the title was to be "Theory of the Earth with Proofs & Illustrations." "These volumes do not contain the whole work," the author cautioned; "there will be another volume, with many plates." This third volume would presumably have included both parts 3 and 4. Though two hundred copies of the sheets and plates were destined for London, moreover, we do not know how many were reserved for William Creech, its Edinburgh publisher, but the total press run could not have been more than five hundred (and might well have been four or even three). Whatever the actual number of copies, there would have been two or three thousand plates to run off; of these, 1,200 would then have to be shipped with the printed sheets of text to London; each copy would be assembled, sewn, trimmed, and bound in boards. Given the tasks and times involved, Hutton's London edition probably appeared in 1796 (which is the date Deluc gave it). If there was no agreement to the contrary, the Edinburgh edition (also in boards, presumably) might have appeared earlier, before the end of 1795.[6]

The title page of volume I (normally the first sheet printed) indicated that Hutton's *Theory* would be in four parts and at least two volumes. Cadell and Davies were listed as the primary publisher, which suggests that the Edinburgh edition may have been smaller than two hundred. Thomas Cadell the younger (1773–1836) took over a flourishing London publishing house from his father in 1793 and ran it until his death. His partner, William Davies (d. 1820), had been especially chosen by the elder Cadell for his already evident business acumen. William Creech (1745–1815) had been the foremost publisher in Scotland since 1773, with Henry Mackenzie and Robert Burns among his authors. *Theory of the Earth* was Hutton's fourth book in as many years; though his works

[6]Hutton to Cadell and Davies, 28 September 1795, was first published by Eyles (1950, 386); the 6 October letter, by Craig (1978, 6). Deluc's 1796 date is in *British Critic* 8 (October 1796): 337. Philip Howard (1797) said that his book was in press when Hutton's appeared. Playfair (1805b) specifies 1795. But Erasmus Darwin was discussing the book with S. T. Coleridge in January 1796, and a passage of general complaint that month by William Smith (the stratigrapher; Phillips 1844, 18) has strong Huttonian overtones. On the whole, then, January 1796 is likely.

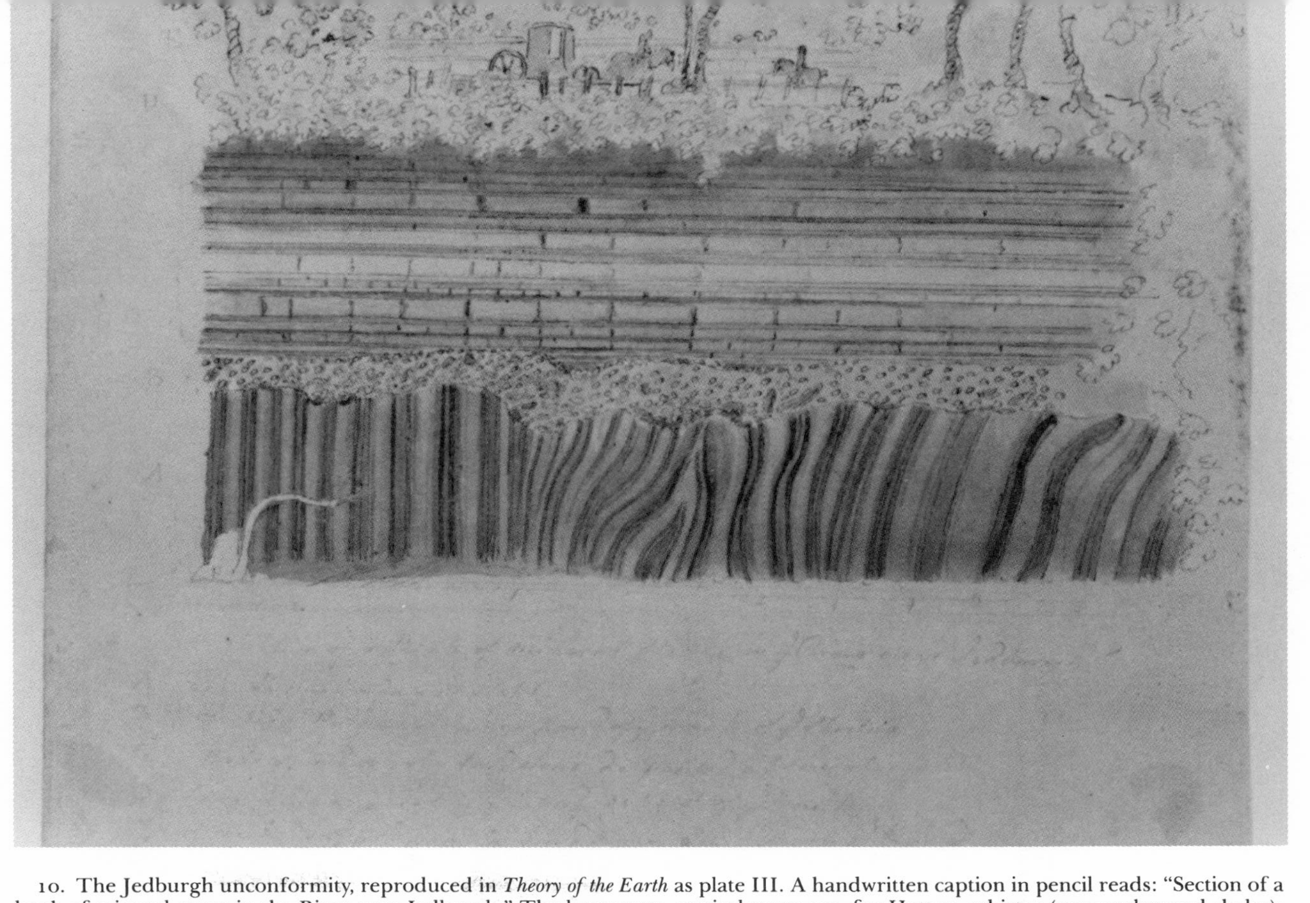

10. The Jedburgh unconformity, reproduced in *Theory of the Earth* as plate III. A handwritten caption in pencil reads: "Section of a bank of mineral strata in the River near Jedburgh." The lowermost, vertical strata are, for Hutton, schistus (greywackes and shales); atop them are conglomerate, sandstone, and marl. Drawing by John Clerk of Eldin, 1787. (Sir John Clerk of Penicuik)

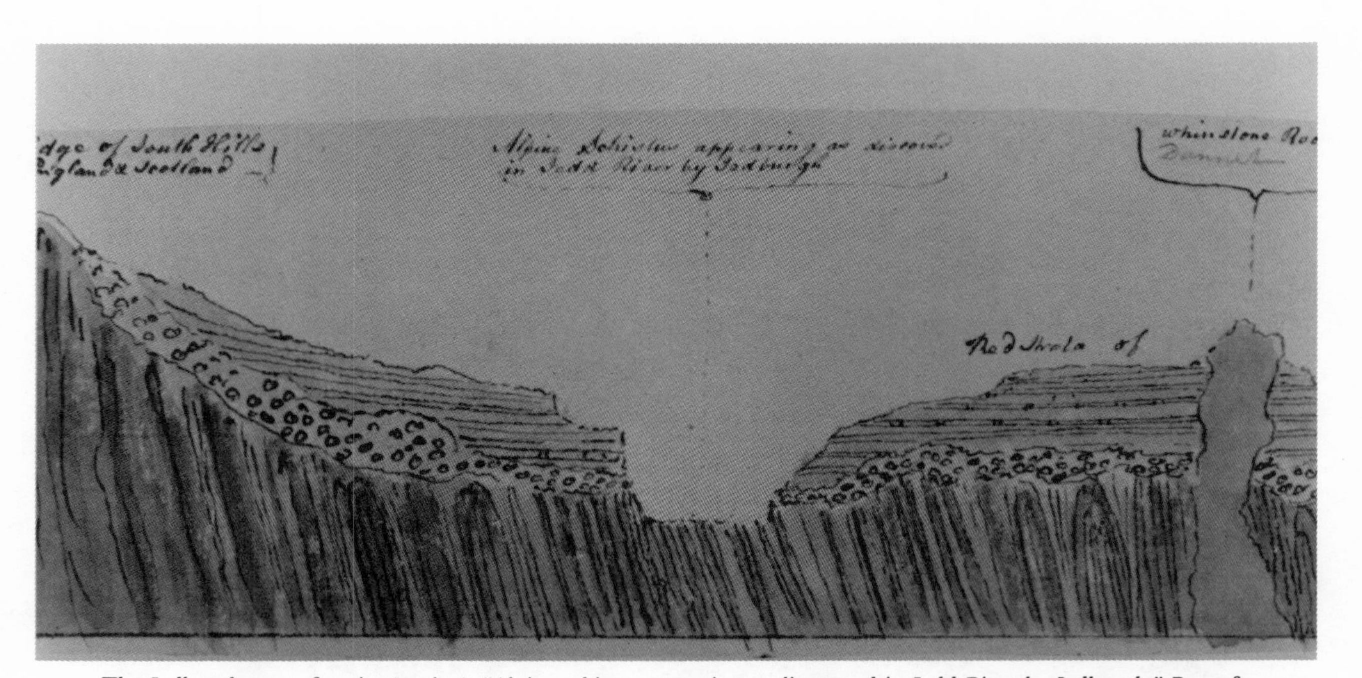

11. The Jedburgh unconformity (section). "Alpine schistus appearing as discovered in Jedd River by Jedburgh." Part of a panoramic section extending some forty miles from the English border to Edinburgh. Drawing by John Clerk of Eldin, perhaps 1787. (Sir John Clerk of Penicuik)

were not for the multitude and uniformly hard to read, he must have been regarded by the knowledgeable as a prominent living author.

His table of contents revealed that all 620 pages of volume one belonged to Hutton's "Part I"; was the complete work, then, to be in four volumes and about 2,500 pages? Part I had eight chapters, three of which were subdivided into sections. Except for a standard heading and occasional footnotes (sometimes lengthy), each page consisted of twenty-five 3-1/4" lines, the generously leaded Roman type of which was reinked as necessary, displaying some irregularities of impression as the edition progressed. Hutton's volume as a whole was made up in eight-page gatherings, beginning with an unlettered one and then A-I, K-U, X-Z, Aa-Ii, Kk-Uu, Xx-Zz, 3A-3I, 3K-3P, 3O, 3R-3U, 3X-3Z, 4A, 3B, 4C-4I, the latter being a half-gathering of four pages. Writing and printing—as was not unusual—overlapped, so that Hutton could refer to page 111 of his printed text while writing page 323. Similarly, page 247 cites the already printed page 139, and 253 cites 68. Volume I ended with four plates, the first three of which were quarto. Plate I depicted a specimen of septaria; plate II, one of graphic granite from Portsoy. Plate III, by Clerk of Eldin, strikingly illustrated Hutton's discussion of Jedburgh in chapter vi, section 1. Cleaving deep within the earth, we see the charming landscape of a country road (with chaise and pair and an attendant horse and rider) underlain first by fertile soil, then by a complex series of perfectly horizontal strata, then by an irregular layer of fragmented deposits, and finally (in irregular union with the last) by vertical and sharply distorted basement strata. Plate IV, folio, profiled a grim hillock, the warped strata of which bend first upward and then downward in concise sweeps but have no relation to the overall silhouette of the hill. Like both of those in volume II, plates III and IV were new to this edition.

Volume II (completing our physical description of the book before proceeding to its contents) begins with a large folio plate redrawn from Saussure's *Voyages dans les Alpes*. In it, we see an imposing, highly serrate Mont Blanc dominating not only lesser eminences but the sky itself; at the very bottom, two minute Swiss climbers with staves emerge from behind a rock to explore the awesome scene. Beyond the title page (unchanged, except for the volume number), a second table of contents announced "Part II," its belated introduction, and fourteen numbered chapters, none of them subdivided. The volume as a whole, comprising 568 pages, is gathered like volume I, from the unlettered preliminaries and A to 4B, with no mistakes. Tipped in at the end of volume II is another large folio plate borrowed from Saussure. Viewing Mont Blanc more distantly now, from high across the Vale of Chamonix, we are equally impressed by its jagged ruggedness and by the immensity of its glaciers.

VOLUME I (PART I): THEORY OF THE EARTH, WITH THE EXAMINATION OF DIFFERENT OPINIONS ON THAT SUBJECT

The reprinting of Hutton's 1788 "Theory," now chapter i of volume I, differed from the original in several respects. For one thing, it was divided into sections rather than parts (though the divisions and their titles were not changed). The small capitals accompanying paragraph indentations were eliminated; quotation marks became more elaborate; numerous small changes in other punctuation appeared; "-ize" words became "-ise" and "enquire," "inquire." Except where wording changed a bit, there were no deletions. But Hutton added several weighty footnotes to augment his previous remarks and, in a few significant cases, expanded the text itself. Because we are tracing the development of Hutton's thinking, it is necessary to review his additions briefly:

1. On page 213 of his 1788 "Theory" Hutton added "subterraneous heat or mineral fire" to electricity and magnetism as actuating powers of the globe. He then added to the second line on page 214, calling "subterraneous fire" the most conspicuous power in the operations of the world but "still less understood, whether with regard to its efficient or final cause" (*Theory*, I: 11–12). Later remarks on the same page were softened with the addition of "perhaps" and "may."
2. To page 215, he added a note replying to Deluc (14).
3. To page 221, he added a note from Saussure (26).
4. On page 222, regarding granite, he changed the "very rarely" of line 6 to "never" (27).
5. His most extensive revisions affected page 225, which was also the beginning of Part/Section II, "An Investigation of the Natural Operations Employed in Consolidating the Strata of the Globe." Here the 1795 version added two introductory paragraphs (33). In 1788, the first sentence of paragraph 2 read, "Thus, fire and water may be considered as the general agents in this operation which we would explore." In 1795, "Thus" became "Therefore" and a long section on subterranean fire immediately followed "explore" (34–41). In it, Hutton called additional attention to "subterraneous fire, or heat, as a powerful agent in the mineral regions, and as a cause necessarily belonging to the internal constitution of this earth" (34). Though much remained to be known regarding it, Hutton praised "the profound philosophy of Dr. Black, in relation to the subject of *latent heat*" (38).
6. To page 231, line 3 ("*sulphureous*"), Hutton added "or *phlogistic*" (51).
7. To page 232, he added two notes regarding siliceous solutions, citing Black, "An Analysis of the Waters of Some Hot Springs in Iceland"

(1794), and Dolomieu, "Memoire sur les pierres composées et sur les roches" (1791–92). Hutton was surprised to find so enlightened a naturalist as Dolomieu supporting the aqueous origin of mineral substances (53–57).[7]

8. On page 234, he joined the last two paragraphs; changes of wording regarding phlogiston followed on page 235, lines 23 and 24 (61, 63), and page 236, lines 3 and 4 (63–64).

9. Minor verbal changes affected pages 248–49 (86–87).

10. Major changes took place on page 250; Hutton added a substantial discussion of agates following line 5 (88–93). He also added a lengthy note—perhaps the most important of all—on compression and consolidation: "The effects of increasing degrees of heat are certainly prevented by increasing degrees of compression," he argued, "but the rate at which the different effects of those powers proceed, or the measure of those different degrees of increase that may be made without changing the constitution of the compound substance, are not known" (94–96).

11. To page 259, Hutton added a long note (largely derived from Saussure 1779–96, vol. 2) regarding Alpine strata and the formation of mineral veins; in the latter, "we see that power by which the strata have been raised from the bottom of the sea and placed in the atmosphere" (112–117).

12. To page 271 (his only addition to Part/Section III), Hutton added a long note on the effects of subterraneous heat inspired by and rebutting Dolomieu in the *Journal de Physique* for May 1792 (138–140; see item #7 above).

13. To page 289, in Part/Section IV, Hutton added a long note responding to Deluc regarding the origin of sand (172–174).

14. On page 291, he cited his own *Dissertations* of 1792 (177).

15. The unnumbered "Explanation of Plate I" was dropped.

These, then, were Hutton's alterations to his "Theory" of 1788, the more important of which affected his Part/Section II especially. All the changes were probably made no later than 1794.

In neither the 1788 "Theory" nor its 1795 revision did Hutton ever mention Richard Kirwan. Since it was Kirwan's criticism that prompted him to take up a book-length version, however, Hutton next turned to confront his adversary directly. At the beginning of chapter ii, "An Examination of Mr. Kirwan's Objections to the Igneous Origin of Stony Substances," he candidly acknowledged the opposition engendered by his theory, among both the learned and the general public. Citing Kirwan's essay specifically, Hutton endeavored to meet all the major objec-

[7]For Dolomieu, see Hooykaas 1970, 281–283.

tions to his thinking its author had raised. Among them, Kirwan had disagreed regarding the origin of soil and the efficacy of fluviatile erosion. He also thought granite the dominant rock in Scotland and elsewhere, though Hutton had stressed sediments and schists. Kirwan, moreover, regarded granite as primary (existing from the Creation); he therefore opposed Hutton's immensities of geological time and the whole mechanism of renovation.

Seemingly accusing him of atheism, Kirwan represented Hutton's succession of worlds as a potentially infinite series. In reply, Hutton emphasized his deistic theology. He had not said that the world is eternal but only that our ability to reconstruct its past is limited. "My principal anxiety," he protested, "was to show how the constitution of this world had been wisely contrived; and this I endeavored to do, not from supposition or conjecture, but from its answering so effectually the end of its intention, viz. the preserving of animal life, which we cannot doubt of being its purpose. Here, then, is a world that is not eternal, but which has been the effect of wisdom or design" (223). Kirwan also opposed the central concept of Hutton's theory, subterranean heat. Throughout the latter half of his reply, Hutton steadfastly supported consolidation by fusion, nervously affirming "an internal heat, a subterraneous fire, or a certain cause of fusion by whatever name it shall be called, and by whatever means it shall have been procured" (239). Subterraneous fire, he guessed hastily, might be fueled by coal produced from the burial of a previous world's plant life (243). As this desperate expedient suggests, Kirwan's attack had been a telling one.

Having replied to Kirwan's specific objections, Hutton went on immediately to chapter iii, "Of Physical Systems and Geological Theories in General." In a rare acknowledgment of predecessors, he quickly reviewed and dismissed the theories of Burnet, Maillet, Buffon, and Deluc, all of whom attempted to explain the present earth through some kind of devastating catastrophe. According to Hutton's deistic conception of nature, such accidents were impossible. A theory of the earth, he knew, "should bring the operations of the world into the regularity of ends and means, and, by generalizing these regular events, show us the operation of perfect intelligence forming a design." If we believe that almighty power and supreme wisdom have been employed to sustain that beautiful system of plants and animals so interesting to us, we must certainly conclude that "the earth, on which this system of living things depends, has been constructed on principles that are adequate to the end proposed, and procure it a perfection which it is our business to explore" (275). Reliance on accidents, he believed, entirely contradicted that purpose.

An accident commonly invoked by theorists was a great deluge of water (perhaps the flood of Noah). But Hutton once again dismissed the agency of water and defended subterranean fire against the charge that it was purely destructive. Yet he also clearly distinguished his own views from those of the volcanic philosophers, who, regarding basalts as lavas, accounted for the nodules occurring within them by aqueous infiltration. They had evidently never considered subterranean heat a general principle in the theory of the earth. Leaving the volcanists, Hutton moved on to consider what had so often led philosophers in general to reason wrongly about the earth. (One should recall his previous chapter at this point; he is actually replying, more philosophically and more calmly, to issues Kirwan had raised.) After dismissing such philosophers as a group—Kirwan is never named—Hutton described what a genuinely scientific theory of the earth would include: natural fact, known laws of nature, and dichotomous exclusion ("if one theory explains natural appearances, then the opposite to that theory cannot be supposed to explain the same appearances," 299). Thus, if fire is right, water must be wrong. Any theory, he believed, should be tested against nature.

Having established his basic criteria, Hutton next included a series of essays, most of them previously written, dealing with the "modern theory of primitive mountains." Because they represented successive stages in his understanding, he inserted them "nearly in the order in which they occurred, or had been written" (310). Having been composed during the 1780s, they are essays with which we are already familiar. Chapter iv, "The Supposition of Primitive Mountains Refuted," begins with his early fragment "On Granite" from 1785; in the latter, Hutton had denied that granite was a primitive rock. But to this he added much more, the remainder of his chapter being mainly a long retrospective emphasizing his own experience. Three types of rock had most often been regarded as primitive: massive, unstratified granite; vertical strata; and unfossiliferous limestones associated with primitive mountains (as described, for instance, by Saussure). Hutton did not dispute that some rocks are older than others; rather, he denied that some rocks originated in a unique way. The critical argument, he believed, is that regarding the alleged absence of fossils. They are not absent, however—only rendered indistinguishable through subsequent mineral operations. A so-called primitive rock found by himself in Wales (1774), he revealed, contains clearly discernible shells. Thus, Hutton concluded, "I had formed my opinion with regard to this alleged fact [the existence of primitive mountains] long before I had seen any description either of the Alps or Pyrenean mountains, and now I have no reason to change that opinion" (326–327). After analyzing reported examples from the Alps, Hutton recalled his own Lake District explorations of 1788 and those of Playfair in 1791. A letter from Sir James Hall in June 1792 provided Scottish examples. All such primitive strata, Hutton believed, would eventually yield fossils also, though they might well be rare.

For the remainder of this essay, Hutton turned on Pierre Simon Pallas, whose *Observations sur la formation des montagnes* (Observations on the formation of mountains, 1777) he quoted at length, confounding the author's major arguments on behalf of primitive mountains and the evolution of life. Hutton's incisive remarks, prompted by suppositions that he totally opposed, are among his most revealing. However much the contents of one stratum may differ from another, he insisted, "there is nothing formed in one epocha of nature but what has been repeated in another" (364). Whereas Pallas had accounted for mammoth remains in Siberia by supposing the elephantine creatures transported from southern Asia by a deluge, Hutton assumed that they had formerly lived where their remains were found. "Thus," he concluded, "may be removed the necessity of a general deluge or any other great catastrophe, in order to bring together things so foreign to each other" (369n). It had not yet been definitively ascertained, as would happen within a few years, that the mammoth was extinct.

Chapter v, "Concerning That Which May Be Termed the Primary Part of the Present Earth," continued his major argument. As in chapter iv, Hutton maintained that all parts of the present earth have had the same origin, the detritus of previous continents "being collected at the bottom of the sea and afterwards produced as land, along with masses of melted substances, by the operation of mineral causes" (371). Hutton then elaborated his concept of metamorphic rocks, reinforcing an emphasis on the more general topic of incessant terrestrial vicissitudes with lengthy quotations from Deluc and Saussure. A long chapter vi, "The Theory of Interchanging Sea and Land Illustrated by an Investigation of the Primary and Secondary Strata," followed; in it Hutton recalled his dramatic fieldwork at Jedburgh, Siccar Point, and Dunglass Burn.

In chapter vii, "Opinions Examined with Regard to Petrifaction, or Mineral Concretion," Hutton reviewed the unsatisfactory attempts of Buffon and Deluc to explain how fossils were formed. For him, petrifaction was only a more specialized form of consolidation, and therefore a topic central to his theory. Naturalists who cited the analogy of stalactites forming in caves to support the aqueous theory of consolidation, he believed, were entirely misguided. Hutton did not doubt that water can infiltrate mineral bodies and fill their interstices with deposits of various kinds, but he stressed that heat is then required also. It was also a mistake to regard consolidation as a form of crystallization. A monograph of 1783 by Jan Filip Carosi, who discarded infiltration but failed to appreciate the necessity of heat and fusion, exemplified the inconclusive reasoning that prevailed in mineralogical writings on the subject. As Hutton himself remarked in a note, "I do not pretend that we understand mineral fusion, but only that such mineral fusion is a thing demonstrable upon a thousand occasions, and that thus is to be explained the petrification and consolidation of the porous and naturally incoherent strata of the earth" (517). In a further section, once more responding to Carosi, Hutton affirmed the process of flintification, which had aroused his interest so early as 1754.

"Petrifaction," he feared, "is a subject in which mineralogists have perhaps wandered more widely from the truth than in any other part of natural history." For Hutton, the reason was plain.

> The mineral operations of nature lie in a part of the globe which is necessarily inaccessible to man, and where the powers of nature act under very different conditions from those which we find take place in the only situation where we can live. Naturalists, therefore, finding in stalactical incrustation a cause for the formation of stone in many respects analogous to what is found in the strata of the earth, and which had come from the mineral region in a consolidated state, have, without due consideration, attributed to this cause all the appearances of petrification or mineral concretion (527).

It was one of Hutton's major objectives to prove them wrong. Once he has done so, only the fusion theory remains. "This," Hutton concluded, "has been the doctrine which I have held out in my 'Theory of the Earth,' and this will be more and more confirmed as we come to examine particular mineral appearances" (557).

The eighth and last chapter of volume (and part) I dealt with a body of evidence not previously utilized; called "The Nature of Mineral Coal, and the Formation of Bituminous Strata, Investigated," it was divided into three sections, the first of which explained Hutton's purpose. In his "Theory" of 1788, Hutton believed, he had established consolidation by fusion beyond any doubt. As he noted ruefully, however, few mineralogists had agreed. The truth of fusion, Hutton maintained, can be seen in *any* mineral body; coal shows us the impossibility of the aqueous alternative. Now, bituminous strata are a worldwide occurrence, and they usually alternate with other strata obviously formed at the bottom of the sea. Theorists disputed, however, what happened thereafter. According to Hutton's opponents, the bituminous matter infiltrated existing strata, transforming them into coal. But as Hutton remarked indignantly, "It was thus that natural philosophers reasoned before the age of science; the wonder now is how men of science, in the present enlightened age, should suffer such language of ignorance and credulity to pass uncensured" (563). The close association of pyrites with coal, among other arguments, proved consolidation by fusion.

Section two of chapter viii elaborated Hutton's views on the origin of coal. According to Hutton, dead plants formed combustible turf, called peat. Streams and rivers then carried this peat to the ocean, where it was precipitated into strata and distilled by subterranean heat. Less pure forms of coal resulted from admixtures, and we have only to examine particular specimens to confirm the reasonability of these suppositions. Hutton then located what coal there was throughout Scotland; why, he asked, was it found only in the flat country? The coal regions of Scotland, he realized, were those that had been invaded from below by masses of basalt. Thus, he could explain both the disordered, broken appearance of coal strata and the particular hardness of the best coal, which was immediately adjacent to the basalt and has been most thoroughly hardened by it.

Hutton's theory of fossil coal appeared in section three as an illustration of mineralogical processes. In both his *Dissertations upon Different Subjects in Natural Philosophy* (1792) and his *Dissertation upon the Philosophy of Light, Heat, and Fire* (1794), he had attempted to distinguish inflammable bodies from combustible ones. All animal and vegetable substances, for him, include both qualities, so it is only through distillation by heat that coal becomes solely combustible. Because such coals do exist in nature, they furnish "one of the absolute proofs of the igneous theory" (613). Certain coal strata, moreover, afforded "the clearest proof of the efficacy of compression" (614). With this seemingly definitive conclusion, Hutton arrived at the end of his lengthy *Theory of the Earth*, volume I. He then devoted the second volume to a masterful discussion of the evolution of landforms, in effect founding the science of geomorphology (Dean 1989).

VOLUME II (PART II): FARTHER INDUCTION OF FACTS AND OBSERVATIONS RESPECTING THE GEOLOGICAL PART OF THE THEORY

In Hutton's theory, according to his three-page "Introduction" (for which there is no equivalent in volume I), earth's present continents were formed originally as horizontal strata at the bottom of the ocean and then elevated. On emergence, they must have looked very different from today's. Rivers, for example, had yet to excavate their own channels. All in all, three causes have determined the form of present land masses: the regular stratification of materials; operations of the mineral region; and operations proper to the surface of our globe, including the influence of the sun and atmosphere, wind and water, and rivers and tides. Hutton explained the effects of these three interacting causes in volume II.

Chapters i and ii: Facts in Confirmation of the Theory of Elevating Land above the Surface of the Sea. In mountainous countries, strata once plain and horizontal have been changed in posture and shape—regularly bent and broken by the same force that had elevated them. Whether primitive or secondary, mountains of various materials show the same essential characteristics. Thus, as lengthy quotations from Saussure established, the Alps and the Jura mountains are basically alike with regard to the position of their strata. Though Saussure had not been able to derive a satisfactory theory of the earth from his own observations, Hutton once more confirmed, he described geological phenomena impartially and in a wholly praiseworthy manner. In chapter ii, Hutton added quotations from Deluc and other sources as well.

Chapters iii and iv: Facts in Confirmation of the Theory Respecting Those Operations Which Redissolve the Surface of the Earth. Once strata have been elevated, they are immediately assaulted by natural forces intended to decompose them into soil. The chief agents through which this decay is effected include the sun, the atmosphere, water, and an immense period of time. (Saussure had sought to explain many of the same phenomena through a brief, catastrophic debacle. Yet quotations from him helped to establish the efficacy of rivers.) In the Alps, enormous masses of solid material have been wasted by the slow effects of air and water, sun and frost—even glaciers—hollowing out barren valleys of immense extent and transporting vast quantities of soil. "Now, whatever may be our theory with regard to the origin or formation of these solid masses of the globe," Hutton asserted at the end of chapter iii, "this must be concluded for certain: that what we see remaining is but a specimen of what had been removed, and that we actually see the operations by which that great work had been performed. We need only to join in our imagination that portion of time which, upon the surest principles, we are forced to acknowledge in this view of present things" (*Theory*, II: 116–117).

Chapter iv, "giving still farther views of the dissolution of the earth," added quoted descriptions of the St. Gothard Pass, the Pyrenees, and elsewhere in the Alps. In a footnote, Hutton denied that gravels had been rounded by rivers (they were shaped by waves on some former coast). He also speculated about the formation of lakes (Lake Geneva, in particular), called attention to the vast amount of substance removed from the surface of the earth over a vast period of time, and then concluded eloquently:

> Whether we examine the mountain or the plain; whether we consider the degradation of the rocks, or the softer strata of the earth; whether we contemplate nature, and the operations of time, upon the shores of the sea

> or in the middle of the continent, in fertile countries or in barren deserts, we shall find the evidence of a general dissolution on the surface of the earth, and of decay among the hard and solid bodies of the globe. And we shall be convinced, by a careful examination, that there is a gradual destruction of everything which comes to the view of man, and of everything that might serve as a resting place for animals above the surface of the sea (157).

No geological theorist was ever more insistent as to the ubiquity of erosion.

Chapter v: Facts in Confirmation of the Theory Respecting the Operations of the Earth Employed in Forming Soil for Plants. Insofar as gravel and stones are worn by attrition, Hutton argued, the farther they travel, the more nearly spherical they will be. Around London, however, we find vast quantities of flint gravel—in the very center of the chalk country. This gravel must therefore have been formed under water, requiring immense destruction of the chalk. (Some of the flint gravel later reconsolidated as puddingstone.) The chalk regions of France, Flanders, and especially England suggest that the sea once covered the land. Having made a particular study of gravels, Hutton could identify at least ten different British types, all of which must have traveled. In a similar problem of larger scale, Saussure described granite boulders that traveled considerable distances from their origin in the strata of Mont Blanc; he attributed their transportation to a vast debacle or flood. Hutton, on the other hand, accounted for them through the carrying power of water and ice.

Chapters vi and vii: A View of the Oeconomy of Nature, and Necessity of Wasting the Surface of the Earth in Serving the Purposes of This World. For Hutton, it was important to establish an apparent paradox—that the land is naturally, but economically, wasted. (The word "economy," which he spelled either with or without an initial o, emphasized wise, proper, and efficient management. "Waste" means erosion.) After reviewing the fundamentals of his theory, Hutton opposed one of Deluc's best known ideas—that erosion will stop once a desired contour or form has been achieved. Nature, he stressed once again, is a "wise system . . . in which nothing is done in vain, and in which everything tends to accomplish the end with the greatest marks of economy and benevolence" (203n). Hutton then demonstrated this system with astonishing and unprecedented brilliance by chronicling the progressive changes made in a fertile plain by an encroaching river that eventually destroys it (205–211).

This grandeur of vision continued into chapter vii, where Hutton again contemplated drifted Alpine boulders while imagining stop-camera changes in a landscape through time. Once beyond some quotations from Deluc on the Alps, Hutton also theorized more specifically than ever on the transporting power of glaciers, "immense valleys of ice sliding down in all directions towards the lower country, and carrying large blocks of granite to a great distance. . . . Such are the great blocks of granite which now repose upon the hills of Saleve" (218). Deluc's belief that erosion will stop with the perfection of form, moreover, was senseless because such a system would endanger the fertility of an earth designed to support life. "What a reverie, therefore, is that idea of bringing the earth to perfection by fixing the state of its vegetable surface!" (220). The system of nature was not to be found in the surcease of natural powers but in their proper balance.

Chapter viii: The Present Form of the Surface of the Earth Explained, with a View of the Operations of Time upon Our Land. Having in his previous chapter opposed Deluc's erroneous view of an eventual stability, Hutton next further illustrated his own theory of perpetual decay. Pebbles and gravels, he observed, amaze us by their sheer quantity. Similarly, we find large accumulations of kaolin, which derive from the decay of granite. Leached mineral veins produce extensive placer deposits (as of gold). All three examples attested to the power and scope of erosion. Turning from the manufacture of soil to the solid rock from which it comes, Hutton recalled the wearing of coastal rocks by waves. He then compared the high, rugged west coast of Britain with the lower, less stark east coast. The chalk cliffs of England and France seemed to be wearing away at the rate of one foot per year. Since they were obviously once joined, it might even be possible to determine how long ago that was. (Hutton did not further pursue the calculation, but if one takes the Channel to be somewhat more than thirty miles wide—and weathering on both sides—the answer would be 80,000 years.) Marine erosion of the land is naturally progressive; an encroachment becomes first an indentation, then a peninsula, islands, and finally barren rocks. "We do not see the beginning and ending of any one island or piece of country," Hutton emphasized, "because the operation is only accomplished in the course of time, and the experience of man is only in the present moment. But man has science and reason, in order to understand what has already been from what appears, and we have but to open our eyes to see all the stages of the operation, although not in one individual object" (267). Amid some further examples (from Italy and Ireland), Hutton cautioned that "Our present inquiry is only with regard to the operation of those causes which we now perceive to be acting upon the coasts of the land, which must be considered as having been operating for a long time back, and which must be considered as continuing to operate" (275–

276). Putting together land masses now separate, Hutton perceived that Ireland had formerly been joined with Britain, the Orkneys with Scotland, the Shetlands with Norway, and (as noted earlier) England with France.

Chapter ix: The Theory Illustrated with a View of the Summits of the Alps. Land may be diminished in height as well as breadth. Having considered coastal encroachments by the sea in his previous chapter, Hutton next examined the diminution of mountains. The strata of the earth, he recalled, are not in their originally horizontal position. Though we cannot know their history fully, it is obvious that an immense amount of sediment has been removed from atop what remains and that an immense period of time has been required to bring about all these alterations, which "are not done by violent changes, but by slow degrees" (289). He could not imagine a better example than the Alps. On whatever side one approached them, some great river could be found discharging both waters and sediments derived from those lofty masses of decaying rock. Such rivers generally run through valleys proportioned to them, so Hutton digressed meaningfully to affirm that the valley was made by the river—in fact, by a system of rivers. Yet, above these river systems—so efficacious in removing solid debris—Hutton detected others connected with them. These are "valleys of moving ice, instead of water," and they move pondrously slow, but with extreme power (296).

Alpine summits often include pyramids and needles of granite that were surely detached from larger masses (as with coastal islands and rocks) by slow-acting forces. "It is true indeed," Hutton observed, "that geologists everywhere imagine to themselves great events or powerful causes by which these changes of the earth should be brought about in a short space of time. But they are under a double deception—first, with regard to time, which is [un]limited . . . [and,] secondly, with regard to operation, their supposition of a great debacle [being] altogether incompetent for the end required. . . . But this is only one of a thousand appearances that prove the operations of time and refute the hypothesis of violent causes" (328–329).[8] Within a human lifetime, he reiterated, we can see only a small part of the larger operation. And what else is required? Nothing but time.

Chapter x: The Theory Illustrated with a View of the Valleys of the Alps. As the mountain degrades, Hutton continued, valleys are formed along its slopes: valleys of ice and valleys of water. Once again, Hutton empha-

[8]The necessary restoration "[un]limited" was first pointed out by A. C. Ramsay in 1846. Hutton had straightforwardly affirmed "infinite space and endless time" in his *Principles of Knowledge* (1794, 3: 645).

ized how much detritus is removed by mountain rivers and over how vast a time. After lengthy examples from the Alps, he unexpectedly quoted Jefferson on the Natural Bridge of Virginia. In both the Old World and the New, Hutton concluded, water and other natural forces incessantly change the appearance of the earth without defacing it. "At all times," he insisted, "there is a terraqueous globe for the use of plants and animals; at all times, there are upon the surface of the earth dry and and moving water, although the particular shape and situation of hose things fluctuate and are not permanent, as are the laws of nature" (378–379).

Chapter xi: Facts and Opinions Concerning the Natural Construction of Mountains and Valleys. Once again Hutton returned to the Alps, attempting to determine whether the great massif had been formed as it is or was shaped by the power of running water. He agreed with Saussure that a mountain's structure influences the placement of its valleys but again insisted that the valleys themselves are then excavated by running water. The development of valleys, moreover, changes the shape of the mountain (which is also influenced by the kind of rock from which it was made). It was likewise important to distinguish associated mountains from isolated ones. Associated mountains, like the Alps, are formed by attrition and tend to resemble one another. Isolated mountains, however, are usually volcanoes and grow upward from a plain. Having never seen a volcano, he described other mountains that were eruptive but not volcanic. Common in the Lowlands of Scotland, they were generally of basalt, Arthur's Seat being a particularly good example. As for the Alps, Hutton quoted further from Saussure, establishing that strata exposed on either side of a river valley corresponded and must once have been contiguous.

Chapters xii and xiii: The Theory Illustrated by Adducing Examples from the Different Quarters of the Globe. Hutton expected his theory to be valid worldwide. "The system which we investigate is universal on this earth," he postulated; "it hangs upon the growth of plants and life of animals." Though certain plants and animals might be found in one climate rather than another, we cannot have one geology for Europe and another for India. Thus, "the operation of a central fire, in making solid land on which the breathing animals are placed, and the influences of the atmosphere in making of that solid land loose soil for the service of the vegetable system, are parts in the economy of this world which must be everywhere distinguishable [i.e., evident]" (433). His next examples, all of them lengthy quotations, invoked Sumatra, Peru, Spain, Columbia, France, and Italy.

Chapter xiii is particularly important because it has given rise to several mistaken interpretations of Hutton's position on catastrophic geological changes (which, despite a few concessions, he steadfastly opposed). I cite the relevant passage in full:

> Naturalists who have examined the various parts of the earth almost all agree in this, that great effects have been produced by water moving upon the surface of the earth. But they often differ with respect to the cause of that motion and also as to the time or manner in which the effect is brought about. Some suppose great catastrophes to have occasioned sudden changes upon the surface, in having removed immense quantities of the solid body, and in having deposited parts of the removed mass at great distances from their original beds. Others again, in acknowledging the natural operations which we see upon the surface of the earth, have only supposed certain occasions in which the consequences of those natural operations have been extremely violent, in order to explain to themselves appearances which they know not how to reconcile with the ordinary effects of those destructive causes.
>
> The theory of the earth which I would here illustrate is founded upon the greatest catastrophes which can happen to the earth—that is, in being raised from the bottom of the sea and elevated to the summits of a continent, and in being again sunk from its elevated station to be buried under that mass of water from whence it had originally come. But the changes which we are now investigating have no farther relation to those great catastrophes, except insofar as these great operations of the globe have put the solid land in such a situation as to be affected by the atmospheric influences and operations of the surface (444–446).

Utilizing an ironic redefinition of the word, Hutton took "catastrophe" to mean the greatest imaginable change in situation, a requirement that going from sea bottom to mountain top met precisely. As he had so often stressed, however, such change was not the brief, extremely violent occurrence other theorists had imagined but an immensely slow and entirely orderly process.

Chapter xiii—and Hutton's discussion of landforms—ends with a celebration of rivers and their central role in preserving the habitable earth.

Chapter xiv: Summary of the Doctrine Which Has Been Now Illustrated. Hutton summarized his work by enumerating seven important points any theory of the earth ought to explain: that the greater portion of the earth has been formed of transported materials collected in the sea; that these loose, incoherent materials have been consolidated; that degrees of consolidation differ; that consolidated strata are always fractured and veined; that the fractures and veins are filled with some other mineral material; that the strata of the earth appear to us as they do—"broken, bended, and inclined" (544); and that some powerful force elevates the strata into place.

But even explaining these observations was not enough. "We live in a world where order everywhere prevails and where final causes are as well known, at least, as those which are efficient. . . . Thus, the circulation of the blood is the efficient cause of life, but life is the final cause, not only for the circulation of the blood but for the revolution of the globe" (545–546). Appropriately, then, Hutton had laboriously completed a coherent intellectual endeavor begun almost half a century earlier with his medical dissertation (on the circulation of the blood) in 1749.

Long as it is, Hutton's *Theory* as published in 1795 still represented only parts 1 and 2 of a projected four-part work. In volume III he hoped to "examine facts, with regard to the mineralogical part of the theory, from which, perhaps, light may be thrown upon the subject, and to endeavor to answer objections, or solve difficulties, which may naturally occur from the consideration of particular appearances" (II: 567). We know that he prepared at least nine chapters (three of them now lost) toward that end. The fourth and final part, by all indications, would have been one more (and even fuller) justification of his theory and God's universe through an extended discussion of final causes.

Theory of the Earth Criticized

On the whole, Hutton's two-volume version of his geological theory attracted less attention than did the earlier version of 1788, in part because Hutton was known to be seriously ill but also because relatively few previously unconsidered issues were involved. Nevertheless, substantial notices did appear, the first of them coming from an old adversary. Jean André Deluc's anonymous review of Hutton in the *British Critic* for October 1796 was the main article, covering sixteen pages. But this was only the first of three parts, the second adding twenty pages and the third eight, bringing the total to forty-four, an unusual length. Deluc began with an epigraph from Pope's Homer: "The fame of one shall proudly reach the skies; / The next, by cold neglect, forgotten dies." In view of Deluc's subsequent reputation, we easily detect an irony that he did not. Yet this self-assured critic undoubtedly believed that his remarks would demolish Hutton for all time.

Deluc's summary of his adversary's theory reduced it to seven propositions:

1. Our continents are composed of strata, which have been formed in the sea.
2. These strata have been produced by the accumulation of substance proceeding from other continents, which, by the action of the atmosphere and the streams of rain water, have been gradually demolished; the materials of these continents were similar to those we observe on our shores.
3. At the same time that the materials of decaying continents are thus carried to their shores, they are there taken by waves, tides, and currents and spread over the whole bed of the sea.
4. Under the water of the ocean reigns an excessive heat, by which the loose materials successively arriving from the shores are melted and changed into a new stony strata.
5. By the time a set of continents is nearly worn out on our globe, the materials proceeding from another set (which, long before, have been delivered into the sea) are consolidated into stony strata and then the same heat that thus prepared them for new continents elevates them in the place of land.
6. This alternate operation, of continents disappearing by being wasted and new continents appearing above the level of the ocean, has already been innumerable times repeated on our globe, at intervals of millions of ages.
7. Our continents are the latest of that series of operations alternately producing sea and land in the same parts of the globe. These continents are in a state of decay; their materials are successively spread, first over the lower parts of the lands, for the purpose of a soil, then over the bed of the ocean, there to be melted and reduced again into stony strata for new continents to come. And that wasting operation has already lasted millions of ages (1796, 338).

Deluc remarked that this summary would undoubtedly appear to some readers "like the outlines of an oriental tale" (i.e., extremely fanciful); its author, however, "seems to be in earnest." The consequences of Hutton's theory were so serious that Deluc believed it his duty to provide a full analysis, which he offered point by point, quoting frequently (and accurately) from Hutton.

Proposition 1 being substantially conceded, Deluc alloted no fewer than twelve pages to the second, which he called "the fundamental proposition of the system." Throughout an eleborate opposition, he often quoted Kirwan's objections to Hutton and Hutton's responses to them, invariably supporting Kirwan. Rocks such as conglomerate and breccia, for example, proved that they were formed from strata broken up and then recemented by the sea itself; Hutton had also failed to consider many relevant phenomena, especially fossils. In his analysis of proposition 3, Deluc doubted that sediments from the shore could be as widely dispersed beneath the sea as Hutton necessarily believed. Temporarily concluding, he promised continued attentiveness to Hutton, for "Visionary theories of the earth are never without their reference to matters of a higher import" (352).

Deluc's second installment analysed propositions 4, 5, and 6, the latter at length. He rejected Hutton's views on subterranean heat, fusion, and consolidation and declared his connection of eroding continents with forming ones tenuous. If the heat that raises continents escaped through volcanic activity, furthermore, would not the continents then submerge back to their former elevation? The theorist himself, Deluc perceived, sensed a difficulty at this point, as he was honest enough to admit. Most important, Hutton had postulated "a succession of worlds" and found "no vestige of a beginning, no prospect of an end." But his theory depended on an analogy between past and present. Where in the present did one see this succession of worlds? "Only in the imagination of this author, who tries to conceal, under the name of the wisdom of nature, what he had himself devised against the Mosaic account of the earth" (474). While never arguing that Hutton was an atheist, Deluc did accuse him of failing to counter atheism sufficiently. He then went on to an analysis of Genesis, "the only source from which man *could* receive a knowledge of the origin of things" (477, emphasis mine), and with which his own geological theory accorded. Like many other writers of his time, Deluc was amused by the contemptible pretensions of those who would prefer their own wisdom to the revealed word of God.

Deluc's final installment was totally devoted to proposition 7, the immense age of the earth, an idea he thought common to all theories that disregarded or opposed the Mosaic history. Hutton's concern with final causes and the wisdom of design was proper, but his arguments were wrong. Deluc's, on the other hand, were based on the formation of soil and the proximity of beds of sea shells to the surface of the continents. Analyses of landforms, moreover, seemingly supported catastrophism and therefore the Mosaic chronology. This lengthy review, representative in many ways of the cumulative opposition Hutton was subjected to, ended with the admonition that geology must never dissociate itself from sacred history.

Beginning on 19 November 1796, Richard Kirwan read to the Royal Irish Academy a series of three essays collectively entitled "On the Primitive State of the Globe and Its Subsequent Catastrophe." "In the investigation of past facts dependent on natural causes," he began plausibly, "certain laws of reasoning should inviolably be adhered to. The first is that no effect shall be attributed to a cause whose *known* powers are inadequate to its production. The second is that no causes should be adduced whose existence is not proved either by actual experience or approved testimony. . . . The third is that no power should be ascribed to an alleged cause but those it is known by actual observation to possess" (1797, 233). He regretted that some recent speculations about the history of the globe had tended toward atheism or infidelity, for if properly pursued "geology naturally ripens, or (to use a mineralogical expression) *graduates* into religion" (234).

In his first essay, "On the Primeval State of the Globe," Kirwan suggested that the globe must originally have been in a soft or liquid state. This liquidity must have been caused either by igneous fusion or solution in water, the latter being far more probable, as Kirwan had shown in 1793. The present materials of the earth must have existed in the original chaotic fluid, proportioned generally as they are now. The late Dr. Hutton, however, had excluded limestone from the list of primeval rocks, believing it formed by shellfish. He also had supposed the present earth to have arisen from the ruins of a former one—but that could lead only to an infinite regression. Granite and gneiss, claimed Kirwan, had formed first, followed by other primitive rocks. Deposited on the interior nucleus of the earth, these had formed the primitive mountains. As the chaotic fluid diminished and recessed (aided in part by volcanoes), the continents gradually emerged. The retreat of the sea continued for several ages, during which the stratified secondary mountains were formed. (Many phenomena once attributed to a universal deluge, he tells us, actually originated in this way.) Hutton to the contrary, stratified hills have always been found reposing on primary rocks or surrounding primary mountains. The succession of geological events detected in the earth by modern observers was, for Kirwan, precisely that recorded by Moses in Genesis.

Since marine life did not exist when the primitive mountains first emerged, fossil shells found on them must have been deposited by a later inundation. Similarly, the bones of elephants and other tropical animals had been found in Siberia, to which they must have been transported by a deluge. After reviewing and dismissing several deluge theories (including Deluc's) in his second essay, Kirwan argued that the deluge of which Noah had written began in the Southern Hemisphere and rushed northward. The British Isles had probably not become insular at that time, but basaltic masses on the Scottish and Irish coasts were shattered into pillars.

Such a violent and universal shock as the diluvian revolution must have effected innumerable changes in the surface of the globe, not only at the time but for centuries thereafter, as Kirwan's third essay explained. Among these were the separation of Asia from America, the

Caspian Sea from the Black Sea, Sicily from Italy, Ireland from England, and Britain from the Continent. The English Channel had probably been created by an earthquake, then gradually widened by tides and currents. Though they had sometimes been mistakenly regarded as evidence for an extended age of the earth, volcanoes such as Vesuvius and Etna were actually recent phenomena.

The first book-length source to mention Hutton's fuller theory was Philip Howard's *Scriptural History of the Earth and of Mankind* (1797), another defense of Genesis. Objecting to Hutton's infinite chronology and unending processes, Howard could not conceive of an earth devoid of man or of geology independent of the Noachian deluge. He also attacked Hutton's views on subterranean fire, fusion, and erosion. For him, the eventual effect of geological forces would be to reestablish the antediluvian configuration of the continents. *Monthly Review* praised Howard's book, noting his remarks on Hutton.[9]

The battle to win acceptance for Hutton's theory had, therefore, still only just begun.

[9]On Howard, see *Monthly Review* 25 (1798): 241–254. For early foreign criticism of the Huttonians (complained of by Playfair), see Carozzi 1990, 113–119.

[4]

Hall, Werner, and Jameson

Throughout Hutton's lifetime, geological questions were often closely associated with chemical ones. We know, for example, that Joseph Black's investigation of lime as affected by heat influenced Hutton fundamentally. Black also experimented with the fusibility of basalt. During the earlier stages of his theorizing, Hutton derived important stimuli from Cronstedt's chemical analyses of minerals, which led to his own analyses of zeolite and geyserite.[1]

Several eighteenth-century experimenters investigated the nature of crystallization. Being primarily interested in the manufacture of porcelain, they dealt particularly with what the Chinese—or so Jesuit missionaries reported—called petuntse (feldspar) and kaolin. Early in the century Réne de Réaumur studied examples of both substances sent him from China. Because petuntse melted easily whereas kaolin did not, he saw correctly that porcelain consisted essentially of a fine-grained infusible earth mixed with a glass. Largely through his work, the chief characteristics of both petuntse and kaolin became well known. By 1768 Black and Watt were experimenting with these and related substances, often melting them in crucibles. "Petuntse" also became a common mineralogical term. Hutton, for example, named the substance repeatedly in his manuscripts of the latter 1780s (which became *Theory*, vol. III [1899])—and was obviously aware of experiments regarding its fusibility. John Williams, in his *Natural History of the Mineral Kingdom* (1789, 2: 15) also mentioned petuntse. Once basic techniques for fusion experiments had

[1]For Black and Hutton, see Chapter 1 above and Appendix 1.

been developed, it was inevitable that further substances would be investigated. Thus, in his *Lithogeognosia* of 1746, J. H. Pott reported some ten thousand experiments of his own on the fusion of various minerals. Twenty years later, Jean Darcet described a large number of similar experiments. Among these, he had reduced vitrified Auvergne basalt and a variety of lavas, all of which fused easily into glasses. But because the amount of heat necessary to fire porcelain was much greater, Darcet hesitated to equate conditions in his laboratory with those apparently less powerful ones existing in nature.[2]

Like Darcet, other eighteenth-century figures were also applying chemical techniques to geological problems. As part of his advocacy of the igneous origin of basalt, for instance, Desmarest (1774) attempted to melt specimens of granite. After succeeding in part, he became convinced that granite and basalt were similarly composed. (Though not chemical, observations by John Strange in Italy, 1775, also suggested a common origin for the two rocks.) In May 1776 James Keir recounted some experiments on the crystallization of glass to the Royal Society of London. "Does not this discovery," he asked, "of a property in glass to crystallize, reflect a high degree of probability on the opinion that the great native crystals of *basaltes*, such as those which form the Giant's Causeway or the pillars of Staffa, have been produced by the crystallization of a vitreous *lava*, rendered fluid by the fire of volcanoes?" (1776, 539). Keir and his associates obviously thought so. Three years later, Saussure reported a series of personal experiments made to determine whether certain kinds of rocks could have derived from the melting of others. Neither granites nor porphyries, he discovered, could be reduced by fusion to anything resembling basalt. Through a letter of his to Josiah Wedgwood, we know that Joseph Priestley was conducting comparative experiments on the fusion of basalt and lava in 1781. Others were regularly reducing various kinds of granite to black glass. Even the accidental discovery that the baked clay sometimes used in England to repair roads regularly broke into hexagonal prisms resembling those of basalt seemed important enough to be called specifically to Hutton's attention. On returning from an extensive tour of Italian volcanoes in 1788, Lazzaro Spallanzani subjected the specimens of lava and other volcanic rocks he had collected to experiments within the glass furnace. "In my volcanic travels," he wrote, "I have been obliged to take upon myself the parts both of naturalist and chemist. The natural history of fossils is so closely connected with modern chemistry, and the rapid and

[2]Smith 1969 and Geikie 1905 (on Desmarest and Saussure) review early experimentation. Kaolin, a fine white clay, is basically a hydrous aluminum silicate; petuntse is of similar composition and likewise derives from decomposed granitic feldspars. For Black and Watt, see Robinson and McKie 1970, 8–13 (mentioning Pott).

prodigious progress of the one so exactly keeps pace with that of the other, that we cannot separate them without great injury to both" (1798, 1: xxv). Experimental geology, then, became an international, well-developed, and productive field of inquiry during the 1770s, with sustained development thereafter.[3]

Hall and Others on Fusion

Sir James Hall (1761–1832), perhaps still properly regarded as the "father of experimental geology" despite that field's preceding history, was the son and heir of Sir John Hall of Dunglass, whom Hutton had known since the 1750s. At age fifteen James succeeded to his father's title, baronetcy, and fortune. Continuing an education begun at Cambridge, Paris, and Geneva, he entered Edinburgh University in 1781, and took courses from Black, Walker, and others. Having developed interests in both geology and chemistry by then, Hall devoted three years (1783–86) to a grand tour of Europe, during which (though in absentia) he was elected a fellow of the Royal Society of Edinburgh (1784). In 1785 Hall investigated the major volcanoes of Italy, including Vesuvius (in mild eruption), Etna, and Stromboli. Returning through Paris, he studied chemistry there for several months under the aegis of Lavoisier. Once back in Edinburgh, Hall came to know Hutton, whose old-fashioned advocacy of phlogiston he opposed.[4]

Initially, Hall also rejected Hutton's theory of the earth. After numerous conversations with Hutton regarding it, however, he changed his mind, the two men becoming good friends thereafter. In June 1788, just before discovering the famous unconformity at Siccar Point, Hutton and Playfair visited Hall at Dunglass. As part of the visit, Hutton probably described for Hall the intrusive granitic veins he had found two years before at Galloway. Hall then went to observe the same phenomena for himself. After finding the junction of granite and schistus Hutton had advised him of, Hall followed it from the banks of Loch Ken to the valley of Palnure. Wherever their junction was visible, Hall found veins running from the granite into the schistus. As he then reported to the Royal Society of Edinburgh in January and March 1790, in a paper called "Observations on the Formation of Granite" (Hall 1794), this example

[3]Priestley to Wedgwood, 8 August 1781, in Smith 1969, 329. *TRSE* 2 (1790): 22–23.

[4]Hall was among the first persons in Britain to endorse the new oxygenic (as opposed to phlogistic) chemistry of Lavoisier. On 4 February, 3 March, and 7 April 1788, the Royal Society of Edinburgh heard him read the three parts of his "A View of M. Lavoisier's New Theory of Chemistry." On 5 May, Hutton read a paper on phlogiston in reply. There was then an extraordinary (i.e., specially scheduled) meeting on 12 May, at which Hutton read some further observations on phlogiston and Hall read another paper in reply to Hutton's first. None of these papers, or abstracts of them, was ever published (*TRSE* 2 (1) [1790]: 26–27). Remarkably, this major difference of opinion was conducted on both sides without enduring acrimony.

(like many others he had observed in Scotland and Italy) firmly established the truth of the Huttonian theory.

Nonetheless, some difficulties still remained. In granite, for instance (a rock composed of quartz, feldspar, and mica), feldspar is usually well crystallized whereas quartz is massive and irregular, having formed subsequent to the feldspar. If all granites were formed by fusion, the very opposite would seem more likely, as feldspar melts easily whereas quartz does not. Hall attempted to obviate this difficulty by supposing the feldspar a flux, with the two minerals thereafter crystallizing simultaneously. Another problem was that previous experiments on granite had shown its fusion to result in an amorphous glass. At one of the Leith glass factories a few weeks earlier, however, Hall had seen a glass that crystallized. Were fused granite allowed to cool slowly enough, he supposed, it might well crystallize also, thereby producing a rock similar to the original. Other unstratified rocks might be similarly explained. But in deference to Hutton, whose fuller statement of his theory was forthcoming, Hall declined to publish this suggestive paper in full.

Thomas Beddoes's "Observations on the Affinity between Basaltes and Granite," read before the Royal Society of London on 27 January 1791, was then highly relevant but only marginally chemical. Despite a recent effort by Abraham Gottlob Werner to the contrary (in 1787), Beddoes believed the igneous origin of basalt to have been thoroughly established by a host of authorities, including Hutton. He also believed granite and basalt to be naturally connected, even changing into one another—with varieties of porphyry in between. "A chemical examination of the basis of a number of these porphyries would be very interesting," he thought, "yet I would not rest the theory of their formation altogether on the result of analysis" (1791, 52). Since rocks were not chemically uniform throughout, "The sensible qualities, the style of fissure, the accompanying fossils, and the form of whole rocks, when surveyed by an experienced eye, are as good criterions of basaltes as a certain proportion of iron, and the black glass which it yields on fusion" (53). Field evidence affirmed that granite and basalt have the same origin; if basalt derives from fusion, therefore, granite must also. The same menstruum yields different rocks because of differences in cooling time, a fact easily substantiated by the manufacture of porcelain and iron. Granite and basalt are so often contiguous and intermeshed, moreover, that they must have formed at the same time. A further extension of this insight implied that the common division of mountains into primary and secondary was erroneous, as only Hutton had foreseen. Thus, "The chains of granite, schistus, and limestone must be all coeval, for if the central chain of the Alps burst as a body expanded by heat from the bowels of the earth, it reared the bordering chains at the same effort" (68). Beddoes's concept of mountain building was clearly more catastrophic than Hutton's.

In his "Examination of the Supposed Igneous Origin of Stony Substances" (1793), Richard Kirwan took issue with Beddoes as well as with Hutton. After citing the work of several chemists to prove that water *can* dissolve siliceous earths, Kirwan dismissed Hutton's insistence on the igneous origin of granite as a "peculiarly unhappy" aspect of his theory. Though Beddoes had supposed changes of texture dependent on time of cooling (as in the manufacture of porcelain), Kirwan produced alternative data to support his contention that intensity of heat was far more important. Glass resulting from the additional fusion of granite differed so much from any form of basalt as to prove beyond question that the two rocks must be entirely distinct. In any case, granite was clearly aqueous in origin.

The next comments pertinent to this controversy were Hutton's own, in his *Theory* of 1795. Early in chapter i (his "Theory" of 1788 reprinted), Hutton added a long note to oppose Dolomieu's erroneous but "ingenious theory for the solution of siliceous substances in water" (I: 54n). A second important note (94–96) then emphasized the significance of compression. In referring once again to Dolomieu, Hutton defended his theory against all chemical evidence thus far: "Nature applies heat under circumstances which we are not able to imitate; that is, under such compression as shall prevent the decomposition of the constituent substances, by the separation of the more volatile from the more fixed parts. This is a circumstance which, so far as I know, no chemist or naturalist has hitherto considered, and it is that by which the operations of the mineral regions must certainly be explained" (140n). Negative evidence from the laboratory, then, was seemingly irrelevant.

Hutton's chapter ii responded directly to Kirwan's paper of 1793. "In the 'Theory of the Earth' which was published," he noted testily, "I was anxious to warn the reader against the notion that subterraneous heat and fusion could be compared with that which we induce by our chemical operations on mineral substances here upon the surface of the earth; yet, notwithstanding all the precaution I had taken, our author [Kirwan 1793, 66–70] has bestowed four quarto pages in proving to me that our fires have an effect upon mineral substances different from that of the subterraneous power which I would employ" (251). As he would emphasize repeatedly, Hutton had only scorn for those who "judge of the great operations of the mineral kingdom from having kindled a fire, and looked into the bottom of a little crucible" (251).

Hall's "Experiments on Whinstone and Lava"

A year after Hutton's death, Hall revived the issue of fusion, this time in a paper entitled "Experiments on Whinstone and Lava," read before the

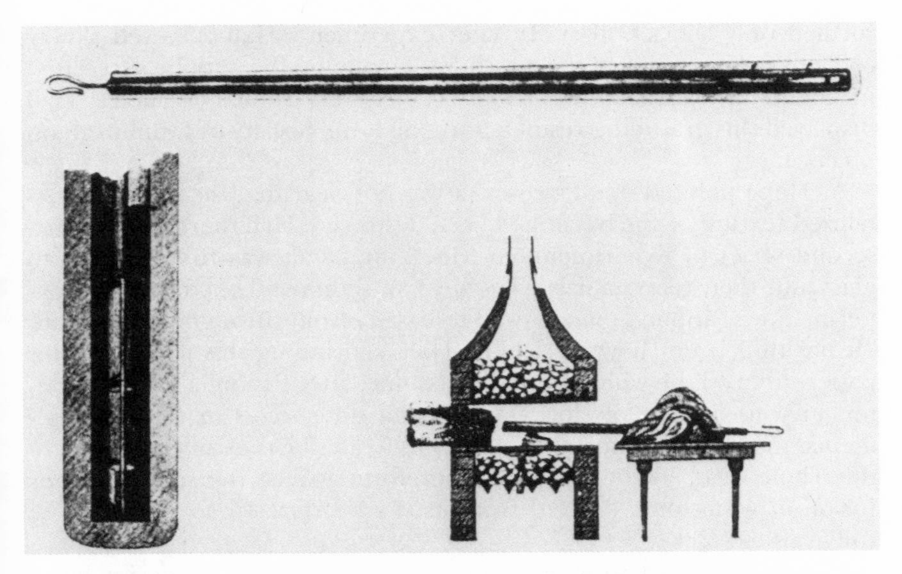

12. Apparatus used by Hall in some of his earliest experiments to determine the effects of heat and compression. From *Transactions of the Royal Society of Edinburgh*, 1805.

Royal Society of Edinburgh on 5 March and 18 June 1798.[5] "The experiments described in this paper," he began, "were suggested to me many years ago, when employed in studying the geological system of the late Dr. Hutton" (1805, 43). Though Hutton had supposed granite, porphyry, and basalt to have formed through fusion, their actual structure in the rock is very different from that of the glassy fusions achieved in the laboratory. Hutton himself explained the differences by emphasizing compression. In his paper of 1790, however, Hall had thought slow cooling a more adequate explanation. After giving it, he had begun to experiment along that line; on learning of his intentions, Hutton offered Hall "but little encouragement," believing that the immense scale of subterranean operations could not be duplicated in the laboratory. Hall then good-naturedly accepted the "little crucible" passage as a censure on experimenters like himself. Even so, he did not agree. But Hutton's disapproval and other "peculiar circumstances" turned Hall's attention elsewhere until the winter of 1797, when his experiments resumed. At the suggestion of Dr. Thomas Hope, he concentrated first on basalt. On 17 January 1798 Hall filled a black lead crucible with fragments of Edinburgh basalt and placed it within the furnace of an iron foundry, where the basalt fused within fifteen minutes; allowed to cool rapidly, it

[5]For Hall, see especially V. A. Eyles 1961; 1963; and in Gillispie 1970–80; and Laudan 1979; 1987.

formed only a black glass. In later experiments Hall achieved slower coolings, which reduced the glassiness. Finally, on the 27th he succeeded in obtaining a crystallization that seemed very similar to basalt. Hall displayed this gratifying result before the Royal Society of Edinburgh on 5 February.

As Hope pointed out, however, it was not clear that the original crystallized texture of the basalt had been destroyed. Hall therefore began a second series of experiments in which the basalt was first reduced to glass and then recrystallized. His success, soon confirmed by other experimenters, applied equally well to basalts from throughout Scotland. "It has thus been shown," claimed Hall, concluding this portion of his paper, "that all the whins employed assume, after fusion, a stony character, in consequence of slow cooling; and the success of these experiments, with so many varieties, entitles us to ascribe the same property to the whole class. The arguments, therefore, against the subterraneous fusion of whinstone, derived from its stony character, seem now to be fully refuted" (56).

Hall then turned to a series of experiments on lava, specimens of which he (accompanied by Dr. James Home and, for a time, by Dolomieu) had collected in 1785 from Vesuvius, Etna, and the Lipari Isles. In their chemical aspects, these Italian lavas closely resembled Scottish basalts. According to Dolomieu and Kirwan, however, lavas must never have been completely fused, for only some kind of glass could have resulted. Hall and Beddoes (in 1790 and 1791, respectively) had argued contrarily that both granites and basalts crystallized through fusion and slow cooling. Seeking to resolve the issue, Hall experimented on specimens of six different lavas. One of these had been collected at Vesuvius during a mild eruption; while there, Hall had also experimented with molten lava actually flowing from the volcano. Combining his Vesuvian observations and experiments with the series just completed, Hall established that "the stony character of a lava is fully accounted for by slow cooling after the most perfect fusion; and, consequently, that no argument against the intensity of volcanic fire can be founded upon that character" (66). Volcanic heat, he believed, has often been intense.

Having experimented at length with both rocks, Hall concluded that basalt and lava are virtually identical. "So close a resemblance," he thought, "affords a very strong presumption in favor of Dr. Hutton's system, according to which both classes are supposed to have flowed by the action of heat" (67). Even so, veins and nodules of calcite occur frequently in basalts but are never found in lavas (the heat of which would not permit them to exist). As basalts formed long ago and at a great depth within the earth, however, the weight of overlying strata was sufficient to confine the carbonic acid and lime from which calcite derived. Such modifications of the action of heat by pressure, noticed only by Hutton, distinguish his theory from all other igneous ones.

But if basalt originally formed far underground, why do we find it in masses on the surface? According to Hutton, Hall reported, the overlying strata had been gradually worn away, during an immense period of time, by the same erosive forces always at work around us. These sediments were then transported to the bottom of the ocean to be deposited as sand and gravel. If exposed to heat, during some subsequent revolution, they might again be converted into stone (Hutton, of course, was not in doubt). But this was the one major aspect of Hutton's theory with which Hall continued to disagree. Like Pallas, Saussure, and Dolomieu, he preferred to believe that "at some period very remote with respect to our histories, . . . the surface of the globe has been swept by vast torrents flowing with great rapidity and so deep as to overtop the mountains. . . . By removing and undermining the strata in some places, and by forming in others immense deposits," he continued, these torrents "have produced the broken and motley structure which the loose and external part of our globe everywhere exhibits" (68n). Hall based his opinion on geological phenomena he had personally observed in Sicily, the Alps, and Scotland. (Hall's friend Lord Daer, who joined in "agreeing with Dr. Hutton in almost every article but this," had some further weighty arguments and observations regarding the Highlands of Scotland. A joint paper by the two of them was now announced, but it never appeared.) Admittedly, Hutton himself, in the second volume of his *Theory* (chapter xii), had straightforwardly opposed such torrents. In Hall's opinion, however, "their existence is not only quite consistent with his general views, but seems deducible from his suppositions, almost as a necessary consequence. When the strata, according to his system, were elevated from the bottom of the sea, the removal of so much water, if not performed with unaccountable slowness, must have produced torrents in all directions, of excessive magnitude and fully adequate to the effects I have thus ascribed to them" (68n). Though Hall purported to endorse Hutton's theory, he denied its central rhythm of complementary cycles, abridged Hutton's concept of geological time, and (while remaining actualistic) for normally gentle uplift substituted periodic upheavals with concomitant torrents. Large parts of Hutton's theory, moreover, clearly did not interest him.

Working closely with Hall, Robert Kennedy presented his "Chemical Analysis of Three Species of Whinstone and Two of Lava" before the Royal Society of Edinburgh on 3 December 1798. "It is well known among the friends of the late Dr. Hutton," Kennedy recalled, "that he made some experiments on zeolite, by which he concluded that soda entered into the composition of that substance" (1805, 87). Though

Hutton had not mentioned these experiments in any of his works, Black regularly cited them in his chemical lectures. Kennedy's analyses now established that "whins, and a certain class of lavas, taken from remote quarters of the globe, consist of the same component elements united in each, nearly in the same proportion" (94), all of them containing soda. He then went on to prove that other kinds of rocks contained soda as well.

Kirwan in Context

Though the papers of Hall and Kennedy would not appear officially before 1805, full texts of both were in circulation well before then, as separates. Having received one of the former's, Richard Kirwan was able to present his "Observations on the Proofs of the Huttonian Theory of the Earth Adduced by Sir James Hall, Bart." as early as 8 February 1800, before the Royal Irish Academy in Dublin. Kirwan dealt first with Hall's "Observations on the Formation of Granite" (1794), which had supported fusion. "Sir James," he concluded after several pages of refutation, "has since very wisely declined justifying his theory of the formation of granite by fusion, and by the advice of Doctor Hope very properly applied himself to experiments on various species of whin, a denomination which in Scotland comprehends grunstein, basalt, trap, wacken, and porphyry" (1800, 13). Hall's experiments with basalt (he noted Kennedy's work also) likewise failed to support Huttonian consolidation, whether of granite, basalt, or any other stony substance. Kirwan acknowledged that the "stony appearance which lavas after cooling exhibit," as discovered by Hall, was of great importance to geology, but the latter's experiments afforded "no confirmation of the high degree of heat attributed to volcanoes, and still less to the many hypotheses gratuitously heaped on each other by Doctor Hutton, or to the volcanic origin of whins" (27). On 28 April and 5 May 1800, Kirwan read two additional papers before the Royal Irish Academy, each of them designed to confirm the reality of the Noachian deluge.

In proposing and elaborating his own geological theory, Hutton had tacitly accepted the opposition of almost every other school claiming to deal with the same subject. As a Plutonist emphasizing subterranean but unerupted rocks, he clashed with those extreme Vulcanists who championed the efficacy of more purely volcanic processes. As a fluvialist emphasizing the erosive powers of running water over long periods of time, he conflicted with all those who still supported geologically effective sudden floods, whether one or more. As a uniformitarian who relied solely on observable present-day forces (except in the bowels of the earth, where we cannot see them), he similarly opposed widespread catastrophes of any kind while postulating an earth much older than traditional estimates allowed. It is not surprising, then, that almost everyone who attempted geological theorizing at all found himself seriously at odds with Hutton in one or more respects.[6]

During the 1770s, before Hutton's theory appeared, geological theorists throughout Europe had divided themselves into two fairly well defined camps, depending on whether they emphasized fire or water more strongly. The fire geologists, called Vulcanists, included all those who, like Desmarest, Strange, Hamilton, Raspe, and Saint-Fond, attributed widespread, significant effects to both ancient and present-day volcanoes. As we know, Hutton associated with all of those Vulcanists just named, being particularly concerned with their observations and opinions concerning basalt. They, in turn, were among a fairly small number of thinkers outside Scotland who noticed Hutton's theory and considered it seriously.

The opposing, Neptunist school generally ignored all but the most immediate effects of volcanoes, which they regarded as recent and largely inconsequential. From their point of view the action of water was far more significant, as it could explain not only the deposition of practically all the earth's crustal components but their subsequent shaping as well. To account for both rocks and landforms in this way, Neptunists postulated some kind of universal ocean, which existed either at the beginning of earth's history (and from which the continents then gradually emerged) or at some point in the recent past, within human history. By Hutton's time, the former mechanism had seemed more plausible, because it was known that marine fossils occurred not only at the tops of mountains but at several other stratigraphic levels. This fact embarrassed both Neptunist explanations to some extent, but was more easily reconciled with the eddies of a retreating ocean over time.

Inevitably, several who supported Neptunist theories stressed the compatibility of their views with the Creation and Flood passages in Genesis. In its earlier development, Neptunism had been even more closely associated with biblical criticism; Thomas Burnet and other seventeenth-century speculators aroused the ire of the more orthodox by freely reinterpreting Scripture to agree with their ingenious but heterodox geological mythologies. Such latter-day Neptunists as Kirwan seemingly derived their geological beliefs on purely empirical grounds but were then pleased to demonstrate how their findings accorded with Genesis. On the whole, it is not unfair to suspect the objectivity of their enterprise to a certain extent. Still, Scriptural corroboration should not

[6]The terms "fluvialist" and "uniformitarian" are anachronistic here; they had not yet been coined.

be equated with Scriptural literalism. By the latter eighteenth century it was no longer common to accept the literal meaning of biblical passages uncritically; on the contrary, almost everyone who mattered agreed that geological arguments needed to be resolved on geological grounds. Those controversialists who knew *only* the Bible came gradually to be excluded from serious geological discussions, and biblical evidence, if used at all, served primarily as confirmation (i.e., a second opinion) rather than proof.

Because the geological arguments supporting Neptunism were occasionally strong, it was quite possible to support aqueous theories without alluding to the Bible at all. Deists, for example, characteristically avoided biblical evidence of any kind in expounding their gradualistic views of nature. But Neptunism as such by no means excluded deistic interpretations. By the end of the eighteenth century, in fact, evidence cited on behalf of the biblical Deluge was generally seen to be of the most recent sort (phenomena, for the most part, that would later be explained by the Ice Age).

Werner

A more fundamental question concerned the origin of the earth's rock masses, some of which were regarded as Primeval, Primary, or Primitive. Though he did not originate this inquiry, Abraham Gottlob Werner (1749–1817) made it his own.[7] An immensely influential figure throughout his adult years, Werner attended both the Mining Academy at Freiberg and the University of Leipzig. While yet a student at the latter, he published *Von den äusserlichen Kennzeichen der Fossilien* (On the external characteristics of minerals, 1774). Though Werner relied on neither the chemical composition of minerals nor on their forms of crystallization, he paid close attention to such other external characteristics as color, luster, fracture, streak, and hardness. Because his methods were the best yet devised for distinguishing one mineral from another, Werner's book quickly superseded Cronstedt's. The first French translation, *Traité des caractères exterieurs des fossiles* (1790), contained numerous corrections and additions. A second French translation (1795) and two English ones followed. Though each of these editions differed from the others in certain details, however, none included the geological theorizing that made Werner and his followers the chief opponents of Hutton's.

After the original publication of his book in 1774, Werner was invited to return to Freiberg as a member of its faculty. Being professor of

[7] Until quite recently, the historiographical treatment of Werner remained far from adequate. Ospovat (in Gillispie 1970–80 and elsewhere) has done much to redress the balance.

mining and mineralogy, he had the cabinet of natural history in his care—like the library, it soon expanded into a major collection—and he was free to teach whatever relevant subjects he wished. Perhaps no other academic appointment in the history of the earth sciences has ever had such momentous consequences. As the scope and power of Werner's splendidly convincing lectures became famous, once-somnolent Freiberg was soon regarded as not only the world's most prestigious mining academy but also its center for geological education. So long as Werner remained at the podium, students came from everywhere to experience at first hand the stimulating vigor and authority of his wide-ranging intellect, which moved effortlessly from such basic subjects as mining machinery and law to vastly suggestive inquiries about the role of the inanimate world in shaping the history and destiny of mankind. For a generation or so, most of the significant geologists in Europe (together with important philosophers and writers) were former students of Werner's, who, though not altogether one himself, reliably transformed them into discriminating field investigators. Werner's immediate impact on the half century of geological endeavor beginning in 1774 was unquestionably greater than Hutton's.

Unlike Hutton, who published only three major geological works (a monograph, a paper, and a book), Werner contributed more than two dozen titles, though all of them after 1774 were relatively short. His second important publication, for instance, "Kurze Klassifikation und Beschreibung der verschiedenen Gebirgsarten" (Short classification and description of the various rocks, 1786) was but twenty-eight pages long and covered only thirty-one types. Many historians, however, regard it as the foundation of modern petrology, comparing Werner with Linnaeus and other systematizers of natural history. Through the efforts of Werner and his students, certainly, the previously inadequate definitions by which rocks were identified, related, and distinguished improved considerably. Offered a common and functional vocabulary, geologists throughout Europe and the Americas gratefully adopted Wernerian terminology and standards, thereby making widespread regional comparisons at least theoretically possible and stimulating immediate interest in more precise stratigraphy and mapping.

In the "Kurze Klassifikation" Werner also asserted unequivocally, on the basis of local examples, that basalt is of aqueous origin. His opinion was widely noticed, largely because previous fieldwork among extinct volcanoes in the Auvergne region of central France (evidence that Werner never saw) and elsewhere had resulted in fairly widespread agreement to the contrary. Following his boldly pronounced contradiction, in the spring of 1787, Werner then examined Saxon basalts at Scheibenberg, in the Erzegebirge, where he found sedimentary layers of sand, clay, and wacke immediately below basalt and naturally regarded all of them as aqueous in origin. He then defended himself against objectors in no fewer than eight brief publications (1788–89). By the time they appeared, a number of adherents had come over to his side.

Five years after the "Kurze Klassifikation," Werner published his *Neue Theorie von der Enstehung der Gänge* (New theory of the formation of veins, 1791), of which there were then augmented French (1802) and English (1809) translations. His major assertion now was that all true mineral veins, including basaltic ones, originated as open rents only subsequently filled—and then from *above*—by aqueous agencies. As with his more general opinion regarding the origin of basalt, Werner's theory of veins provoked yet another major geological controversy, which, with much associated fieldwork, lasted a quarter of a century.

Despite all the notice accorded his other pronouncements, Werner achieved his most pervasive impact on the science of geology (which he preferred to call "geognosy") through a comprehensive scheme of petrological chronology that he took for granted in his "Kurze Klassifikation" and other publications but never wrote out in detail. Though largely a synthesis of earlier ideas, Werner's subsequently notorious theory of geological succession established itself through his teaching and explications derived from class notes by his former students. Werner did not, however, in any way originate the concept of geological succession and, as Alexander Ospovat has noted, the so-called law of superposition was already an eighteenth-century commonplace.

Adopting observations to like effect by some German and French predecessors, then, Werner not only acknowledged a broad concept of regular and predictable stratigraphic succession but agreed as well that certain strata could be grouped into larger units called formations. Initially, he discerned four of these units, each of which corresponded (in his mind) to a distinct epoch in the history of the earth. A Deist somewhat like Hutton, Werner accepted the immensity of geological time while ignoring the Creation narratives in Genesis. Perhaps a million years ago, he guessed, a universal ocean covered the earth. Its chemical precipitates, crystalline and devoid of any fossils (life having not yet begun), became the *Primitive* rocks, of which granite was the most basic. As the universal ocean gradually drained away, its increasing turbulence and the advent of life together created a second, less orderly formation, the *Floetz* (layered) rocks, including distorted fossiliferous limestones among others. The Primitive formation and some Floetz strata were worldwide in extent. His last two formations, *Volcanic* and *Alluvial*, were only local occurrences and relatively unimportant. Far from affirming Hutton's critical belief in a central fire or heat, Werner held that volcanoes originated from the spontaneous combustion of shallow-seated coal deposits. His Alluvial rocks resulted from local floods and similar conditions, but not from the biblically affirmed Deluge of Noah.

Throughout his lifetime, Werner's basic formational scheme underwent considerable elaboration. A most striking addition, announced in lecture, took place around 1796, when he introduced a third universal formation intermediate between the other two and supposedly originating from mechanical (rather than chemical) deposition. These were his *Uebergangsgebirge* (*Transition*) rocks, of which greywacke was a frequent example. Regarding both his Primitive and Floetz formations, moreover, Werner found it necessary to complicate the essential harmony of his scheme by postulating some interruptive floods, in which the subsiding universal ocean had temporarily risen again to inundate land already exposed. Later stratigraphical distinctions by Werner and his followers further elaborated (but did not outrightly contradict) the established five-part progression. Though destined not to achieve permanence, Werner's secular theory of the earth's history was the first to be widely accepted.

Jameson

Some of Werner's students at Freiberg went on to become courageous geological explorers of unknown and even exotic regions; others followed his own example more closely. Like him, these latter consistently preferred teaching and pretty much limited their subsequent fieldwork to a search for examples that would confirm Wernerian theories. Such, at least, was for a time the case with Robert Jameson (1774–1854), whose broad interests in natural history made him a favorite pupil of John Walker's at the University of Edinburgh.[8] Through Walker's influence, Jameson was placed in charge of the university museum, where geology and mineralogy particularly attracted him. In 1794 he spent three months exploring the geology, mineralogy, zoology, and botany of the Shetland Islands.

Having been apprenticed to a surgeon originally, Jameson was elected to the Royal Medical Society of Edinburgh (a student organization) in 1795, just before Hutton's *Theory* appeared. In 1796 he read before that society two papers controverting Hutton's opinions: "Is the Volcanic Opinion of the Formation of Basaltes Founded on Truth?" and "Is the Huttonian Theory of the Earth Consistent with Fact?" In the first of these, published only in 1967, Jameson noted that the volcanic theory was first proposed in 1763 by Desmarest, who affirmed that basalt had

[8]Jameson, like Werner, has been poorly treated by earlier historians, and outrightly ridiculed by some. One should prefer J. M. Eyles in Gillispie 1970–80, and other articles on him by Sweet.

the same external character as lava. On this gross error, Jameson observed, "succeeding geologists have founded a most extraordinary chimera, attributing the formation of basaltes to fire!" (1967, 93) Incomprehensibly, many prominent geologists had adopted this position, though it could be refuted by common observation, as he proceeded to argue in a series of numbered points. Among his more important assertions were that basalt and lava are chemically distinct; that basalts contain zeolites, whereas lavas do not; that lavas do not become, or grade into, basalts; and that the pumice often associated with basalt is not volcanic in origin. He claimed as well that the supposed extinct volcanoes found adjacent to basalts are chimerical; though basalt occurs frequently in volcanic countries, it also appears in Scotland, where no such volcanoes have ever been detected. Observations of the crystallization of lava—and glass—seemingly favorable to the vulcanist position, he suggested, could be otherwise explained.

"Upon such a basis," Jameson concluded scornfully, "is the famous Volcanic theory founded, which for many years consigned three-fourths of our globe into the hands of Pluto—until the immortal Werner, from a careful examination of nature, declared the absurdity of such a hypothesis (if it can be so called)" (94). Kirwan too, in his admirable *Elements of Mineralogy* (2d ed., 1794), had ably defended and further substantiated the Neptunist hypothesis. Jameson himself now stepped forward to present seemingly conclusive arguments: (1) Sand, clay, wacke, and basalt gradually pass into each other; since the first three are obviously aquatic in origin, all of them must be. (2) Basalts, similarly, pass into syenite (a granitelike rock), as Jameson had observed on Arthur's Seat and Salisbury Crags. (3) Basalts have been observed to contain organic remains, a fact sufficient of itself to refute the vulcanist theory. (4) Serpentine, marl, granite, porphyry, and indurated clays are sometimes columnar in form, like basalt, yet all of them are aqueous in origin. He then concluded the paper without further comment.

Jameson's second paper, "Is the Huttonian Theory of the Earth Consistent with Fact?" remained unpublished until 1892. "The celebrated theory of Doctor Hutton," it began, "has, for several years, attracted the attention of geologists, not more from the ingenuity with which it is supported than the vast collection of facts which it contains, rising in this respect far superior to all former conjectures" (1892, 37). Because a thorough review of the theory would require inordinate time, Jameson proposed to examine only its basis: Hutton's assumption that all the strata of the globe have been consolidated by heat from an original state of fusion.

Hutton's theory assumed both insuperable compression (necessary to explain how various quartz rocks were formed) and slow cooling. Regarding the first, Jameson disputed Hutton's beliefs that quartz was insoluble in water and must therefore have been crystallized by fusion; that flint nodules had been formed by fusion; and that wood petrified when liquid flint was injected into it under immense pressure. The idea of insuperable compression, he argued, should therefore remain unacceptable until proven otherwise. Though this rejection was of itself sufficient to demolish Hutton's theory, he thought, further corroboration could be found in the mistaken concept of slow cooling, which originated not with Hutton but with Sir James Hall (in his 1790 observations on granite). Whereas earlier experiments with the fusion of granite had reduced that rock to an amorphous glass, Hall had pointed out, slow cooling might well have produced crystallization, resulting in a rock much like the original. Jameson, however, believed that such cooling constituted chemical change. Dramatically, he then reported a series of his own experiments—but on glass rather than granite. These experiments proved that changes in glass did not depend on slow cooling but on the degree of heat applied, which freed the glass from effects of included alkali. "It appears, then," Jameson concluded, "that immense compression and slow cooling are, as yet, wanting of proof; consequently the pillars of the Huttonian conjecture are not well founded" (38).

In July 1797 Jameson visited Dublin and Hutton's most vociferous opponent, Richard Kirwan. During their conversation, the latter proposed several examples of natural phenomena that would apparently refute the Huttonian theory. For instance, when the wooden foundations of an old Roman bridge across the Rhine were recently exhumed, they were discovered to be completely silified—in such a short time. Sandstone in a quarry was known to renew itself after cutting, presumably leaving the Huttonian doctrine of denudation in doubt. Granite and gneiss, moreover, prove to be quite distinct once they both decompose. Jameson also saw part of Kirwan's geological collection and was admitted to his laboratory, where numerous experiments on the fusibility of stones were being conducted. From Dublin, Jameson went on to the Hebridean island of Arran; expending much effort, he filled an extensive journal with geological details (Sweet 1967b).

The next year, Jameson published *An Outline of the Mineralogy of the Shetland Islands, and of the Island of Arran* (1798), which mentioned Hutton in several places and was the first book-length attempt to controvert his theory on the basis of field evidence. (Hutton himself, we recall, visited Arran in 1787 but did not publish his results.) Unlike other of Hutton's critics, and himself in other places, Jameson did not here attempt to consider Hutton's theory at length. Most of his remarks therefore appear as footnotes or asides. In the first of them, Jameson chastized an unnamed "late celebrated Theorist" for grouping gneiss and several varieties of schist under the general name "schistus." "This want of accuracy is to be regretted," he observed, "for it is plain, if such confusion be sanctioned, that the science of mineralogy will revert to the unmeaning jargon of a Sir John Hill" (1798, 18n). Jameson then controverted Hutton's revelation that granite is not necessarily a primeval rock and can form later than schist (27). He noted that Huttonians had fallaciously denied the solubility of silex (53–54n) and then cited and seemingly refuted Hutton's central contention regarding heat, fusion, and consolidation (60–61n) and soon thereafter denied the igneous origin of granite (72n). A further note on the origin of coal (100–101n) preferred the Neptunist position, and another denied the evidence from septaria (102n). In his most extensive note (110–111n), Jameson questioned Hutton's belief that the matter of this world had been formed by the decay of an earlier one, implying that Hutton resembled but went beyond Buffon in the extent of his religious heterodoxy. On more strictly geological grounds, Jameson found no proof that Huttonian detritus actually reaches the bottom of the ocean; on the contrary, land *forms* along coasts. Ten pages later (121–122n), Jameson rejected Hutton's assumption that granite, porphyry, and basalt are only varieties of each other. A final observation contradicted Hutton regarding the formation of peat (164). Evidently, then, Jameson had proclaimed himself the ally of Kirwan, Deluc, Howard, and Walker.

Jameson spent the summer of 1798 exploring the Hebrides; in 1799 he visited the Orkneys and returned to Arran. The literary result of these excursions was *Mineralogy of the Scottish Isles* (1800), which followed Walker (to whom it is dedicated) by regarding geological theorizing as premature. In an apparent hit at Hutton, Jameson wrote, "I fear that the theories of the formation of the earth, interesting as they are, often mislead the mind and pervert the understanding, and those who yield to them become so involved in delusive speculations, so blind to fact and experience, that, like Archimedes, they find but one thing wanting to raise worlds" (1800, I: vii). The thing wanting, for Hutton, would be subterranean heat, but Jameson's charge that theories delude the senses would eventually be turned against him.

Though not a conventional travel book, Jameson's *Scottish Isles* alternates chapters of narrative and mineralogical analysis. In volume 1, chapters 1–8 take us from Edinburgh to Arran. Among these, chapters 3–8 are only slightly revised versions of his chapters 1–6 from 1798, though some of the appended references to Hutton have been deleted or changed. Some new observations on the mineralogy of Arran consider the possible stratification of basalt. Chapters 9–15 then deal with the Hebrides and related geological problems. Volume 2 continues with Hebridean islands, the Shetlands (from 1798), the Orkneys, and the return to Edinburgh, with additional chapters on peat and kelp (both 1798, with additions). Unexpectedly, three pages of brief observations by Lord Webb Seymour and John Playfair on the mineralogy of the country between Fort Augustus and Inverness are included as an appendix.

Most of Jameson's strictures against Hutton in *Scottish Isles* reprinted or restated those of 1798, but there was one important addition. Regarding the island of Bute and geological phenomena found there, Jameson quoted at length from Hutton's *Theory* (II: 265–267), the memorable passage (in Part II, chapter viii, "The Present Form of the Surface of the Earth Explained, with a View of the Operation of Time upon Our Land") in which Hutton described the gradual evolution of a coastline as erosion forms peninsulas that eventually become islands, the western islands of Scotland being his most conspicuous examples. Jameson then replied: "This is very probably a correct delineation of the mode which nature follows in altering the land, in some few instances; but it cannot be general, as it would give an age to the world quite inconsistent with the Hebrew chronology; we must therefore consider it as untenable. I am fully persuaded that any chain of reasoning that does not coincide with that chronology is false" (1800, I: 138). His argument against Hutton continued, on geological grounds, for another couple of pages.

[5]

Playfair

Repeatedly attacked by increasingly vociferous opponents, and with its most prominent defender now unable to respond further, Hutton's geological theory might well have sunk into oblivion had its cause not been hazarded by some of the dead author's friends, chief among whom was John Playfair (1748–1819).[1] The son of a Forfarshire minister, Playfair was educated at home by his father until age fourteen, then sent to the University of St. Andrew's to qualify himself for the church. Mathematics, however, soon proved to be his real field. From 1769 to 1773 he was in Edinburgh, associating with Robertson, Smith, Black, and perhaps Hutton (though Playfair himself later dated their real friendship from about 1781). When his father died in 1772, Playfair succeeded to livings at Liff and Benvie—but only because of help from Robertson. While at Liff, he corresponded regularly with the latter about philosophy, physics, anthropology, and other learned matters. Besides making occasional visits to Edinburgh, Playfair explored Perthshire in 1774 with Dr. Nevil Maskelyne (the Astronomer Royal) and was stimulated by him to submit a successful paper on mathematics to the Royal Society of London.

Between 1782 and 1787, having resigned his church livings, Playfair tutored for a family near Edinburgh and was therefore able to participate in the city's intellectual life as a winter resident. Appointed joint

[1]Our primary source for Playfair's life is the biographical sketch of him written by his nephew and included in the Playfair *Works* of 1822. More details are now available, but they have never been comprehensively assembled. In addition to later references in the present book, see also John Challinor in Gillispie 1970–80, and Dean 1983.

professor of mathematics at the University of Edinburgh in 1785 (the year in which Hutton's theory was first made public), he then moved to Edinburgh two years later, joining his mother and sisters. From 1787 to 1795, Playfair published on various topics in the *Transactions* of the Royal Society of Edinburgh and (in 1795) his popular *Elements of Geometry*, which remained standard for half a century.

Though Playfair had probably met Hutton by 1769 (both were then in Edinburgh, enjoying the same friends) and had presumably developed some kind of rapport by 1781, the most meaningful part of their association took place during the last decade of Hutton's life. We know, for example, that Playfair had not been privy to the details of Hutton's theory before its presentation in 1785. On hearing that theory, moreover, he failed to see how it could explain the oblate shape of the earth—for him a necessary condition—and therefore remained skeptical of its validity.[2] Yet, by 4 June 1788, Playfair and Hutton were visiting Siccar Point together. That year or the next, they were in Ayrshire. Sometime after 1791, Playfair wrote a geological letter about the English Lake District to Hutton, who published the major portion of it four years later. Though there must have been a good deal in Playfair's extensive papers and correspondence regarding Hutton and his theories, we have lost this evidence entirely.

Illustrations of the Huttonian Theory, 1802

According to Playfair's nephew, James G. Playfair, his uncle John devoted the five years from Hutton's death in March 1797 to the publication of his *Illustrations of the Huttonian Theory* (1802) in preserving his friend's thought and life. Having at first intended to summarize Hutton's geological theory as part of a more general biographical memoir of him (he had earlier memorialized the life of his mathematical predecessor at Edinburgh, Dr. Matthew Stewart), Playfair began once again to read through Hutton's *Theory*, only to be appalled at how verbose and impenetrable it was. Hutton's stylistic inadequacy, he soon came to believe, had been in large part responsible for the failure of his thinking to achieve broader acceptance. Knowing himself to be a much more gifted writer (and perhaps urged on by Robertson, who was equally aware of Hutton's verbal deficiencies), Playfair set aside his memoir of Hutton for a time in order to explicate the latter's theory.

On the evening of 1 July 1799, at the Royal Society of Edinburgh

[2]John Whitehurst's theory (1778) began with an oblate earth and may well have seemed preferable to Playfair for that reason.

Francis Horner heard Playfair read the first part of his analysis of Hutton's theory, "a very distinct and luminous deduction from as powerful a train of arguments as ever was given in favor of a mere hypothesis" (Horner 1853, 1: 86, emended). Playfair read his second part at the next meeting. As we have every reason to suppose, the two halves of Playfair's presentation later became the 140-page main section of his book, which, except for one addition, must therefore have been completed by June 1799.

With wonderful economy, Playfair emphasized the fundamental simplicity of Hutton's argument by presenting it under only three major headings. A masterful introduction began by citing three examples of fossils to prove that present land was once immersed beneath the sea; evidence derived from broken strata similarly established that "the earth has been the theatre of many great revolutions, and that nothing on its surface has been exempted from their effects" (1802, 2). To trace, explain, and connect these revolutions, he believed, is the proper object of a theory of the earth. Though geology was the youngest of the sciences, many theories had already been proposed, most of them stressing either fire or water; Hutton's, however, utilized both.

Playfair first divided the mineral kingdom into stratified and unstratified substances, discussing each, then both. In the chapters that followed, each paragraph is consecutively numbered and brief footnotes cite Playfair's own notes (to follow) or Hutton's *Theory* or some other source.

SECTION I. OF THE PHENOMENA PECULIAR TO STRATIFIED BODIES

1. Materials of the Strata. Though most theories of the earth agree that much of the earth's rock consists of strata once deposited at the bottom of the sea, Dr. Hutton's uniquely stresses that the materials so deposited came from previously existing rocks. Calcareous, siliceous, argillaceous, and bituminous examples follow, but granite and other so-called primitive rocks (except schistus) are excluded. Hutton calls such rocks "primary" rather than "primitive."

2. Consolidation of the Strata. Only two agents, water and fire, can possibly effect consolidation, but water does so very imperfectly. Igneous consolidation also requires pressure; it therefore differs considerably from any process observable at the earth's surface. Specimens of fossil wood, flint, puddingstone, and sandstone, among others, support the Huttonian theory. Septaria, coal, rock salt, and trona are also excellent examples.

3. Position of the Strata. Mere diminution of the sea (as in the theories of Buffon and Werner) cannot explain the earth's exposed strata, which have evidently been uplifted by some internal force. Such uplifts (marks of which are especially evident among the primary rocks) are not the result of disorderly violence but rather part of "a regular system essential to the constitution and economy of the globe" (49). In some cases, the primary strata were twisted into an almost vertical position and significantly eroded before younger strata were deposited on top of them. Such alternate elevations and depressions of the sea bottom, however extraordinary they may seem, are part of the system of nature. The same powerful subterranean heat that consolidated the strata at the bottom of the sea also raised and tilted them.

SECTION II. OF THE PHENOMENA PECULIAR TO UNSTRATIFIED BODIES

1. Metallic Veins. That unstratified minerals in veins formed later than the strata they traverse is obvious. Their fluidity must have resulted from simple fusion and slow but forcible injection. Not all veins, however, are of the same age. They may be found in both stratified and unstratified rocks but are especially common in primary strata, a fact according well with Hutton's theory.

2. Of Whinstone. An unstratified rock, whin (basalt) exists either as veins or in irregular masses. Though very much resembling lava, whin differs in being unerupted. Occasional columnar form, the presence of pyrites, included fragments from the surrounding rock, and frequent agates all prove whin to be igneous, as does other evidence.

3. Granite. For Hutton, granite is always unstratified (but Playfair qualifies this, distinguishing gneiss from granite). Though granite usually underlies stratified rocks, it is actually younger, having been injected while fluid, like basalt. Basalt and granite, moreover, "seem to be united by a chain of insensible gradations" (83).

The three parts of this section, Playfair concludes, prove that all mineral substances, whether stratified or not, have been formed by fluidity, consolidation, and uplift. But that "fire, or more properly heat, was the cause of the fluidity of these mineral bodies, and also of their subsequent elevation," he admitted, "is not perhaps to be considered as a truth so fully demonstrated . . . ; it is, no doubt, a matter of *theory*" (90). Yet hot springs, volcanoes, and earthquakes all support the concept of a subterranean heat.

SECTION III. OF THE PHENOMENA COMMON TO STRATIFIED AND UNSTRATIFIED BODIES

Once rocks have been elevated above sea level, Playfair continued, they are constantly subjected to other forces—both mechanical and chemical—intended to reduce them. This "system of universal decay and degradation . . . may be traced over the whole surface of the land, from the mountain top to the sea shore" (100). In particular, we may consider the nature and economy of rivers:

> Every river appears to consist of a main trunk fed from a variety of branches, each running in a valley proportioned to its size, and all of them together forming a system of valleys, communicating with one another, and having such a nice adjustment of their declivities that none of them join the principal valley either on too high or too low a level, a circumstance which would be infinitely improbable if each of these valleys were not the work of the stream that flows in it. (102; since called "Playfair's law of accordant junctions")

It is often possible to trace successive river platforms rising above each other; they strikingly suggest "an antiquity extremely remote" (103). River valleys bisecting mountains are not created by great convulsions of nature but by simple streams working gradually over a long period of time. The same natural forces that take soil away from the surface of the earth, however, are simultaneously restoring it to virtually the same degree.

Among several examples of the extent to which this wasting occurs, one may cite immense deposits of gravel and flints, the diffusion of gold and other metals from mineral veins to rivers, and especially the highest mountain peaks on earth, those of the Alps. The structure of mountains is largely determined by the rock from which they are made. Their valleys relate to one another as did the streams that formed them. Thus, "Through them all, this law is in general observed, that where a higher valley joins a lower one, of the two angles which it makes with the latter, that which is obtuse is always on the descending side. . . . This alone is a proof that the valleys are the work of the streams" (113–114). Both the mountains and the valleys have been formed by the "operation of rains and torrents, modified by the hardness and tenacity of the rock" (114). The same forces have hollowed out the valleys, detached the separate mountain peaks, reshaped their slopes, and transported their materials into distant valleys—"hence the granite of Mont Blanc is seen in the plains of Lombardy, or on the sides of Jura" (115). These, then, are the inexorable processes taking place daily throughout an indefinite period of time.

We cannot doubt that new strata are even now forming and will someday emerge, continuing a cycle of vicissitudes of which "we neither see the beginning nor the end," as in the motions of the planets, which reveal "no mark either of the commencement or the termination of the present order" (119). It is, indeed, unreasonable to suppose that any such marks exist, because the divine laws by which nature works will never destroy themselves. As Playfair insists

> the Author of Nature has not given laws to the universe which, like the institutions of men, carry in themselves the elements of their own destruction. He has not permitted in His works any symptom of infancy or of old age, or any sign by which we may estimate either their future or their past duration. He may put an end, as He no doubt gave a beginning, to the present system at some determinate period, but we may safely conclude that this great *catastrophe* will not be brought about by any of the laws now existing, and that it is not indicated by anything which we perceive. (119–120)

We know, however, that even the present order of things began an extremely long time ago. Yet Hutton's theory in no way contradicts the chronology of Genesis, which deals only with the origin of mankind, an event that must have taken place no more than six or seven thousand years ago.

In conclusion, Playfair argues splendidly, the Huttonian theory is both novel and beautiful. Unlike others, it "presents us with a system of wise and provident economy, where the same instruments are continually employed, and where the decay and renovation of fossils [mineral bodies] being carried on at the same time in the different regions allotted to them preserve in the earth the conditions essential for the support of animal and vegetable life" (128). Like the water cycle, it seems to demonstrate beautiful contrivance in nature. Among his many novel facts and observations, Playfair notes, Hutton calls attention to granitic veins, irregular junctions of primary and secondary strata, the distinction of whinstone from lava, the local changes in strata surrounding veins of whin, and the affinity of whinstone and granite; the universality of unlimited decay; and the sufficiency of present-day causes to explain geological phenomena (130–131). Even were a comprehensive theory of the earth beyond human capacity, these would still be valuable additions to our knowledge. Because no other includes these points, we may justly call that now before us the Huttonian theory (135).

This was, in all probability, Playfair's original ending, as read by him before the Royal Society of Edinburgh in 1799. Once the success of his presentation (by which I mean respect for the quality of his work rather than conversions to the theory) had become obvious, Playfair—like Hutton before him—soon began to elaborate his own remarks with a far longer section called "Notes and Additions." The form itself—theory, then more particular discussions—had been followed by Hutton also, though a good deal less elegantly. It had probably been suggested to both writers by Robertson. In any case, Playfair seemingly began writing the latter half of his now-anticipated book during the fall of 1799.

Kirwan's *Geological Essays*, 1799

To the author of the *Illustrations*, Robert Jameson was little more than an aggravating nuisance. The real opponent, for Playfair as for Hutton, was at this time Richard Kirwan, who had only recently published a much-noticed book in which Hutton was severely criticized. (From a standpoint quite independent of Hutton's, Kirwan regarded Johann Gottlob Lehmann, who had first clearly distinguished primary from secondary mountains, as the founder of modern geology.) Though reprinted within it were the three essays previously appearing in 1797, Kirwan's *Geological Essays* (1799) was now more often concerned with mountains: primary, secondary, and volcanic. "Geology," he suggested, "is the science that treats of the various relations which the different constituent masses of the globe bear to each other. It at once unfolds and shows how to read the huge and mysterious volume of inanimate nature, of which mineralogy supplies the alphabet" (1799, iii). The same strictures he had applied to Hutton in 1797, therefore, remained unchanged. To them, however, he now added an entire essay, "On the Huttonian Theory of the Earth" (433–499). In his previous memoir of 1793, "Examination of the Supposed Igneous Origin of Stony Substances," Kirwan had, he thought, already established the inconsistency of Hutton's theory with natural phenomena. Since its appearance, Hutton had replied to Kirwan's strictures "with much acrimony" in his *Theory* of 1795. A detailed refutation of all he presented there would not be required. "It will be sufficient," Kirwan held, "to subvert the fundamental principles upon which his system is constructed, and occasionally to point out the absurdities that follow from it" (433); for this, a brief examination would suffice. Kirwan then reduced Hutton to four propositions: (1) that the destruction and renovation of soils and lands are natural operations of the globe; (2) that land formed in the ocean, from the materials of a previous habitable earth (and, more specifically, that limestone originates from marine animals); (3) that the loose materials thus collected were fused and consolidated by heat, rather than accretion or crystallization; and (4) that the consolidated substances were then elevated and expanded by extreme heat, thus forming our present continents. Kirwan then examined each proposition together with the major points of Hutton's 1795 rebuttal. To propositions 1 and 2, for instance, Kirwan objected that sediments derived from continental erosion are deposited at the mouths of rivers or nearby, not in the middle of the ocean. He also affirmed that limestones must have preceded the origin of life, for the globe could not have been habitable without them. As for proposition 3, petrified wood and other fossils disprove the possibility of fusion. Beyond that, Hutton had apparently supposed that "at some past period a degree of heat prevailed under the ocean superior to any that has ever been known to exist, and which must have taken effect in circumstances the least favorable to its production and exertion" (452). Through this supposed heat, immense masses of calcareous matter were fused and crystallized. Yet no experimenter thus far (including Lavoisier, Saussure, and Ehrman) had succeeded in producing a laboratory heat sufficient to achieve the like. From what was known also of volcanoes, no such heat existed. And where did Hutton suppose it to exist? "Under the ocean in the bowels of the earth, where neither a sufficient quantity of pure air nor of combustible matter capable of such mighty effects can with any appearance of probability be supposed to exist; and without these, such degrees of heat cannot even be imagined without flying into the region of chimeras" (454). Many rocks and minerals, of course, clearly did *not* originate from fusion, coal being the most obvious example. Finally, regarding proposition 4, "it were idle to inquire how our continents could be raised by a cause of whose existence we have no proof" (476), so Kirwan devoted his few remaining pages to rebutting Hutton's assertions of 1795 regarding himself.

Playfair's "Notes and Additions"

The long series of notes Playfair added to his forthcoming *Illustrations* was designed primarily to counter Kirwan, with no more than cursory notice of the less imposing Jameson. There are twenty-six notes of about 15 pages each, or 385 pages in all. As modern editions of the *Illustrations* are readily available, and Playfair a clear writer, the following comments do not constitute a full summary.

Note i: Origin of Calcareous Rocks. Playfair attempts to refute Kirwan's charge that Hutton believed all calcareous rocks to be of animate origin (but Hutton had said exactly that).

Note ii: Origin of Coal. Hutton was only one of several theorists who believed coal to be of vegetable origin. Kirwan thought otherwise, but his "incongruous and ill-supported hypotheses" (1802, 160) are not convincing.

Note iii: Primitive Mountains. The concept was developed by Lehmann, Pallas, Deluc, and Saussure.

Note iv: Primary Strata Not Primitive. Playfair cites fossil shells found by himself in supposedly primary limestones of Cumberland and Devonshire; "hence, the existence of shellfish and zoophytes is clearly proved to be anterior to the formation even of those parts of the present land which are justly accounted the most ancient" (165). Schistus, also considered primary, contains sand, proving its derivation from former rocks as well. Werner and others to the contrary, "there is *no order of strata yet known* that does not contain proofs of the existence of more ancient strata. We see nothing, in the strict sense, primitive" (170).

Note v: Transportation of the Materials of the Strata. Citing mountains throughout Britain and Europe, Playfair defends Hutton's concept of the formation of secondary strata from primary against the objections of Deluc and Kirwan. In the chalk beds and limestones of England and France, we find large numbers of fossilized tropical animals, which must have been transported from equatorial regions.

Note vi. Mr. Kirwan's Notion of Precipitation. Kirwan's explanation of how strata precipitate from a solution is absurd.

Note vii. Compression in the Mineral Regions. Huttonian terrestrial compression is similar to Newton's stellar compression (*Opticks*, Query 11). Kirwan's objections to Hutton's advocacy of subterranean fire are irrelevant because "it is not *fire*, in the usual sense of the word, but *heat*" that the theory requires, "and there is nothing chimerical in supposing that nature has the means of producing heat, even in a very great degree, without the assistance of fuel or of vital air" (186). Playfair concedes that he cannot explain the earth's heat, though he knows it to exist.

Note viii. Sparry Structure of Calcareous Petrifactions. Fossils in limestone have been created by heat.

Note ix. Petroleum, etc. These naturally occurring organic substances provide more evidence of heat within the earth.

Note x. The Height above the Level of the Sea at Which the Marks of Aqueous Deposition Are Now Found. Strata and shells high in the mountains (attributed by Kirwan and other Neptunists to the Deluge) attest to Huttonian uplift.

Note xi. Fracture and Dislocation of the Strata. Slips, like veins, have been filled from below.

Note xii. Elevation and Inflexion of the Strata. Huttonian unconformities and the double raising up and letting down of the ancient strata attest to the high antiquity and great revolutions of the globe. (Playfair alludes to significant original fieldwork undertaken on behalf of the theory by other Huttonians and himself.) Examples include Siccar Point, the west coast of Scotland, Arran, Jedburgh, Yorkshire, and Cumberland. These and further instances of elevated and disturbed strata prove that the strata must have been pliant and soft when acquiring their present form, and that their curvature has been vertical only. Such curvature can be explained only by supposing that strata once flat and horizontal were impelled upward while still capable of flexibility (Saussure's theory of crystallization in place is therefore highly improbable). "Dr. Hutton's theory," Playfair concludes, "is nowhere stronger than in what relates to the elevation and inflexion of the strata, points in which all others are so egregiously defective" (234).

Note xiii. Metallic Veins. Specimens of native metals establish their igneous origin. The Huttonian theory holds that metallic veins were filled by the *violent* injection of fluid matter from below. The opposite belief, of filling from above, is refuted.

Note xiv. On Whinstone. Whinstone is neither volcanic nor aqueous, but rather of igneous origin—a truth discovered independently by John Strange and Hutton. Many supposed extinct volcanoes are really basaltic. Vulcanist arguments by Saint-Fond and Dolomieu, though faulty, are still infinitely closer to the truth than Neptunian alternatives. Still, Neptunism "has become the prevailing system of geology" (274), largely because of Werner. The latter's arguments on behalf of the aqueous origin of basalt (1791) are examined and refuted (but Playfair emends one of his own earlier statements) with evidence from Salisbury Crags, Arthur's Seat, and elsewhere.

Note xv. On Granite. Though Hutton was the first geologist to explain granitic veins, others had noticed them; they occur throughout Britain, and at Glen Tilt especially. Graphic granite can be seen at Portsoy. In an explicit disagreement with Hutton, Playfair asserts on his own authority that "granite does form strata where it has no character of gneiss" (327).

Though some examples cited by the Neptunists remain illusory, Saussure's observations are authoritative. The existence of stratified granite in no way embarrasses Hutton's theory, however. Though he responds to the anti-Huttonian arguments of Robert Jameson (*Mineralogy of the Scottish Isles*, 1800), Playfair regrets the necessity of controversial writing. He concedes to Kirwan that Hutton greatly understated the role of granite in forming the earth's crust. Regarding Scotland particularly, "Dr. Hutton has certainly erred considerably in defect" (346); as simple mathematics will establish, Scotland actually contains a large tract of granite (but Playfair allows Kirwan no credit for his correction of Hutton's mistake).

Note xvi. Rivers and Lakes. As erosive forces, rivers are capable of producing great effects. They have, in general, hollowed out their valleys, indeed whole systems of valleys. Some rivers were once chains of lakes; a lake, therefore, is "but a temporary and accidental condition of a river" (357). Examples include the English lakes and Lake Superior; waterfalls such as Niagara are constantly diminishing and will someday disappear. Polished rocks in waterfalls attest to constant wear. Lake Geneva should long ago have been filled up with sediments from the Rhone; one can only guess why it has not. Rivers have shaped our seacoasts also.

Note xvii. Remains of Decomposed Rocks. Extensive plains throughout Europe (formed from the debris of higher ground) demonstrate the effectiveness of erosion.

Note xviii. Transportation of Stones, etc. "It is a fact very generally observed that where the valleys among primitive mountains open into large plains the gravel of those plains consists of stones evidently derived from the mountains" (381). Saussure, thinking that gravel the product of a debacle, underestimated "the effects of action long continued" (383). "For the moving of large masses of rock," Playfair believes, "the most powerful engines without doubt which nature employs are the glaciers" (388). Huge fragments of rock may have been carried a great distance. Next in importance to the glaciers are the torrents. Fragments of rock that oppose the torrent are "rendered specifically lighter by the fluid in which they are immersed, and lose by that means at least a third part of their weight: they are, at the same time, impelled by a force proportional to the square of the velocity with which the water rushes against them, and proportional also to the quantity of gravel and stones which it has already put in motion" (390). To explain some instances of transportation, however, we must suppose that the mountains were higher and the ground smoother than they are now. Causes other than glaciers and torrents may also have been involved, but never a debacle. As part of a lengthy refutation, Playfair unexpectedly concedes that the great Scottish valley stretching from Inverness to Fort William was certainly not excavated by the rivers that now flow through it. He also notes Sir James Hall's support for the debacle concept in 1798.

Note xix. Transportation of Materials by the Sea. "When the detritus of the land is delivered by the rivers into the sea, the heaviest parts are deposited first, and the lighter are carried to a greater distance from the shore" (413). The mass of detritus brought seaward is so great that several narrow seas have shallowed. Currents distribute detritus from the land over the bottom of the sea. "Amid all the revolutions of the globe," Playfair continues, "the economy of nature has been uniform, in this respect, as well as in so many others, and her laws are the only thing that have resisted the general movement. The rivers and the rocks, the seas and the continents, have been changed in all their parts, but the laws which direct those changes, and the rules to which they are subject, have remained invariably the same" (421–422; later, the epigraph to Lyell's *Principles of Geology*, vol. I). Kirwan's objections to Hutton's concept of sea bottom deposition are then refuted (including an unusual estimate of the time required for continental erosion).

Note xx. Inequalities in the Planetary Motions. Elaborating another argument of his own, this time regarding Hutton's "no vestige of a beginning," Playfair discusses stability and wise design in planetary motions, as discovered by Lagrange and Laplace.

Note xxi. Changes in the Apparent Level of the Sea. "While the land has been gradually worn down by the operations on its surface, it has been raised up by the expansive forces acting from below. There is every reason to think that the elevation has not been uniform, but has been subject to a kind of oscillation, insomuch that the continents have both ascended and descended, or have had their level alternately raised and depressed, independently of all action at the surface, and this within a period comparatively of no great extent" (441–442; an important modification of Hutton's actual beliefs, but not a contradiction of them). Numerous examples prove how much sea level has changed; it was the land that moved.

Note xxii. Fossil Bones. Mammoth bones and others belong to animals now extinct. "The inhabitants of the globe, then, like all the other parts of it, are subject to change. It is not only the individual that perishes but whole species" (469; later, with additions from this same note, Lyell's epigraph to *Principles of Geology*, vol. II). The animals now found as fossils in Siberia must have lived there, and during a period which, "though beyond the limits of ordinary chronology, is posterior to the great revolutions on the earth's surface and the latest among geological epochas" (476; these ideas go far beyond Hutton and specifically, though not explicitly, contradict him).

Note xxiii. Geology of Kirwan and Deluc. Regarding Hutton's two most outspoken opponents, Deluc is a poor theorist but good observer while Kirwan is a cabinet-bound Wernerian who has never looked on nature with his own eyes.

Note xxiv. System of Buffon. Though Hutton's theory has been compared with Buffon's (by Kirwan and Jameson), they are actually very different.

Note xxv. Figure of the Earth. The earth, a solid spheroidal body compressed at the poles, has nearly the same shape it would have if liquid. To assume that it once *was* liquid, however, involves the Neptunist in many difficulties. A Vulcanist theory such as Buffon's is also inadequate. Hutton himself (a Plutonist) did not feel required to explain the oblate figure of the earth. For Playfair, on the other hand, this originally seemed a serious objection to the theory, and the purpose of this long note is therefore to reconcile that theory with the astronomy of Laplace and other moderns. (Playfair had published a related paper in 1797.)

Note xxvi. Prejudices Relating to the Theory of the Earth. To many, such a comprehensive geological theory as Hutton's will seem premature. After all, there have been a great many theories, most of them soon discredited. But Hutton's is founded on a solid body of established facts and observations, many of which others have also acknowledged. Had Saussure, for example, understood his own observations, he would have become a Plutonist. Dolomieu also approached, but likewise failed to achieve, Huttonian insights. Adequate materials for proposing a theory of the earth are now at hand. No natural discovery was more important to this endeavor than Black's discovery of the causticity produced in limestone by exposure to fire; before that, Hutton's concept of consolidation could not have been supported (though his theory was already partly formed by then). Similarly, other discoveries may also throw new light on geology.

Playfair's Modifications of Hutton

Of Playfair's notes—often, short essays—intended to illustrate Hutton's theory of the earth, the first twelve (excepting the problem in note 1) were pretty much orthodox attempts to explain and corroborate what Hutton actually thought. Note xiii, however, added an insistence on violence not found in the original. Note xiv, similarly, extended Hutton's thinking on basalt, particularly in the reply to Werner, an opponent whom Hutton never mentioned. In note xv, on granite, Playfair twice specifically contradicted Hutton but claimed that the corrections made no difference to the theory. Notes xvi and xvii are substantially Playfair's own work. In note xviii, Playfair's understanding of glaciers goes beyond Hutton's. His mathematical argument for the carrying power of streams, moreover, has no Huttonian equivalent. Hutton had supposed detritus from the land to reach the middle of the ocean primarily through the action of waves (*Theory*, II: 562, I: 14). Like Deluc in 1790 Kirwan saw the problem with this (potentially a serious one), so in note xix Playfair stressed the more plausible concept of distribution by currents. Note xx is entirely his. Note xxi, regarding oscillating continents, changed Hutton's theory fundamentally, immensely complicating its basic rhythms (as if adding epicycles to circular planetary motions). Likewise, note xxii, on fossil mammoth bones, seriously considered some recent data that Hutton had steadfastly opposed; Playfair's arguments straightforwardly contradict Hutton's. Finally, note xxv, on the figure of the earth, is again Playfair's own and places Hutton's entire theory within a context that Hutton himself had not considered. By 1802, in other words, we must clearly distinguish Hutton's theory per se from Playfair's modifications of it.

A comparison of Playfair's *Illustrations* as a whole with Hutton's *Theory* reveals several significant departures:

1. Playfair regards Hutton's theory as a human construct, not as a discovered system of nature.
2. More so than Hutton, Playfair regards the theory as open to and capable of improvement.
3. Because the theory belongs to Newtonian science rather than to philosophy or theology, Playfair assumes scientific presuppositions, without having to establish them.
4. Playfair assumes the existence of a science called geology, which is for him an aspect of cosmology.
5. Playfair normally regards Hutton's theory as nothing more than a description of physical processes, some of which are expressible mathematically or geometrically as laws.

6. Incessant processes, for Playfair, replace Huttonian cycles. The alternation of land and sea, for example, is no longer a matter of simple replacement.
7. Heat, rather than fire, is definitely the subterranean agent.
8. Heat causes depression as well as elevation.
9. Playfair is more prone than Hutton to ascribe violence to geological processes.
10. Playfair accepts extinction and a unique prehistoric past (compare his understanding of fossil Siberian mammoths in note xxii with Hutton's in *Theory*, II, chapter iv.)

Playfair also disagreed with Hutton about granite; distinguished more formally between stratified and unstratified rocks; abandoned Huttonian emphases on divine purpose, agricultural fertility, and methodology; and rejected his predecessor's theories of mountain building and a new Pacific continent. Finally, Playfair regarded science as in part a branch of literature and was therefore concerned (like the French) to write elegantly and well.

Before leaving Playfair's *Illustrations*, we should also notice the few parts of it not yet discussed. Among these, the most important is pages 135–140 (paragraphs 132–134), which follow the rhetorical climax that presumably ended Playfair's oral presentation in 1799. The original thinking presented in this section went well beyond mere exposition and must have been added later, after the notes were done. Both the last paragraph of Playfair's text and the first of his notes are numbered 134, indicating a last-minute addition.

Playfair argued in these added remarks that what gave Hutton's theory its unique character and exalted it infinitely above all others was its "principle of pressure, to modify the effects of heat when applied at the bottom of the sea" (135). Through this principle Hutton seemingly reconciled the appearances of fiery action with the apparent impossibility of subterranean fire. Hutton himself never saw his own theory in this light, however, and Playfair had not done so at any other time. Since the phrase "principle of pressure" is new at this point, one would expect a note, but there is none. It was Playfair rather than Hutton, moreover, who specified heat rather than some unknown subterranean force that Hutton often called fire. Playfair's second argument at this point, again entirely his own, was that Hutton's theory remained the only one that could satisfactorily explain the spheroidal figure of the earth. Though John Whitehurst had based his theory of 1778 on that very fact, Hutton neither explained nor mentioned it. Playfair's exposition then concluded with a splendid peroration affirming the progress of science, as future times would "fill up the bold outline which Dr. Hutton has traced with so masterly a hand" (139). The last part of the book to be written was an "advertisement" dated 1 March 1802 in which Playfair represented himself as nothing more than a humble expositor but admitted that "some arguments may have taken a new form, and some additions may have been made to a system naturally rich in the number and variety of its illustrations" (iii–iv). In actuality, Hutton's version of his own theory had now been consigned to a dormancy approaching oblivion.

Playfair's "Biographical Account of the Late Dr. James Hutton," 1805

On 10 January 1803, when the fortunes of Plutonist geology seemed even bleaker than the weather, John Playfair presented his long-delayed "Biographical Account of the Late Dr. James Hutton, F.R.S. Edin." before the Royal Society of Edinburgh. In so doing, he at last discharged an obligation of almost six years' standing and of which his respected but as yet less than triumphant *Illustrations* had been a by-product.

Following Hutton's death in March 1797, Playfair began almost immediately to collect information and reminiscences concerning him. He consulted Black, for example, but the elderly chemist had by that time given up his teaching post and would die himself within two years. Sir James Hall provided some early letters from Hutton to his father. John Clerk of Eldin, perhaps together with his son, must have volunteered details. Isabella Hutton, the one surviving sister, obviously told Playfair a good deal about her brother's life and habits. She also allowed him considerable access to the mound of unpublished material that Hutton had written and accumulated throughout his adult life. Playfair himself had significant firsthand knowledge of Hutton from about 1785 onward. Though they had at least been introduced well before then, it was surely their prolonged geological discussions following the presentation of Hutton's theory in March and April 1785 that drew them together. (Throughout the relationship, however, a certain deference to the older man was apparently required.) Yet, for all these witnesses, the "Biographical Account" derived largely from Hutton's own writings, whether published or not.[3]

We may presume that the meeting of 10 January 1803 was attended in large part by those members of the Royal Society of Edinburgh who had themselves known Hutton and were sympathetic to his memory; they had come to hear a eulogy. This Playfair surely intended to provide, but two distinct obstacles stood between him and easy success. The first was

[3]These inferences derive from the work itself.

that the intellectual accomplishment for which Hutton had been primarily remembered was his geological theory. The reputation of that theory, notorious but in no sense obscure, had fallen to its nadir and gave every indication of obsolescence, like the phlogiston theory in chemistry (which Hutton had also championed). The second obstacle to be overcome was that everyone who had known Hutton fully appreciated his eccentricity. To much of Edinburgh society his unfashionable dress and queer behavior had made him something of a joke. For years, visitors who knew of Hutton's interests and deliberately sought him out did so with the understanding that he was an oddity. Except for his sisters, the mature Hutton generally avoided women and their cutting displeasure throughout his life. Virtually all his socializing took place in men's clubs or at the houses of male friends, who were usually bachelors like himself. When Playfair rose to honor his friend's memory, therefore, it must have been with a subdued mixture of bravado and trepidation.

With customary directness, Playfair began by recalling what was known of Hutton's first quarter century, from his birth and education to his agricultural apprenticeship in Norfolk. Playfair had often heard Hutton reminisce about his two years in East Anglia, which were probably the happiest of his life. Except for some notes made by Hutton during his Highland excursion of 1764, however, Playfair had little specific information about his subject's activities away from Edinburgh. After 1768 (as Isabella recalled and manuscript evidence probably indicated), Hutton spent a great deal of time in private, reading, writing, and conducting chemical experiments. In many respects, however, science for him at this time was primarily an amusement. (Because he failed to appreciate the amount of fieldwork Hutton had already accomplished, Playfair almost certainly represented Hutton's geological and perhaps chemical researches as being more random than they were.) For the years from 1777 on, Playfair relied heavily on Hutton's publications, beginning with his pamphlet on coal and culm; he also noticed several tracts and essays now lost.

From 1785 onward, Playfair could speak with more authority, not only because his closer friendship with Hutton began then but also because the latter's announcement of his geological theory made him a bit more definable than previously. The "Biographical Account" (52–55) epitomized Hutton's theory by reducing it to five major assertions: (1) present-day rocks and strata have been formed out of earlier ones; (2) they were consolidated and lithified by subterranean heat; (3) these new strata were then elevated and distorted by the same force; (4) basaltic, granitic, and porphyritic veins were subsequently injected into these strata by the same force; (5) once elevated, all mineral bodies are subjected to incessant decay (thereby returning to step 1). Though this cycle ("revolution") had a beginning, and will no doubt have an end, we are unable to discern indications of either.

Playfair then summarized the earliest surviving version of Hutton's geological theory and speculated on the development of his thought, emphasizing the apparently central role of a chemical experiment by Black. In an appended note (which may have been added after January 1803), Playfair expressed considerable skepticism about the adequacy of Hutton's proof regarding subterranean compression, a matter presumably verifiable through experiment.[4]

Once Hutton's theory of the earth had actually appeared, Playfair continued, "several years elapsed before anyone showed himself publicly concerned about it, either as an enemy or a friend" (1805b, 61). Two causes, he thought, contributed to this neglect: the glut of unsuccessful theories, and Hutton's serious defects as a writer. Playfair then turned directly to Hutton's theory of rain (also published in 1788), which seemingly attracted more attention than his geological theory had. In a long and valuable note, Playfair recorded Hutton's further efforts in meteorology. He then sketched Hutton's fieldwork in 1785 (Glen Tilt), 1786 (Galloway), 1787 (Arran), and 1788 (Siccar Point; Isle of Man). The account of Glen Tilt corrected a mistake in Playfair's *Illustrations* regarding granitic veins; it and others also recalled drawings of geological phenomena made on these trips by John Clerk of Eldin at Hutton's request.

Playfair's Account of the Expedition to Siccar Point

Playfair's concise narrative of the Siccar Point adventure by Hutton, Hall, and himself and his description of the unconformity discovered there is the most famous field report in the literature of modern geology. Though its eloquent language is obviously Playfair's, the imaginative reconstruction of past geological events (resembling those in *Theory*, II, chapters vi and vii) remains Hutton's own, as Playfair himself affirmed.

> The ridge of the Lammermuir Hills, in the south of Scotland, consists of primary micaceous schistus and extends from St. Abb's Head westward till it joins the metalliferous mountains about the sources of the Clyde. The seacoast affords a transverse section of this alpine tract at its eastern extremity and exhibits the change from the primary to the secondary strata, both on the south and on the north. Dr. Hutton wished particularly to

[4]This note (59–60), like the added remarks in Playfair's *Illustrations*, (1802, 135), reflects the influence of Sir James Hall. Presumably, Hall read both the *Illustrations* and Playfair's life of Hutton in proof.

13. The Siccar Point unconformity. Though not at all concerned with landscape aesthetics (unlike the drawings by Clerk), this diagrammatic representation vividly portrays the stark incongruities of time and process that created one of the most important chapters in Hutton's "bible" (his term) of geological evidence. "Drawn by Sir James Hall," probably 1788. (Sir John Clerk of Penicuik)

14. The folded strata at Lumesden burn, reproduced in *Theory of the Earth* as plate IV. In his same conceptual, if unaesthetic, style, Hall again documents the existence in nature of a central Huttonian concept—in this case, the flexibility of strata under the influence of heat, pressure, and time. "By Sir James Hall," probably 1788. (Sir John Clerk of Penicuik)

examine the latter of these, and on this occasion Sir James Hall and I had the pleasure to accompany him.

We sailed in a boat from Dunglass, on a day when the fineness of the weather permitted us to keep close to the foot of the rocks which line the shore in that quarter, directing our course southwards in search of the termination of the secondary strata. We made for a high rocky point or headland, the Siccar, near which (from our observations on shore) we knew that the object we were in search of was likely to be discovered.

On landing at this point, we found that we actually trode on the primeval rock which forms alternately the base and the summit of the present land. It is here a micaceous schistus, in beds nearly vertical, highly indurated, and stretching from southeast to northwest. The surface of this rock runs with a moderate ascent from the level of low-water, at which we landed, nearly to that of high-water, where the schistus has a thin covering of red horizontal sandstone laid over it; and this sandstone, at the distance of a few yards farther back, rises into a very high perpendicular cliff. Here, therefore, the immediate contact of the two rocks is not only visible but is curiously dissected and laid open by the action of the waves. The rugged tops of the schistus are seen penetrating into the horizontal beds of sandstone, and the lowest of these last form a breccia containing fragments of schistus—some round and others angular—united by an arenaceous cement. Dr. Hutton was highly pleased with appearances that set in so clear a light the different formations of the parts which compose the exterior crust of the earth, and where all the circumstances were combined that could render the observation satisfactory and precise.

On us who saw these phenomena for the first time, the impression made will not easily be forgotten. The palpable evidence presented to us, of one of the most extraordinary and important facts in the natural history of the earth, gave a reality and substance to those theoretical speculations which, however probable, had never till now been directly authenticated by the testimony of the senses. We often said to ourselves, what clearer evidence could we have had of the different formation of these rocks, and of the long interval which separated their formation, had we actually seen them emerging from the bosom of the deep?

We felt ourselves necessarily carried back to the time when the schistus on which we stood was yet at the bottom of the sea, and when the sandstone before us was only beginning to be deposited (in the shape of sand or mud) from the waters of a superincumbent ocean. An epocha still more remote presented itself, when even the most ancient of these rocks, instead of standing upright in vertical beds, lay in horizontal planes at the bottom of the sea and was not yet disturbed by that immeasurable force which has burst asunder the solid pavement of the globe. Revolutions still more remote appeared in the distance of this extraordinary perspective. The mind seemed to grow giddy by looking so far into the abyss of time; and while we listened with earnestness and admiration to the philosopher who was now unfolding to us the order and series of these wonderful events, we became sensible how much farther reason may sometimes go than imagination can venture to follow.

> As for the rest, we were truly fortunate in the course we had pursued in this excursion; a great number of other curious and important facts presented themselves, and we returned, having collected in one day more ample materials for future speculation than have sometimes resulted from years of diligent and laborious research. (71–73)[5]

In further sections of the "Biographical Account," Playfair devoted a surprising amount of space to Hutton's philosophizing on physics and metaphysics. He had, Playfair stated, completed manuscript treatises on both sometime prior to 1781. Those on physics were apparently collected in Hutton's *Dissertations on Different Subjects in Natural Philosophy*, a portion of which Playfair summarized. In remarks original to himself, he also compared Hutton's thought in this area with that of Roger Joseph Boscovich, a Jesuit writer who had interested Dugald Stewart. Playfair insisted, however, that Hutton's physical theory was both independent and better. He then disposed quickly of Hutton's *Dissertation upon the Philosophy of Light, Heat, and Fire* in order to explore the ponderous three volumes of *An Investigation of the Principles of Knowledge*, emphasizing Hutton's acceptance of conventional morality, the existence of God, and the immortality of the soul. Playfair compared his thought with that of Berkeley.

From 1794 onwards, according to Playfair, Hutton was plagued by poor health and required some dangerous surgery. In this state he was attacked strongly by Kirwan for his theory of the earth. As a result, Hutton resolved immediately to enlarge that theory into a book, the first two volumes of which appeared in 1795. A third volume, necessary to complete the work, remained behind and was still in manuscript. Meanwhile, Hutton had gone on to write and nearly complete another work, tentatively called "Elements of Agriculture"; it remained in manuscript also. Hutton had then been seized once more by his recurring illness and confined at home, where in the winter of 1796–97 he received and read the last two volumes of Saussure's *Voyages dans les Alpes*, which had just appeared. On Saturday, 26 March 1797, Hutton wrote down some remarks on new attempts at the classification of minerals, experienced a spasm of shivering that evening, and died.

The final ten pages of Playfair's account characterized Hutton as an eminently well-qualified, if unconventional, natural philosopher, remarkable alike for observation and deduction. "There may have been other mineralogists who could describe as well the fracture, the figure, the smell, or the color of a specimen," Playfair supposed, "but there have been few who equalled him in regarding the characters which tell not

[5]Besides modernizing his accidentals, I have also revised Playfair's paragraphing. Shakespeare alluded to the "abysm of time" in the *Tempest* (I, ii, 50) and thought of looking into the "seeds of time" in *Macbeth*, a play any educated Scot would know.

15. Philosophers. James Hutton (left) and Joseph Black in 1787. From John Kay, *A Series of Original Portraits* (1837), I.

only what a fossil *is*, but what it *has been*, and declare the series of changes through which it has passed. His expertness in this art, the fineness of his observations, and the ingenuity of his reasonings were truly admirable" (89). The specimens in Hutton's own mineral cabinet, otherwise unsystematic, were all chosen for this purpose. Regrettably, a great deal of Hutton's knowledge perished with him.

As Playfair went on to recall, Hutton's unusually isolated life and neglect of self-conscious behavior sometimes led him into amusing eccentricities. His conversation, whether serious or jocular, was lively, original, and well informed. His friendship with Joseph Black was highly valued by them both. John Clerk of Eldin, another valuable friend (and surely among Playfair's audience), not only accompanied Hutton on his geological fieldtrips but undoubtedly contributed significantly to his theory. After commenting further on Hutton's social peculiarities and virtues (recalling the Oyster Club in particular), Playfair lauded his subject's "upright, candid, and sincere" character. His rock collection, presented by Isabella Hutton to Black, had since been given to the Royal Society of Edinburgh, to be arranged there and kept forever separate as an illustration of the Huttonian theory of the earth. (In actuality, the collection soon disappeared.)[6]

The Friday Club

Playfair was such an agreeable man that anything he wrote was certain to attract an audience. Though his resume of Hutton's life and affectionate evocations of his personality were no doubt well received, some surely wondered whether they had come to praise Hutton or to bury him. As if in answer to that dilemma, the most concerned Huttonians of Edinburgh (and some others, less involved) met in June 1803 to found the Friday Club, a social group originally meeting once a week at Bayle's Tavern in Shakespeare Square. Among the founding members were Playfair, Sir James Hall, Dugald Stewart, Walter Scott, Francis Jeffrey, Henry Brougham, Francis Horner, and Henry Mackenzie. Lord Webb Seymour joined the following year. As Henry Cockburn, another founding member, would later recall, "John Playfair was the person who, from our very first meeting till the summer in which he died, was unquestionably at our head. For sixteen years the Friday Club, both for the pleasure which he conferred and which he enjoyed, was the favorite scene of that inimitable person" (Cockburn 1910, 116). To a remarkable extent, these few persons—or rather a selection of them—maintained Huttonian positions against a formidable opposition lasting a quarter of a century.

[6]For Hutton's rock collection and what became of it, see J. Jones 1984. Playfair's biographical sketch of Hutton was apparently in print by 2 March 1805 (Robinson and McKie, 1970, 394).

[6]

Huttonian Controversy

The Advertisement prefacing Playfair's *Illustrations* is dated 1 March 1802. By around then, he must have returned to his printer the last proofs of his text and some final manuscript preliminaries. No index was planned, so only a few days more would have been required for publication, which took place in Edinburgh, presumably sometime in March. At this time, when the war against France had been temporarily suspended, Edinburgh was stirring with renewed cultural and literary energies, many of which were critical rather than creative. Certainly this was the case with Francis Jeffrey (1773–1850), a conspicuously unsuccessful young lawyer who had already contributed to *Monthly Review* and would later become well known (if not outright notorious) for his scathing critiques of major Romantic poets. Together with Sidney Smith, Henry Brougham, and other reform-minded contributors, Jeffrey helped to found the impressive *Edinburgh Review* (1802–1929), which from 1803 to 1829 he would edit.

Jeffrey and Murray on Playfair's *Illustrations of the Huttonian Theory*, 1802

In October 1802, the first month of its existence, *Edinburgh Review* published a fifteen-page discussion of Playfair's *Illustrations* by Jeffrey. "No species of physical research, perhaps," he proposed boldly, "is involved in greater intricacy, or productive of greater ambiguity of deduction, than that which respects the constitution and history of our globe"

(1802, 201). There were so many hypotheses, and they differed so fundamentally, that "attempts to form a theory of the earth may still be considered rather as exercises for fanciful and speculative minds than as sources of improvements to useful science" (201). Though Hutton could not be considered either a Vulcanist or a Neptunist, such Neptunists as Deluc, Kirwan, and Werner had strongly opposed him.

The very basis of Hutton's theory, Jeffrey proposed, was its assumption of a perpetual central heat within the earth—for which Playfair could not account. Yet even if one granted the Huttonians this important supposition, numerous difficulties remain. If there were such a heat, for example, it must either be decreasing through time or at best remaining uniform, since one could not imagine it increasing. According to Hutton, however, the degree of heat fluctuates—withdrawing, for instance, as continents are allowed to subside. More damagingly, Hutton apparently supposed that this central heat operated only under ocean basins. Once a new continent has emerged, Hutton seemed to say, it would not be further elevated to compensate for the downwearing so important to his theory. "It is difficult to conceive anything more fantastic and improbable," Jeffrey declared, "than those laws of action; and yet, without supposing that the central heat is subjected to them, Dr. Hutton can never make out his leading proposition, that the whole of the present land has been formed out of the ruins of that which had disappeared, and that the next generation of continents will be preceded by the submersion of the present" (203). As Jeffrey acknowledged, Playfair had asserted in note xxi that the expansive power acts under dry land as well. But Hutton himself did not propose this and could not logically have done so.

Since Hutton's theory now seemed inadequate overall, Jeffrey continued, its more specific inconsistencies were of lesser consequence. If the continents really were crumbling into the sea, however, and the sea bottom simultaneously elevated from beneath, sea levels would be rising worldwide. But in northern Europe they had been falling for many centuries. Second, it seemed to follow from Hutton's theory that the diameter of the earth is increasing—and would continue to do so until it reached the moon! Although Huttonian continents came and went, each was built on a crust more distended than the last.

By far the most objectionable parts of the Huttonian theory for Jeffrey were those regarding the separation, consolidation, and disturbance of the strata. If all our strata had really formed at the bottom of the sea, as Hutton supposed, they would probably resemble those heterogeneous mixtures commonly found at the mouths of rivers. In actuality, however, the strata of the earth tend to be consistent within themselves but highly various in sequence. Though Playfair had ably suggested difficulties in the Neptunian idea of consolidation, the Huttonian alternative was little better. Why, for example, do we so often find layers of relatively unconsolidated strata underlying solid limestone? Being nearer to the central heat, such unconsolidated strata should have become even more solidified than the overlying limestone. Furthermore, it was difficult to suppose strata softened by heat, which would more likely liquefy them. If impelled by central heat, mineral veins would probably burst through the strata they intersected. Hutton's arguments regarding granitic veins seemed stronger, but none of his school could account for the stratification of flint nodules, which are not randomly distributed (as the intrusion theory would suggest) but definitely in layers. Flints, moreover, are often associated with fossilized organic remains that would have been destroyed by such intense heat. Finally, Playfair had not responded adequately to an argument of Robert Jameson that the supposed igneous origin of granite was inconsistent with its occurrence in high mountains. If molten like lava, surely granite would have flowed over the surface of the earth.

Despite the unquestionable ability of its most recent expositor, Jeffrey concluded, the Huttonian theory was far from established. "It is a system, indeed, built on postulates so bold, and involving operations so prodigious, so capricious, and so incapable of exemplification from actual experience that we do not conceive it susceptible of any complete or satisfactory defense" (215). Nonetheless, Playfair himself must be highly praised for having constructed a well-informed and beautifully written argument. Whatever its ultimate fate, Hutton's theory was now before the public in its clearest and most skilfully argued form.

A second and longer reply to Playfair, written specifically from the Neptunist point of view, appeared later that same year, or perhaps at the beginning of the next. *A Comparative View of the Huttonian and Neptunian Systems of Geology: In Answer to the Illustrations of the Huttonian Theory of the Earth, by Professor Playfair* (1802) was the product of that John Murray, M. D. (d. 1820), who otherwise wrote and lectured on chemistry and pharmacy. Being a direct response to Playfair, Murray's 256-page book must have been written sometime between March and December 1802. Though primarily concerned with the *Illustrations*, Murray also knew Hutton's *Theory* of 1795 and quoted from it.

In a modest preface dated 8 December, Murray credited Playfair with having given discussion of the Huttonian theory "an interest and form in a great measure new." To Murray, however, the Huttonian doctrines, whatever their ingenuity and novelty, appeared all the more clearly "visionary, and inconsistent with the phenomena of geology" (1802, iii). Murray's book, therefore—though not written in any spirit of hostility—would evaluate the Huttonian and Neptunian systems impartially so as to establish the superiority of the latter.

Following an introduction in which Murray reviewed the vast antiquity and immense turmoil of geological history, the earth's once-fluid state, and the critical question of whether fusion or solution (fire or water) was primarily responsible for its present form, he proceeded in part I to restate the Huttonian and Neptunian theories. Besides its emphasis on regularity and design, Hutton's theory was characterized by imaginative grandeur. Broadly speaking, it consisted of four propositions: that the globe is a uniform system of decay and renovation; that an immense heat fuses and consolidates oceanic deposits derived from continental decay; that the subterranean fire operates under vast pressure; and that this fire's occasionally augmented expansive power elevated the strata. Hutton did not attempt to explain the first formation of the globe and assumed that its present state was not the original one.

The less pretentious Neptunian system, as devised by Werner, accounted for all strata as having been deposited at various periods and in various ways from a fluid. Granite had been deposited first, at a period before the advent of life. Other Primitive rocks (gneiss, schistus, porphyry) then followed, eventually becoming mountains as the earth's universal ocean gradually diminished in depth, draining into our planet's internal cavities. Partly mechanical but mostly chemical in origin, a group of Transition strata (varieties of limestone, schistus, and basalt) were laid down, some of them containing fossils. As the diminution of the waters continued, a suite of mechanically produced Secondary strata followed (sandstone, limestone, puddingstone, some basalt). These could generally be found in horizontal layers rich in fossils, the animal and vegetable kingdoms having by this time come to their full development. Above these, we find only some miscellaneous volcanic productions and alluvium wasted out from the Secondary strata by running water. Such, then, were the Huttonian and Neptunian theories.

In his part II, Murray evaluated the first principles of the two rival explanations. Though Huttonian emphasis on surface erosion appeared to be well founded, rocks such as granite and basalt degrade very slowly. Whether the detritus produced by erosion actually reached the depths of the sea seemed more debatable, and the whole notion of Huttonian decay presupposed an age for the earth that was beyond human conception. He also questioned Hutton's assertion that all strata were derived from the waste of a former world, the Wernerian view being fully as adequate. While generally accepting Hutton's concepts of universal decay and fluvial transport, therefore, we need not agree that these processes give rise to a succession of worlds, including one antecedent to the present.

The alleged principle of subterranean heat was particularly important to Murray because it provided "a direct demonstration of the falsity of the Huttonian hypothesis" (28). If this immense heat was produced by combustion, he argued, what fueled it and how was the necessary oxygen supplied? Who could imagine a heat of the intensity required? (Great mountain chains, for example, must have been uplifted together.) Calcium carbonate, of which limestone, chalk and marble are composed, cannot be fused by any heat *we* can produce artificially. Quartz is nearly as infusible, yet whole mountains are made of it. Any heat centered within the earth, moreover, would eventually diminish. Thus, "The system has within itself a principle of decay; its operations must, of necessity, have an end, and are incapable of producing that indefinite succession of worlds supposed by the Huttonian geologist and inseparably associated with the principles of his theory" (52). Murray then surveyed the Neptunian position, being particularly concerned to refute objections against it made by Hutton in 1795 (*Theory*, I, chap. ii, opposing Kirwan).

In part III Murray compared Huttonian and Neptunian explanations for the distortion of strata. Both sides agreed that stratified rocks could be horizontal, vertical, or at any angle in between. Hutton supposed that deposition had been horizontal in all cases, with subsequent orientations the product of uplift. Neptunists, on the other hand, thought contorted strata either to have been deposited in that state on an irregular base or to have assumed their bizarre forms through crystallization. Surprisingly, Murray favored the Huttonian position (which Saussure had also proposed in his final volume), believing that some recognition of the elevating power of subterranean heat—as a local and temporary force—would actually enhance the Neptunist alternative (112–113). But he rejected Hutton's general account of stratification, which could not explain alternating strata (successive layers of sandstone, limestone, sandstone, limestone, for example) or the occurrence of loosely consolidated layers beneath well compacted ones. In actuality, he thought, the Huttonian theory of stratification (insofar as valid) was the Neptunist one.

The two theories also differed greatly regarding unstratified rocks. Neptunists, for instance, considered granite a sediment and the oldest of all rocks; Hutton, on the other hand, believed it to be the youngest, and of igneous origin. These differences centered on the interpretation of granitic veins. Though Wernerian theory held that granite was Primitive, there were instances in which it was clearly of later formation. As Playfair had shown, therefore, granites could be of different ages; the same must then be said of basalt (126–129). Both granite and basalt, however, had evidently been deposited as veins from the percolation of a fluid. Layers of basalt are sometimes found alternating with those of other sediments; contrary to Playfair, moreover, Werner himself had reported instances of basalts and stratified sediments grading into one another. This phenomenon, one of the strongest proofs of the Wernerian system, virtually destroyed the Plutonic hypothesis (151). As one would expect, both granite and basalt were sometimes found stratified themselves, with the latter even containing fossils. The aqueous origin of basalt therefore was beyond doubt.

In the fourth and final part of his rebuttal, Murray discussed the support given each of the two contending theories by specific minerals and rocks. How could silicified wood, for example, possibly have been formed through fusion without the wood being carbonized? Since petrified wood is actually a form of agate, moreover, we may assume that all agates were deposited from water. To take the argument a step farther: agates often occurred in basalts; since the agates were deposited from water, the basalts must have been also. In a related manner, Hutton's theory of the formation of flint was totally at variance with appearances; the frequent association of these nodules with marine fossils proved their aqueous origin (183). In this whole area, then, the Neptunian theory was obviously superior to the Huttonian. A major area of concern, furthermore, and one almost totally ignored by Hutton, was the matter of animal and vegetable fossils. Given the delicacy of the original fish or plant, it was not conceivable that intense heat could have played any role in the consolidation of such fossils. Even Hutton's favorite ferruginous septaria were sometimes fossiliferous, thereby proving their aqueous origin. That coal, rock salt, and metals were igneous was, of course, highly improbable. And, finally, granite and whin were both clearly the products of water.

With all of the evidence now available, Murray concluded, one can easily see that the Huttonian theory is imaginary, fantastic, absurd, and physically impossible. It should therefore be consigned to the same oblivion that has overwhelmed other equally fanciful speculations from our prescientific past. The Neptunian theory, on the other hand, confines itself to the present world, explains only phenomena that presently exist, and does so through fair and legitimate deduction. Thus, "All the phenomena of geology conspire to prove that water has been the great agent by which minerals have been formed, and the surface of the earth arranged" (255). Despite the best efforts of both Hutton and Playfair to sustain it, then, the Huttonian theory of the earth must fail.

Playfair and Murray on Hall and Kennedy (1802)

As we would expect, both Playfair and Murray mentioned Sir James Hall and Robert Kennedy. "A series of the most interesting experiments, instituted by Sir James Hall and published in the *Transactions* of the Royal Society of Edinburgh," Playfair wrote, "has removed the only remaining objection that could be urged against the igneous origin of whinstone" (1802, 79–80). Those experiments had shown most satisfactorily that melted whin, if cooled *slowly*, does not vitrify but rather hardens into a stony substance hardly distinguishable from basalt or lava. Dr. Kennedy's experiments then proved that whin contains alkali, "by which, of course, its fusion must have been assisted" (80). In a further allusion to Hall's paper, Playfair accepted his proof of the intensity of volcanic fire (91–92); Hall's experiments, he affirmed, "have added much to the evidence of the Huttonian system; and, independently of all theory, have narrowed the circle of prejudice and error" (306–307). His observations on granite were also approvingly cited (316–317). As for Hall's advocacy of torrents, however, Playfair tactfully demurred (412).

Though Murray was professionally a chemist and included chemical arguments throughout his book, he mentioned Kennedy and Hall only in its closing pages. The former's conclusion that basalt and lava are nearly identical suggested to Murray that "lava has been formed from the fusion of trap; and if this be true, the resemblance between them is no proof whatever of their having had a similar origin" (1802: 245). For his part, Hall had clearly shown in several excellent experiments that when basalt is fused and then slowly cooled it assumes a stony character scarcely distinguishable from that of natural basalt. But to regard this evidence as in any way proving that basalt had originated from fusion was evidently a mistake. Having no volatile elements within it that fusion would affect, the basalt under investigation did nothing more than retain its former composition and physical qualities (249–250). Finally, though Kennedy's experiments had indeed shown that whinstone contains alkali (supposedly assisting fusion), such alkali would also have increased solubility. Murray therefore regarded the aqueous origin of basalt as having been indisputably established (251). Thus, without contesting the truth of either Hall's or Kennedy's experimental results, both Neptunians and Huttonians found within them only confirmations of their previous beliefs.

Distorting Mirrors

Various contributors now sought further evidence and arguments of disparate kinds. One such attempt, read (in absentia) before the Royal Society of Edinburgh on 7 March 1803, was "Remarks on the Basalts of the Coast of Antrim" by the Reverend William Richardson. In it, Richardson described some highly fossiliferous layers of what he called "siliceous basalt" to be found on the peninsula of Portrush, about six miles west of the Giant's Causeway in northern Ireland. The siliceous basalt (named by a foreign visitor two years earlier) alternated with layers of regular basalt, a fact affirming the common origin of them both. He was aware, however, that other mineralogists had refused to call the shell-bearing layers basalt. Interestingly, a note appended to the published version of this paper affirmed that several members present when the paper was read agreed with this objection, for cogent reasons: most of the supposedly siliceous basalt did not look like basalt at all; none of the undisputed basalt layers contained any traces of fossils; though veins often issued from the real basalt into the "siliceous," none ever originated in the latter; and, finally, the pseudo-basalt was stratified, whereas real basalt never is. Obviously, then, Richardson had failed to make his case.

He tried again (and in person) before a friendlier audience, the Royal Irish Academy in Dublin, on 2 May 1803, presenting his allegedly scientific "Inquiry into the Consistency of Dr. Hutton's Theory of the Earth with the Arrangement of the Strata and Other Phenomena on the Basaltic Coast of Antrim." Hiding behind a cowardly disclaimer that fooled no one, Richardson virtually equated Hutton with certain "antichristian conspirators" whose unbelief had helped to foment the horrors of the French Revolution. Like too many of them, Hutton seemed but one more example of those infidel geologists who sought to discredit the veracity of Moses. For Richardson, Hutton's conclusion that in the history of the world we can perceive neither a beginning nor an end was tantamount to an admission of atheism. Yet the Scot had apparently been highly esteemed by his contemporaries, and (however regrettable) many British naturalists have since adopted his theory.

After summarizing that theory briefly, Richardson attacked all its major suppositions: the surface of the earth, he declared, is not constantly in decay; strata are not consolidated at the bottom of the sea by heat and fusion; nor are they elevated by the same force. "It is difficult to find patience for the sober discussion of such a suite of extravagancies thus announced with solemn pomp," he fumed intemperately; to this theory, the face of nature in Antrim "gives the most direct and positive contradiction" (1803, 472). For example, all its basaltic strata are horizontal, and some of them include cavities filled with fresh water. Obviously, then, these strata could not have been subjected to fusion or uplift by subterranean forces. Like Kirwan in 1800, Richardson had earlier found marine fossils in the Portrush basalts. On the evidence, therefore, Hutton's pernicious theory—scientifically as well as theologically repugnant—lacked any kind of empirical justification.

Richardson's third and most significant attempt, read 10 December 1804 in Dublin, again before the Royal Irish Academy, was a long, three-part paper "On the Volcanic Theory." Ostensibly a belated reply to Desmarest's famous memoir of 1771, it was anti-Huttonian in design as well. After labeling Desmarest the inventor of the volcanic theory, Richardson went on to repeat the perverse arguments of the Abbé Augustin Barruel (*Histoire du Jacobinisme*, 1797; trans. R. Clifford as *Memoirs Illustrating the History of Jacobinism*, 1797–98) against him. Desmarest's intentions, said Barruel, were to give the lie to Moses regarding the creation of the world—its date, particularly—and, indeed, to engineer the extirpation of Christianity. Having devoted part 1 of his essay to Desmarest and his aims, Richardson examined the arguments of all who supported the volcanic theory in part 2, then advanced seemingly decisive arguments to the contrary in part 3, citing evidence from the peninsula of Kenbaan, near Ballycastle, where basalt and limestone were in frequent contact.

At some later time, Richardson met Sir James Hall, Dr. Thomas Hope, and David Brewster in Edinburgh. Accompanying these avid Huttonians to Salisbury Crags, he was shown the famous junction of sandstone and basalt (later called "Hutton's Section") with its one piece of sandstone entirely surrounded by intrusive basalt. "When Sir James had finished his lecture," Brewster recalled years later, "the Doctor [Richardson] did not attempt to explain the facts before him on any principle of his own; nor did he recur to the shallow evasion of regarding the enclosed sandstone as contemporaneous with the trap; but he burst out into the strongest expressions of contemptuous surprise, that a theory of the earth should be founded on such small and trivial appearances!" Accustomed to finding nature in her grandest manifestations, like the gigantic Irish cliffs, Richardson could not imagine how opinions thus formed might be shaken by such "minute irregularities" as those just shown him. Thoroughly flabbergasted by this reply, the Huttonians turned to other topics of conversation (Brewster 1837, 9–10n).

Parkinson's *Organic Remains*

Written in a charming epistolary style derived from Fontenelle, James Parkinson's *Organic Remains of a Former World* (vol. I, 1804) was primarily concerned, in its first volume, with the world of fossil plants. Despite this unusual context, Parkinson still found it necessary to consider Huttonian geology more than once. Thus, in letter XXIII, he reviewed opinions concerning the formation of coal:

> Dr. Hutton ingeniously labored to establish his hypothesis of the succession of worlds by a system of revolutions occurring at regular periods, each successive period of existence being, according to our measurement of time, of infinite duration; leaving us no vestige of a beginning, nor prospect of an end. Agreeable to this hypothesis, the Doctor imagined that coal

> is formed by the slow deposition of oily and bituminous matters at the bottom of the sea. These bituminous and oily matters he supposes to have originated in the dissolution of the various animal and vegetable bodies, which are continually perishing on the surface of the earth, and in the waters of the ocean (1804, 1: 240)

These various materials, Parkinson continued, "he supposed to subside together in an uniform manner, producing a stratified mass, which, becoming covered by an immense weight of superincumbent earth, must have been thereby exceedingly compressed and condensed, and finally consolidated, by the powerful influence of subterranean heat" (240). Though the matter was still much at issue, many of the French chemists agreed with Hutton in emphasizing subterranean fire and some kind of distillation promoted by it (244). Kirwan's theory, on the other hand, was entirely different.

Both Hutton and Playfair appeared in Parkinson's letter XXVIII, which dealt specifically with petrified woods in which the wood had been replaced by pyrite. Hutton himself had considered such specimens in both the 1788 and 1795 versions of his theory; for him, pyritization "is performed by means of heat or fusion; and there is no person skilled in chemistry that will pretend to say this may be done by aqueous solution" (1788, 235; 1795, I: 63). Playfair strongly affirmed Hutton's position in his *Illustrations* (1802, 33, 58–59), stressing metals and their ores. Parkinson, however, consigned both of them to obsolescence in this respect, their opinions being "no longer tenable, when compared with the ingenious experiments and observations of still more recent date" (1804, 1: 293). Citing contemporary sources, Parkinson upheld the aqueous origin of pyrites.

Letter XXIX then dealt more broadly with theories of petrifaction as such. Here again, proposals differed widely. Hutton, of course, believed siliceous fossil wood to have been formed by an injection of molten flint (1788, 234; 1795, I: 61) and thought he could disprove the aqueous alternative. As expected, Playfair also supported him in this (1802, 25). Utilizing Hutton's own criteria, however, Parkinson again found the opposing opinion stronger (1804, 1: 309–313). For him, wood petrified when saturated by silica-bearing waters effecting precise replacement through crystallization (315). A last controversy, in letter XXXII, regarding the formation of siliceous pebbles, ended similarly. Here again, the Hutton-Playfair assertion that agates and the like derive from heat, fusion, and injection was civily but decisively rejected (325–331). Thus, in three of four specific instances, informed opinion—most of it chemical—had examined aspects of the Huttonian theory and found it wanting.

Watt's "Observations on Basalt"

An altogether more promising effort was Gregory Watt's "Observations on Basalt, and on the Transition from the Vitreous to the Stony Texture Which Occurs in the Gradual Refrigeration of Melted Basalt, with Some Geological Remarks," read 10 May 1804 before the Royal Society of London and soon published in their *Transactions*. Having received an early copy of Hall's 1798 paper on whinstone and lava (Hall, 1805), Watt decided to repeat some of the same experiments, but on a grander scale. He therefore fused about seven hundredweight (approximately 785 pounds) of a local amorphous basalt called Rowley Rag in the furnace of an ironworks, achieving at first only a glass. Watt then traced correlations between crystallization and temperature. According to his lengthily described results with glass and other mineral substances, the formation of basalt by either fire or water seemed equally possible (1804a, 303). Unfortunately, Watt (1772–1804, son of the famous James Watt by his second marriage) died of consumption the same year.

Davy's Lectures on Geology

Between 1797 and 1798 the younger Watt had been closely associated with Humphry Davy (1778–1829) when both were living in Penzance, Cornwall—where Davy was born.[1] During these centrally important years, Davy (whose formal education ended at age fifteen) was serving a medical apprenticeship and developing his interest in chemistry. Watt, a graduate of Glasgow come to Penzance for his health, boarded with Davy's mother while Davy himself lived with his master, Dr. John Borlase. The two young men became fast friends, sharing their strong interests in geology and poetry at every opportunity. In October 1798 Davy moved to Bristol, where for two and a half years he worked closely with Thomas Beddoes at the latter's Pneumatic Medical Institution (which, based on experiments by Beddoes and James Watt, was attempting to treat consumption and other diseases with gases). Beddoes himself remained seriously interested in Huttonian geology, having published a significant, pro-Huttonian paper on the similarities of basalt and granite in 1790 after living in Edinburgh from 1784 to 1787. Under his guidance, Davy read some of the controversy's literature, including in March 1800 Kirwan's *Geological Essays*. The young chemist soon developed ideas of his own.

Following the publication of his *Researches, Chemical and Philosophical:*

[1]For Davy, see Treneer 1963 and David M. Knight in Gillispie 1970–80.

Chiefly Concerning Nitrous Oxide and Its Respiration (1800), Davy moved to the Royal Institution in London, where his reputation speedily increased. He had been recommended for his new position by Dr. Thomas Hope, the Huttonian professor of chemistry at Edinburgh, who had succeeded Black, his former teacher. During the summer of 1804, Davy made a geological tour of Scotland, meeting Playfair and Hall and going with them to inspect Salisbury Crags, Arthur's Seat, and Siccar Point. This tour then became an important influence on the ten-part series of weekly geological lectures undertaken by Davy in 1805, beginning on 7 February. No general summation of the present state of the science had yet been published (and specialized "geologists" scarcely existed), so Davy's efforts toward even a superficial overview were of public interest.

After an introductory lecture and another on geological references in Greek and Roman classics, Davy's third moved rapidly from ancient Alexandria through the Middle Ages, the Renaissance, and the Enlightenment to Buffon and (briefly) Hutton, the latter two having much in common. But the idea that our present land was formed in the bosom of the sea by the mere action of water," Davy assured his listeners, "may be controverted and overturned with ease" (1805, 44). Most of our high rocks and elevated mountains are nonfossiliferous, unstratified masses consisting of granite and other rocks which are "wholly incapable of being consolidated by water." And even stratified rocks, though they seemed to have been deposited by water, were "often blended with rocks of another description so as rather to indicate a great inundation and rapid transit of the ocean over the land than a gradual and quiet deposition of the solid matter of the earth in the bed of the sea" (45). In this respect, then, neither the Huttonian nor the Wernerian hypotheses (though he did not specifically name them) could be regarded as plausible. Davy next argued that, though the lapse of ages might cause some slight diminution of mountains by rivers, no significant changes of that kind had taken place in nearly forty centuries.

Lecture four surveyed geological theorists since Buffon, including Maillet, Whitehurst, Deluc, Wallerius, and Hutton. The appearance of Hutton's theory, Davy held, "may be almost considered as forming an era in the history of geology. Few works have ever excited more attention, few works have been more warmly admired or more vehemently attacked, and it has at least answered the important end of keeping the public interest alive to the facts and discoveries of the science" (54). Many of Hutton's separate ideas could be found in the works of his predecessors, including Hooke, Buffon, and particularly Lazzaro Moro the Vulcanist. But if he borrowed, Hutton also synthesized, adapting parts to the whole in an ingenious and orderly manner.

Davy then summarized Hutton's basic ideas regarding erosion, uplift, fusion, and cyclicity, contrasting them with the opinions of Werner and Kirwan, both of whom he praised for accomplishments other than their geological theories. "Controversies between the theorists who consider the present state of the globe as resulting from the agency of fire and those who regard it as the effect of the power of water have been carried on with great ardor," Davy added, "and the two systems have been named in opposition to each other, the Plutonic and the Neptunian" (57). But Dr. Hutton's theory was no less conjectural than those of Werner and Kirwan, "for as yet there have been no positive proofs that the principal materials of our globe are either soluble in water or capable of being fused by fire and deposited in regular forms in consequence of cooling" (57). Sir James Hall's masterly experiments notwithstanding, their results could not be regarded as analogous with the actual processes of nature. Davy could not close, however, without complimenting Playfair for the verbal excellence of his recent book, in which the questionable ideas of Hutton were developed with precision and beauty.

Lecture five began with a review of the primitive rocks, including occurrences in Cornwall, Scotland, and the Alps. Davy's remarks on granitic veins and similar intrusions favored Hutton somewhat, but he confessed that the general structure and appearance of the primitive rocks offered very few facts favorable to either hypothesis. "If the Huttonian or Plutonic theory of the formation of the Primitive rocks be founded upon conjectures," however, "the Neptunian geognosy of Werner and Kirwan is almost wholly built upon error and chemical impossibilities" (68), for they supposed stones and metals soluble in water yet precipitating from it without any changes in the fluid, which they held to both dissolve and crystallize. At present, there was *no* acceptable explanation for the origin of the primitive rocks.

In lecture six, Davy turned to secondary strata and the fossils they contained. Under this category he included basalt, which, he said, was usually unfossiliferous but could include coal or even the impressions of fishes. Lecture seven then speculated on the origins of various secondary rocks, a topic of great uncertainty. Regarding the origin of limestone, Davy again praised the experiments of Sir James Hall: "I have had the pleasure of examining his specimens and of witnessing his method," Davy acknowledged, "and I have seen pieces of compact stone, even containing crystals, which have been formed by heat under pressure from coarsely powdered chalk" (92). Yet many fossiliferous limestones in nature preserved unaltered casts of shells that could hardly have survived through fusion.

Analyses of calcareous and siliceous sandstones supported the same conclusion. Davy particularly remembered a quarry near Edinburgh in which intact pieces of whin had been found embedded in siliceous sandstone. Basalt is more easily affected by heat than quartz minerals are, so its unaltered presence proved that the sandstone was not consolidated through fusion. Chalk, like sandstone, must be aqueous in origin, and it is obvious that flint strata (whatever their actual origin) could not have been produced by fusion. Coal, despite some recent experiments by Hall, was of vegetable but unfused origins. Finally, Davy considered the problem of basalt. Here he again referred to his experiences in Edinburgh, exhibiting specimens from Arthur's Seat, and to Hall's experiments. Rather than defend a specific position on this most controversial issue, however, Davy alternatively endorsed both the Neptunist and Vulcanist positions, concluding with a eulogy of his departed friend Gregory Watt, whose paper of 1804 (proposing that the formation of basalt by either fire or water was equally plausible) had just appeared.

Lecture eight, on metallic veins, surveyed ancient opinions and Cornish mines before opposing the Huttonian and Wernerian views, both of which Davy believed to be inadequate. Lectures nine, on volcanoes, and ten, attributing volcanic fire to veins of pyrite, did not mention either Hutton or Werner, though Davy closed with an affirmation of divine design. Despite occasional vagueness in his positions, Davy—the most prestigious British chemist of his generation—promulgated an overall impression that the Huttonian theory of the earth (like its rival) had not been founded on an adequate understanding of nature.

The Leslie Affair, 1805

Early in 1805, John Playfair was appointed to the chair of natural philosophy at the University of Edinburgh, replacing John Robison, who had died. Playfair's own chair of mathematics being thus vacant, John Leslie (1766–1832), author of the highly regarded *Experimental Inquiry into the Nature and Properties of Heat* (1804), was nominated to fill it. But the prospect of Leslie's normally routine election to the chair by the town council of Edinburgh provoked unexpected opposition from the presbytery of the kirk, which claimed partial authority over university appointments and put forward another candidate.

Clerical objections to Leslie centered on two notes within his *Experimental Inquiry*. The sixteenth note, provoking a familiar Edinburgh charge of atheism, had endorsed David Hume's reasoning with regard to physical causation and was therefore regarded by some as an affront to natural religion. His twenty-fifth note, endorsing Huttonian geology, seemed to be an attack on the truth of Revelation. "Let any unprejudiced man peruse that note," the Reverend D. Knox proposed to the general assembly on 22 May 1805, "and say whether it be possible to believe that the author entertains any kind of respect for the Mosaic account of the Creation." Leslie's affirmations of immense geological time, vast cycles, slow Alpine erosion, and central fire were all offensive to Knox, who took the opportunity to castigate Playfair's extended chronology as well. Playfair himself responded effectively and bluntly in two pamphlets affirming the necessary freedom of geological inquiry and castigating the clergy for endeavoring to invade chairs of secular learning. Though the controversy continued into 1806, Leslie's appointment stood.[2]

Hall's "Account of a Series of Experiments," 1812

At this critical juncture, Sir James Hall presented before the Royal Society of Edinburgh his lengthy "Account of a Series of Experiments, Showing the Effects of Compression in Modifying the Action of Heat," which he read at a series of sessions, beginning 3 June 1805. "Whoever has attended to the structure of rocks and mountains," he began, "must be convinced that our globe has not always existed in its present state, but that every part of its mass, so far at least as our observations reach, has been agitated and subverted by the most violent revolutions" (1812, 71). But it was impossible to make much progress in geological theory until a developed science of chemistry emerged to assist the endeavor. Though partisan theorists advanced the rival claims of both fire and water, neither agency, as normally employed, could account for known appearances within the mineral kingdom. Until Dr. Black, through his discovery of carbonic acid, explained the nature of carbonates, no rational understanding of limestone—the earth's most important rock—was possible. Since Black proved that lime decomposed when heated, however, his discovery seemed hostile to any igneous theory of its origin. "The contemplation of this difficulty," Hall proposed, "led Dr. Hutton to view the action of fire in a manner peculiar to himself, and thus to form a geological theory by which . . . he has furnished the world with the true solution of one of the most interesting problems that has ever engaged the attention of men of science" (73–74).

Hutton, he said, supposed

> (1) that heat has acted, at some remote period, on all rocks; (2) that during the action of heat all these rocks (even such as now appear at the surface) lay covered by a superincumbent mass of great weight and strength; and (3) that in consequence of the combined action of heat and

[2]The Leslie affair was remembered by Cockburn (1856, 200–211), and by Brewster, (1837, 9n, from which the Knox quotation comes).

> pressure, effects were produced different from those of heat on common occasions—in particular, that the carbonate of lime was reduced to a state of fusion more or less complete, without any calcination.

For Hall, the "essential and characteristic principle of [Hutton's] theory is thus comprised in the word *Compression*" (74).

The theory itself, however, involved so many suppositions contrary to everyday experience that few persons had considered it in detail. Hall himself admitted to a long period of skepticism regarding it, "for I must own that on reading Dr. Hutton's first geological publication I was induced to reject his system entirely" (74) and would probably have continued to do so but for his intimate friendship with Hutton, whose lively, insightful conversation contrasted so strikingly with the obscurity of his writings. "After three years of almost daily warfare with Dr. Hutton on the subject of his theory," Hall continued, "I began to view his fundamental principles with less and less repugnance" (75).[3] He then proposed to Hutton that the latter undertake experiments to prove that the action of heat was indeed modified by pressure. "I urged him to make the attempt," Hall attested, "but he always rejected this proposal, on account of the immensity of the natural agents whose operations he supposed to lie far beyond the reach of our initation; and he seemed to imagine that any such attempt must undoubtedly fail, and thus throw discredit on opinions already sufficiently established, as he conceived, on other principles" (75–76). Being far from convinced of that himself, Hall thought it necessary to prove that compression would actually prevent the calcination of lime at high temperatures. In deference to Hutton's opinion, however, he had abstained from continuing the series of early experiments begun in 1790.

Following Hutton's death in 1797, Hall resumed his experiments on carbonate of lime. There were to be more than five hundred of them in all, culminating on 30 August 1804, when their results were laid before the Royal Society of Edinburgh, and now with the presentation of his lengthy paper describing them. After experimenting with various devices to achieve heat acting under pressure, in January 1798 Hall settled on common gun barrels, cut off at the touch hole and plugged at the breech with iron. Into them he put a paper cartridge filled with lime, rammed the rest full of pounded clay, and then had the muzzle sealed like the breech. When heated, several of Hall's gun barrels failed (at one time blowing up the furnace in which they had been placed), but a series

[3]The "three years" must have fallen somewhere between the latter half of 1786 (when Hall returned from abroad and married) and the first few months of 1790, when, in his paper on granite, he proclaimed himself a Huttonian. Hutton's "first geological publication" was his "Abstract," of which the Halls owned two copies (Eyles 1950).

of suggestive melts resulted. This phase of his research continued (with an interruption) through 1803. For more than a year, he experimented with porcelain tubes rather than gun barrels, and sometimes with oyster shells rather than limestone. Despite many failures, he was able to prove "that by mechanical constraint the carbonate of lime can be made to undergo strong heat, without calcination, and to retain almost the whole of its carbonic acid, which, in an open fire at the same temperature, would have been entirely driven off; and that, in these circumstances, heat produces some of the identical effects ascribed to it in the Huttonian theory" (95). This announcement had immediate repercussions.

Once Hall had presented the results of his experiments before the Royal Society of Edinburgh on 30 August 1804, he was besieged with comments and replies from eminent chemists and mineralogists. One section of his paper was therefore devoted to their objections, which called forth yet another series of experiments in the spring and summer of 1805. A critical issue was the possible influence of impurities. In response, Hall selected unusually pristine specimens (including some pure carbonate of lime prepared in a laboratory) and carefully protected them against contamination. This material proved to be troublesome but eventually yielded to a convincing solidarity. "These last experiments," Hall concluded, "seem to obviate every doubt that remained with respect to the fusibility of the purest carbonate, without the assistance of any foreign substance" (139).

In a concurrent series of experiments pursued throughout 1804, Hall examined the effects of heat and pressure on coal (as well as other organic substances). He did not achieve results comparable to those with lime, but the substance he obtained closely resembled that peculiar form of coal—which Hutton had discussed—sometimes found immediately adjacent to intrusive dykes of basalt. Experiments with horn and sawdust convinced Hall that both animal and vegetable remains contributed to the formation of coal. The severe reduction of both under heat, moreover, suggested that the coal now remaining to us represents but a small portion of the organic matter originally deposited. Hall reported his experiments with coal in Nicholson's *Journal* for October 1804.

In the final section of his voluminous paper of 1805, Hall considered the geological implications of his experimental discoveries. "The most powerful and essential agent of the Huttonian theory," he proposed, "is Fire, which I have always looked upon as the same with that of volcanoes, modified by circumstances which must, to a certain degree, take place in every lava previous to its eruption" (157). Though the actual source of the earth's internal fire was unknown, one needed only to know that it exists ("This topic, however, has of late been much urged against us"). Volcanic fire, like that of Vesuvius, Hall continued, "is subject to perpetual and irregular variations of intensity, and to sudden and violent renewal, after long periods of absolute cessation" (158). The fire in Hutton's theory, he thought, demonstrated exactly the same attributes, sometimes acting repeatedly in the same place and at other times ceasing entirely. Hall's just-concluded experiments on lime, combined with his earlier ones on lava, proved that even the feeblest exertions of volcanic fire were sufficiently intense to agglutinate, and sometimes entirely fuse, lime.

Hall went on to consider the mechanics of volcanic eruption, the related formation of limestone, the intrusion of basaltic dykes, and the elevation of strata. For Hutton, rocks once elevated are slowly eroded by rain, rivers, frosts, and other ordinary surface agents acting with the same force over a very long period of time. But Hall had never been able to agree with him on this, having always preferred the opinion of Saussure and others that "vast torrents, of depth sufficient to over-top our mountains, have swept along the surface of the earth, excavating valleys, undermining mountains, and carrying away whatever was unable to resist such powerful corrosion" (172). He had found evidence of such torrents around Edinburgh, as well as in Switzerland. With this important emendation, Hall believed the Huttonian theory of the earth to be satisfactorily established. Though many parts of it had not yet been fully worked out, the truth of its most questionable principle—the fusion of limestone under pressure—had now been placed beyond doubt by direct experiment.

[7]

Huttonians and Wernerians

By 1805 it was evident that a major theoretical dispute existed within the new science of geology, that the Huttonian and Wernerian hypotheses were mutually contradictory, that the outcome of the dispute would determine the later development of the science (with significant implications for all those taking part in it), and that no resolution was as yet in sight. During the next few years, then, newly gained adherents on both sides came forward to be heard, advancing points and arguments of various kinds.

Greenough and Silliman (1805)

A characteristic response to theoretical dilemma was that of George Bellas Greenough (1778–1856), who, after pursuing mineralogical endeavors in Germany, Cornwall (there meeting Davy), France, and Italy, toured Scotland in 1805 with Hutton and Werner very much in mind. After geologizing his way north from London, Greenough arrived on 1 July at Siccar Point, "an object of considerable interest to the geologist from the grandeur of the junction which takes place here of the primitive with the secondary strata."[1] Three days later, he met Sir James Hall in Edinburgh, where they examined basaltic dykes and Salisbury Crags together. Hall described for Greenough the series of experiments he was

[1]All quotations from Greenough in this section were originally transcribed by Rudwick (1962) from an unpublished diary.

currently presenting to the Royal Society, affirmed that heat and pressure regularly alter limestone to marble, and likewise supported Hutton's distinction between basalt and lava (no zeolite or spar being found in the latter). His own attachment to theory, Hall declared, in no way discredited his evidence but rather gave "strength to his cause and probability to his opinions."

Greenough returned to Salisbury Crags the next day, being particularly interested in contacts between its sandstone and basalt. During their previous visit, Hall had temporarily convinced Greenough that (contrary to Davy and the *Edinburgh Review*) the sandstone was noticeably harder when adjacent to the basalt, by reason of local fusion. On this occasion, however, that analysis seemed less persuasive. Finding some very hard sandstone in a location quite independent of the whin, Greenough instead attributed that hardness to the admixture of jasper and softness elsewhere to simple decomposition. Greenough then maintained this opinion on the 14th, when he examined Salisbury Crags for a third time, now in the company of Dr. Thomas Hope, the professor of chemistry at Edinburgh who was one of Hutton's strongest supporters. The devil himself, Hope argued picturesquely, "with all his omnipotence of confusion," could never contort, dislocate, indurate, or elevate strata by means of water alone. Hope, then, likewise supported the igneous origin of basalt.

Greenough left Edinburgh impressed with the grandeur of its geological setting, which (in contrast to that of London) would inspire even the ignorant to ask basic questions about the formation of rocks. While en route overland to the Western Isles, he found "Dr. Hutton's system of constant though gradual decay and disintegration which our earth is continually undergoing" becoming increasingly plausible to him. At Oban, he doubted the stratification of basalt. Then, on the isle of Staffa (which Davy had visited the year before), Greenough spent four hours perusing wonderful examples of columnar basalt, in which zeolites were often found. Whin dykes along the coast seemed marvelously grand but neither the product of violence nor perhaps of water. He believed himself to have detected "a palpable inconsistency in the Wernerian system," for basalt with the appearance of a primary rock and the occurrence of a secondary one belonged equally to both classes and therefore to neither. On the isle of Rum, where gigantic basaltic dykes were also evident, Greenough realized that it was "utterly impossible to account by the Wernerian theory for the formation of any large mineral veins that have not a perpendicular or at least a position considerably inclined to the horizon" because only gravitation could fill them.

As he immersed himself in the primary rocks of western Scotland and northern Ireland, Greenough became increasingly skeptical—of both the Huttonian and Wernerian hypotheses and of the accuracy of Robert Jameson's published geological descriptions. He objected to the Huttonian theory because he had found a mass of unaltered chalk completely imbedded in whin; why had it not been changed to marble? The Huttonian theory of flints, moreover, seemed highly improbable; if these nodules had really been injected into the chalk beds by subterranean heat, why were they always in horizontal layers? Similarly, Werner had taught that the more ancient rocks are always the more crystallized, but the most highly crystallized rocks that Greenough had seen (the castle hills of Stirling and Edinburgh) were decidedly of secondary origin. By now, therefore, he was prepared to discard at least portions of both theories, remaining diffident toward theory as such.

While traversing the Scottish Highlands in October, Greenough began to speculate on the nature of geological explanation. Amid the glens he found reasons to endorse Saussure's opinion (and Hutton's) that most valleys had been hollowed out by the streams running through them; they had certainly been formed slowly and gradually rather than by some sudden, violent force like the Deluge. The extensive gravels he found throughout the Highlands predisposed him to accept at least one of Playfair's points. "A man must have some credulity to believe in the Huttonian theory," Greenough noted, "but unless he believes it he cannot but see that the world is daily approaching towards its dissolution. . . . everywhere we see marks of decay, nowhere of renovation." From this time onwards, Greenough struggled inwardly to find acceptable proofs of Huttonian cyclicity. By 3 October he was at Blair Atholl and Glen Tilt, near rocks which had first confirmed Hutton's theory of granite. "Though I am by no means a convert to this theory," Greenough acknowledged, "I cannot but be impressed with sentiments of deep respect towards the author of it, a man of original genius who saw with his own eyes, who saw clearly, and knew what was worthy of being seen. If his conceptions were more clear than his language, this is the less to be regretted, since it produced Playfair's *Illustrations*."

In further remarks written the same day, Greenough pursued his Huttonian dilemma at length. Hutton, he mused, maintained that the consolidation of bodies could be effected only by heat under compression. Yet the more usual effect of heat was to dissipate rather than to concentrate. Such organically derived solids as bones, teeth, and shells, moreover, clearly were not formed by heat. The experiments of Sir James Hall, it is true, *had* shown that fused chalk in closed vessels could be turned to marble. But where in nature would the necessary pressure come from? And how might oily and bituminous particles *sink* in water so as to be converted into coal? While exploring Glen Tilt, Greenough was unable to trace the granitic veins there back to their parent rock. On the other hand, he also found objections no less formidable to the Wernerian system. Despite these difficulties, Greenough realized the value of hypotheses ("a system is a good thing—it leads us to the observation of facts and it serves to tie them together") and, at least with regard to basalt, he somewhat favored the Huttonian point of view. After returning to Edinburgh the same month, he geologized further with Sir James Hall and met professors Playfair and Hope before continuing his geological investigations on the isle of Arran—but his journal for that portion has not survived. Had he chosen to complete and publish his observations, Greenough might well have had a major impact on the Huttonian-Wernerian controversy. As it happened, however, he returned in silence to London, where, little more than two years later, he, Parkinson, and Davy, among others, founded the Geological Society, which was specifically committed to the discovery of facts rather than the promulgation of theories.

No such restraint characterized the prevailing milieu in Edinburgh, where the Huttonian-Wernerian controversy proved to be a perfect antidote for the scarcely sublimated feelings of pent up, frustrated aggression that were directed more obviously against France but sometimes more meaningfully against London. In a situation that persisted for several years, one prominent center of debate continued to be the University of Edinburgh, where both sides sought to obtain converts from among the younger generation. As Benjamin Silliman (1779–1864) of New Haven, Connecticut, would soon after record,

> There was no distinct course of geology in Edinburgh in 1805–6. Some dissatisfaction was indeed expressed regarding Professor Jameson—who had then recently returned from Werner's celebrated school of geology at Freiberg, in Saxony, and who was fully imbued with the doctrines of his great master—that he did not commence his course of instruction. He had, however, an able substitute in Dr. Murray [author of *A Comparative View*], who was a well-instructed and zealous advocate of the Wernerian theory on the agency of water while Dr. Hope, on the other hand, was an ardent and powerful supporter of the Huttonian or igneous theory. The discussions on these subjects were held in the midst of the chemical lectures, being introduced in connection with the elementary and proximate constitution of rocks and minerals (Fisher 1866, 1: 167).

During the five months that Silliman (later to be the first professor of geology at Yale) spent in Edinburgh, he tended to side with Murray, who advocated the aqueous origin of granite, porphyry, and basalt.

We have no better testimony than Silliman's regarding the spirit of controversy that then prevailed:

> The followers of Hutton were now organized into a geological phalanx, and my residence in Edinburgh occurred at the fortunate crisis when the combatants of both sides were in the field. . . . Being a young man, uncommitted to either theory, I was a deeply interested listener to the discussions of both the Wernerian and Huttonian hypotheses. From the fierce central heat of the philosophers of fire, and its destructive heavings and irruptions and overflows, I went to bathe in the cool ocean of Werner; and as both views were ably and eloquently sustained, the exercise was to me a delightful recreation and a most instructive study . . . and I was not long in coming to the conclusions that both theories were founded in truth. (169–170)

He therefore accepted the igneous origin of basalt but was slow to prefer other aspects of Hutton's theory and for some years did not.[2]

From the university, the controversy spilled over into a number of popular journals, particularly the *Edinburgh Review*. As we know, that publication had commissioned Francis Jeffrey to review Playfair's *Illustrations* in its opening number—which he did, unsympathetically—and likewise to dismiss Murray's *Comparative View* when it appeared soon afterwards. By 1803, however, Gregory Watt had become the *Edinburgh Review*'s resident critic on matters geological, an opportunity he used to good advantage. Thus, in the issue for April 1804, reviewing Scipio Breislak, and again in July, reviewing Dolomieu, he pointedly included and rejected Wernerian alternatives. After Watt's death in October 1804, James Headrick took over his post and continued to attack the Wernerians. Predictably, then, a lengthy review of Robert Jameson's *System of Mineralogy* (vol. 1, 1804) that month commended the author's previous works while chiding his Wernerian orthodoxy. "Basalt, as might be expected," Headrick wrote, "is considered of Neptunian origin; and Mr. Jameson roundly affirms, 'it is *now universally* admitted to be an aquatic production.' Does he consider his countrymen who follow the theory of Dr. Hutton as nonentities? Or how does he dispose of almost every mineralogist in France and Italy, and, among others, of M. d'Aubuisson, '*a scholar of Werner*,' and once a staunch stickler for the aqueous origin of basalt, but who, since he visited Auvergne, has candidly owned that some basalt at least is volcanic?" (Headrick 1804, 75n). In April 1805 a similarly dismissive review of Jameson's *Mineralogical Description of the County of Dumfries* (1805) by Headrick misspelled the author's name, upbraided his devotion to Werner again, and assured readers that the German's theory "will never apply to Scotland" (1805, 233). Salisbury Crags, Jameson had written, was essentially a coal formation. "Upon the whole," Headrick concluded, "it is perfectly apparent that, in place of a

[2]Silliman then appropriated the Huttonian theory of basalt for himself in a paper on the mineralogy of New Haven (1806; published 1810).

mineralogical account of Dumfriesshire, Mr. Jameson has only been solicitous to find a vehicle for his newly acquired theory from the school of Freyberg" (245). A less indoctrinated disciple might have temporized his theorizing with facts, but Jameson "never deviates into practical utility"; he could just as easily have written his book in Germany, without looking into Dumfries at all.

Finally, in October 1806 Playfair himself ventured into the morass of geological controversy, expending a dozen well-camouflaged pages on a close summary of Hall's "Account of a Series of Experiments" (1805), which had evidently become available as an author's separate, though not published officially until 1812. The accuracy and importance of Hall's conclusions seemed incontestable to Playfair, as did the existence of subterranean fire. Calcareous rocks, he supposed, might well have been consolidated by that fire, but evidence thus far remained problematical. Maintaining his disguise of impartiality, Playfair accordingly declined to endorse Hutton's theory unreservedly.[3]

Jameson and the Wernerian Natural History Society

Jameson himself was not indifferent to the calumnies leveled against him. By all indications, however, his responses to them remained sober and dignified. As both the Royal Society of Edinburgh and the newly formed (November 1807) Geological Society of London—of which he and Playfair were honorary members—leaned more strongly toward Hutton than otherwise, Jameson successfully founded a new learned society specifically intended to apply and promulgate the teachings of Werner while reflecting their German master's broad interests as well. On 12 January 1808 the Wernerian Natural History Society came into being, with Jameson as its president (for forty-six years!) and Werner its first honorary member. (The second, Richard Kirwan, would also be honored by the founding of a Kirwanian Society—Dublin, 1812—but little ever came of it.) Though Jameson soon proved for a time to be almost the only resident member who regularly contributed papers on geological subjects, the Wernerian Society promoted its international influence by extending honorary memberships to such prominent foreign geologists as Leopold von Buch, Alexander von Humboldt, J. F. d'Aubuisson, Louis Necker de Saussure, Jean André Deluc, J. F. Blumenbach, Scipione Breislak, Alexandre Brongniart, and Georges

[3]Attributions of authorship are based on those in Walter E. Houghton, ed., *The Wellesley Index to Victorian Periodicals*, vol. 1 (Toronto: University of Toronto Press, 1966).

Cuvier, with numerous Wernerians from Germany and elsewhere much in abundance. English honorary members included Humphry Davy, G. B. Greenough, and James Watt. Eventually, many of the most prominent naturalists in Europe and America came to be affiliated with the Wernerian Society in one way or another.

A fundamental difficulty attending the Wernerian position was that Werner himself had never published a definitive version of his gradually evolving theory, so that those adhering to it risked some confusion in doing so. The third volume of Jameson's *System of Mineralogy* (1808) then filled a need its author had first acknowledged eight years before, when his introduction to *Scottish Isles* regretted that "we have not, as yet, in any English publication" (1800, xiii) an account of Werner's theories. By 1808 a fairly adequate understanding of Werner's ideas had been available for some time, John Murray's *Comparative View of the Huttonian and Neptunian Systems of Geology* (1802) being essentially a handbook of Wernerian beliefs. Murray's *System of Chemistry* (1806–7), moreover, included a substantial appendix on the formation of minerals. Based primarily on Jameson's own lectures, it criticized the doctrine of subterranean heat and regarded Werner's theory as a much better approximation of observable natural facts than Hutton's. Now that Neptunian theories were coming under prolonged attack, however, Jameson realized that a proper explication of them under his own name was vital.

After giving an elementary description of the earth's surface in chapter 1, Jameson described the effects of water upon that surface. (Some readers may be surprised by the cogent arguments through which he emphasized the efficacy of both water and ice as present-day geological forces.) All valleys, to be sure, are not formed in the same manner, but rain and rivers certainly account for some. Icebergs transport erratic boulders great distances. If water destroys, moreover, it also creates, as new deposits of sediment are laid down daily within rivers and along our coasts. In addition to these mechanical functions, water fosters chemical deposition also.

Chapter 3, on the internal structure of the earth, began with a rapid look at the history of geological speculation ("still very limited and confused") that soon culminated with "the comprehensive mind of Werner," the "illustrious mineralogist to whom we owe almost everything that is truly valuable in this important branch of knowledge" (1804–8, 3: 41). "We should form a very false conception of the Wernerian geognosy," Jameson noted, "were we to believe it to have any resemblance to those monstrosities known under the name of *Theories of the Earth*" (42), an allusion to the as yet unmentioned Hutton. A four-part discussion of geological structures then included mountain rocks, mountain masses, rock formations, and the earth itself. Jameson's petrology was largely classificatory, but under "Mountain Masses" he discussed strata. Though differing considerably in the angle they form with the horizon, he stressed, "all inclined strata, with a very few exceptions, have been formed so originally, and do not owe their inclination to a subsequent change" (55–56). Among the more problematic rocks, gneiss is always stratified; granite, frequently; and syenite, sometimes.

Jameson next cautioned against "micrological" observations that fail to consider the structure of a mountain as a whole. "By not attending to this mode of examination, he continued, "geognosts have fallen into numberless errors" (56). That he had Hutton in mind becomes clear on page 344, where a note to page 56 is entitled "Huttonian Theory." By failing to consider nature on the grand scale, Jameson declared, we remain in such a state of ignorance as to discover in the globe marks of dislocation, contortion, and confusion that exist only in our own minds. "The late Dr. Hutton of this place," he then revealed, "a man of unquestionable ingenuity but very imperfectly skilled in mineralogy, inferred from his observations that the present world has been formed from the debris of two former worlds inhabited by numerous tribes of animals and clothed with a profusion of the most magnificent vegetables. According to this strange hypothesis, even gneiss, mica slate, and clay slate are but mechanical deposits which have been softened by the action of heat, so as to permit their being elevated without breaking from their supposed original horizontal position to their present vertical one" (344–345). If Jameson could establish the real existence of primitive rocks, however—those that had existed since the formation of the earth—"the Huttonian theory, notwithstanding the powerful eloquence used in its support [i.e., by Playfair], must be rejected as groundless" (345). His note attempted to do precisely that.

It is not surprising that the remainder of chapter 3 and all of the next seven deal with Werner's famous sequence of universal formations. These, global (though interrupted) in extent, he claimed, constituted much the greater part of the earth's crust; partial formations occur only here and there. Almost all the Primitive, Transition, and Floetz formations are universal, being products of a universal ocean; a partial formation, on the other hand, would be highly local, caused perhaps by a transient flood. As Jameson argued at length, the Wernerian sequence of universal formations provided a virtually complete geological history of the earth.

The spheroidal figure of the earth, he continued, is a proof of its original fluidity. Assertions to the contrary by Playfair and Hall notwithstanding, only water (rather than fire) could have produced that fluidity. A universal ocean must once have covered the entire earth, as many earlier theorists had dimly realized. Most probably, that ocean then diminished gradually, leaving on its bed a series of chemical deposits, beginning with granite. Because of the universal ocean, no mechanical deposition was possible. Since life had not yet appeared, moreover, Primitive rocks (though sedimentary) do not contain fossils. As the higher parts of the earth rose above the diminishing waters, however, mechanical deposits became increasingly possible; accordingly, they competed with and eventually almost replaced chemical deposition as a source of younger rocks. Transition rocks are therefore partially mechanical in origin; they are also the first to contain fossils (zoophytes). Floetz strata, in which limestones predominate, are even more mechanical and filled with such advanced marine fossils as shellfishes, fishes, and seaweeds. These organic remains were from creatures that differed entirely from those later seen on the earth. In newer formations, however, we find the remains of known genera, and the most recent fossils of all closely resemble species now living. Land plants appeared long after marine ones, and the land animals still later (chapter five). Following the Floetz strata, Werner recognized only partial formations, the Alluvial and Volcanic.

Though the theory of universal formations generally supposed a significant regularity in the order and form of sediments, several undeniable exceptions were known. In particular, what Jameson called the "Newest Floetz-Trap Formation" (including basalt) occurred widely but almost always unconformably, with great disorder and very little continuity. Thus, basalt, a coarse chemical precipitate in Jameson's view, irregularly overlay more mechanical deposits. "It is evident from the nature and position of these rocks," Jameson could only declare, "that they have been formed by a vast deluge." (84). To explain this embarrassing anomaly, he, like Werner, had as it were to run the film backwards. Contrary to its normal, orderly diminution, the world ocean must have risen rapidly, remained calm at its higher level long enough to lay down the new formation, and then swiftly retired to its former level. Heaps of petrified wood and the remains of land animals such as deer showed how recently this untoward event took place; the resulting formation was often found on mountain tops.

Other passages of Huttonian interest appear in chapter 6, where Jameson alluded specifically to Hutton's researches on granite (109–110n, 368) and clay slate (121–122, 350–351, 147, 152). Chapter 8, on Floetz rocks, includes an important discussion of basalt (185–189), in which Werner and his students had unquestionably discovered fossils (187, 204). Hollows on the summits of basaltic hills had sometimes been misidentified as volcanic craters by superficial observers (188). Obsidian and pumice were not volcanic rocks (195–196). As Werner had argued, volcanic fire was but a local phenomenon caused by the ignition of coal beds (199, 219–221).

Finally, having concluded his systematic exposition of the Wernerian formations, Jameson turned to the question of mineral repositories, including beds and veins. His discussion of the latter (for which Werner's book of 1791 was naturally a major source) climaxed with a note on the Huttonian theory of veins. As part of his energetic defense of the Wernerian position (that mineral veins are filled from above, by aqueous deposition; not from below, by subterranean forces) Jameson quoted—and seemingly refuted—Playfair's *Illustrations* no fewer than seven times. As he believed, "we may very safely challenge the supporters of the Huttonian theory to produce a single instance, in the Primitive, Transition, Floetz, or Alluvial rocks, of a vein whose characters show that it has been formed by the injection of fluid matter from below" (367). In retrospect, however, 1808 was perhaps the last year in which Wernerian assertions of any kind might be regarded as safe. Within his book, for example, Jameson also announced a forthcoming volume of *Geognostic Essays* (188); it was destined never to appear.

Miscellaneous Contributors

Though its author certainly did not acknowledge his contribution as such, further evidence of the weakening Wernerian position appeared in a public letter addressed to Humphry Davy by William Richardson and read before the Royal Society of London on 17 March 1808. In his epistle, Richardson recalled field investigations made by Davy and himself, apparently in 1806, among the basaltic phenomena of Derry and Antrim in northern Ireland. Emphasizing the great promontory of Bengore particularly, Richardson detailed each of its sixteen "strata." A characteristic occurrence throughout this region, he claimed, was that strata once obviously contiguous were no longer so. Had the force that separated them come from beneath or from above? The first hypothesis, he argued, could not be correct because the basaltic strata had remained horizontal. Playfair's advocacy of rivers as erosive agencies was likewise incorrect because their depredations had been "comparatively insignificant" (1808, 209) and, like subaerial erosion and internal decay, tended to diminish rather than augment inequalities. The sharply delineated inequalities of the basaltic district in northern Ireland, then, attested to the "performance of operations to which we have seen nothing similar" (210) and to the "former existence of powers of far superior energy to any we have ever known in action" (210–211). These powers must have acted from above. Thus, Richardson denied both Huttonian and Wernerian theories to favor some sort of catastrophism evidently requiring a swift, exceptionally powerful deluge.

Though he was no longer resident within Edinburgh, Francis Horner (who had heard Playfair's epoch-making exposition of Hutton when first presented and subsequently became a founder of the *Edinburgh Review*, as well as of the Geological Society of London) maintained close ties with the city through his friendship with Playfair's favorite companion, Lord Webb Seymour. On more than one occasion, Horner (1778–1817) attempted to cope with the possible implications of Huttonian cyclicity as a theory of history.[4] On 8 July 1808, for example, he wrote to an associate regarding his anxieties about the present warfare in Spain. The outcome of the struggle there (in which British troops were engaged) would, he thought, determine the fate of Europe. A British victory would revive that nation's determination to maintain the cause of freedom, whereas a defeat would lead to widespread despair that either liberty or prosperity would be seen again in any part of Europe. *Was* the history of the world progressive or must we forever regress to barbarism? "And shall we have future theories of the moral history of the earth," he asked, "like Hutton's system of its changes upon the surface, tracing back many former transitions from civilization to barbarism, and presenting, in the future prospect, an endless, irksome succession of the same changes?" (1853, 1: 55).

A letter to Seymour on 27 January 1809 tried to cope with the same problem in a more naturalistic manner. Ever since modern astronomers had completed the Newtonian system of the universe, it was generally assumed that supposed irregularities in the cosmos are periodic and constant, all the revolutions of the system being uniform, orderly, and repetitious. Hutton's theory clearly proposed the same regarding the surface of the earth; for him, change was in no sense progressive. But was there not a chemical objection to this? If every mineral is an oxide—as Horner assumed—then the process of oxydation is progressive and could be traced back (at least in theory) to the first instance. "I have no doubt," he concluded, "that the principle of periodic revolution will be found to prevail throughout nature, but I cannot state to myself in what manner this apparent exception is to be reduced under it" (476).

By late August 1811 Horner had gone beyond these theoretical considerations and was utilizing Playfair's *Illustrations* as his guidebook to the geology of Devonshire. He was particularly interested in seeing the important junction of primary and secondary rocks (schistus and sandstone) that Playfair and Seymour had visited in 1799 and which the former then described in his paragraph 189. For Playfair, such junctions "present to the senses the most striking monuments of the high antiquity and great revolutions of the globe" (1802, 212). Greatly impressed with Playfair's chaste eloquence, Horner readily commended the book as a stylistic model: "Though he has varnished over too partially the imper-

[4]See, e.g., Horner 1853.

fections of Dr. Hutton's reasoning respecting the hypothesis of a central fire," Horner advised his brother Leonard, "and in this particular has probably violated the rules of just philosophy, yet I know no work in which the logic of inductive reasoning may be learned to more advantage than in the *Illustrations*" (1853, 2: 84–85). Though most of "the hypothetical part" of Hutton's theory would probably be discarded as geological knowledge progressed, Playfair's book would remain standard, both for its apt description of the nature and object of geology and for its rules and examples of proper geological thinking.

A very different attitude toward Playfair had appeared two years earlier in Jean André Deluc's *Elementary Treatise on Geology* (1809), which, as *Monthly Review* noted, was "in reality a controversial work, consisting almost entirely of an attack on the hypothesis of the late Dr. Hutton, as given to the world by his friend Professor Playfair."[5] Like Murray in his *Comparative View*, Deluc often followed Playfair in the order of his topics, his two chief arguments being the incompatibility of Huttonian geology with even a liberal reading of Scripture ("days" being periods of time) and the somewhat more effective point that his own firsthand knowledge of Alpine geology gave his views an authority to which Hutton could not lay claim. Besides controverting a series of Huttonian arguments (often by turning them inside out) and denying the applicability of Hall's experiments, Deluc also thought it necessary to reprint anti-Huttonian letters he had originally published during the 1790s.

Once these remarks had been acceptably received, Deluc, now over 80, continued his attacks on Hutton and Playfair in a series of geological travel accounts. To place his intentions beyond doubt, Deluc began the first of these (*Geological Travels, I: Travels in the North of Europe*, 1810) with a detailed comparison of his geological system and Hutton's, arranged under no fewer than twenty-seven headings. Among his major contentions were that rivers have not excavated the valleys through which they flow; that the ocean does not erode its coastlines; that geological changes are primarily catastrophic; and that our continents, which cannot have existed for long, were produced by the subsidence of others. Throughout the work as a whole (including two further volumes of travels in England, 1811), Deluc insisted endlessly on the superiority of his observations to Playfair's. He continued, moreover, to affirm biblical chronology and the Deluge.

Though not specifically intended as such, an effective Huttonian rejoinder to Deluc's old-fashioned but still appealing arguments on behalf of creationism, abridged chronology, and catastrophic forces was Sir George Steuart Mackenzie's *Travels in the Island of Iceland* (1811), record-

[5]*Monthly Review* 63 (1810): 493.

ing a trip the previous summer. "The contest between the Huttonians and Wernerians," Mackenzie (1780–1848) wrote in a chapter devoted to mineralogy, "has become so keen . . . that new facts bearing upon any system must not be received unless they are distinctly unfolded" (1811, 357). Though naturalists had long been fascinated with volcanoes and their productions, he continued, it was not until Playfair's *Illustrations* had dissipated the obscurity of Hutton's prose that they began to understand the importance of compression. Since then, many ingenious arguments had appeared to oppose Hutton and support Werner. Whatever difficulties the Huttonian doctrine of central heat might entail, however, it was surely more reasonable than the vacillating ocean postulated by Wernerians. Specimens brought with him from Iceland, Mackenzie believed, established the igneous origin of obsidian and pumice. Active present-day volcanoes, moreover, were merely the later manifestations of the same subterranean fire that first broke out at the bottom of the ocean, filling its basin with subterranean lavas, some of which then reached the surface to form land. In addition, the same force often elevated rocks of other kinds, creating the presently uneven surface of the globe—including huge mountains that rain and rivers could not possibly have carved out by themselves. Thus, in defending Hutton, Mackenzie jettisoned a great part of his theory.

Scipio Breislak (1748–1826) similarly accepted only portions of Hutton's theory in his *Introduzione alla Geologia* (Introduction to geology, 1811) but strongly criticized Neptunist positions. "I respect the standard raised by Werner," he remarked, "but the flag of the marvellous and mysterious will never be that which I shall choose to follow."[6] On the whole, his Vulcanist objections to Wernerian dogmas proved to be more durable than the alternative theory Breislak proposed, which utilized the expansive power of hypothetical subterranean gases to crack and distort the earth's crust. For him, granite, porphyry, gneiss, and schist were all igneous rocks originating from the cooling of primeval matter. But Hutton's theory seemed to him insufficient, among other reasons, because the bottom of the sea had been found to be as cold as its surface. Nonetheless, he spoke well of the Huttonians and particularly praised Hall for the quality of his experiments. *Edinburgh Review* called Breislak "the most zealous champion of the Plutonic cause" in Italy (148).

Breislak's book, of which there were soon French and German versions, clearly indicated that the Huttonian-Wernerian controversy could no longer be dismissed by Continental geologists as a foreign squabble. Among further signs of increasing cosmopolitanism was the visit to Edinburgh in 1811 of Louis Simond, a French geological traveler who

[6]As translated by Geikie (1905, 257).

then devoted almost twenty pages of his subsequently published journal to the Huttonian theory (an English translation appeared in 1815). Though generally unknown to foreign readers, Simond maintained, Hutton's theory seemed nonetheless "the most probable of any existing explanation of the awful revolutions our world has evidently undergone" (1815, 2: 1–2) and was therefore worth summarizing. Simond's familiarity with the subject included not only Playfair's *Illustrations*, which he praised, but also Jeffrey's review of it, Hall's experiments, Jameson's *Mineralogy*, and Mackenzie's forthcoming book on Iceland. From Hall in person he accepted both the distinction between basalt and lava ("a most curious confirmation of the Huttonian theory") and "overwhelming floods" (2: 8, 15). It was apparently Thomas Allan who took Simond on the inevitable tour of Arthur's Seat and Salisbury Crags.

That March, following Simond's departure, Allan read to the Royal Society of Edinburgh an instructive paper on local rocks. He began by noting the recent surge of interest in geological questions, of which Sir George Mackenzie's hazardous voyage to Iceland was such a fine example. "In the present state of our knowledge," he believed, "to divest geology of theory would be to deprive it of all its interest. . . . A system of geology may yet be formed, founded exclusively on the phenomena of nature, or at least on reasoning much less hypothetical than is now required" (1812, 406). Toward this end, a careful examination of natural facts would seem to be most expedient.

Arthur's Seat and Salisbury Crags, "in all probability, first suggested the theory of Hutton," which, beautiful as it was, Allan could by no means accept entirely. He was "very far indeed from following him through his formation and consolidation of strata or the transportation and arrangement of the materials of which they are composed." The unusually contorted strata of Berwickshire, moreover, while "totally irreconcilable with any other hypothesis, are yet but imperfectly explained by his" (408).

Being primarily concerned to establish a general theory, Hutton did not describe particular districts as such. Neither he nor any other geological writer, amazingly enough, had ever elucidated the geology of Edinburgh and vicinity in detail. The only comprehensive terminology available to the attempt, moreover, was that of Werner. "The Huttonian theory," Allan pointed out, "has no language peculiar to itself" (411). In accord with his position regarding facts, he then presented to the Society an extensive collection of local rocks begun some time ago. A major interest of his in securing specimens was to reinforce the Huttonian interpretation of veins.

No part of Salisbury Crags had attracted more scientific interest than a section in which a mass of sandstone had been totally enclosed by basalt ("greenstone"). "This spot," Allan noted, "has been the scene of much controversy between contending geologists. While the Huttonian considers it as a most incontrovertible proof of violence and heat, the Wernerian contends that there is nothing in the least extraordinary in the appearance" (423). Allan's own remarks solidly affirmed Hutton's earlier interpretation, that the basalt had originally been fluid. He then went on to discuss induration as such, contrasting Huttonian and Wernerian viewpoints to the detriment of the latter.

Hall's "Convolutions" and "Revolutions" (both 1815)

Sir James Hall also thought it necessary to preserve Hutton by emending him. On 3 February 1812, somewhat in reply to Allan, he addressed the Royal Society of Edinburgh "On the Vertical Position and Convolutions of Certain Strata and Their Relation with Granite." All across Scotland—from Galloway to Berwickshire—Hall began, there was an extensive exposure of the rock commonly called greywacke—but which he preferred to call killas (a Cornish term). This formation was everywhere stratified, and the strata lay parallel to each other, but were more likely to be vertical or bent than horizontal. Since the killas was particularly well exposed along the Berwickshire coast (exhibiting a series of regular bendings, some two or three hundred feet in height), Hall investigated this region during the summer of 1811. As he soon ascertained, the killas strata must originally have been deposited horizontally and were then compelled to assume their present contorted shape by the exertions of "some powerful mechanical force" (1815a, 82). This idea had first occurred to Hall in 1788, when visiting the Berwickshire coast with Hutton and Playfair (a famous sketch of that coast by him appeared in Hutton's *Theory*, vol. I.). Specifically, he assumed, these originally aqueous deposits "had been urged, when in a soft but tough and ductile state, by a powerful force acting horizontally" (84), which was then opposed either by the inherent resistance of the rocky mass or by an equivalent force acting in the opposite direction. Overlying strata also weighed down the killas but could be displaced by sufficient force.

Being convinced of this interpretation, Hall decided to illustrate it by experiment. He piled several pieces of cloth (representing the killas strata) on a table, placed an unhinged wooden door and some weights over them (representing the superincumbent pressure), and then moved two vertical boards at the ends of the pile toward each other by hammering them with a mallet. In consequence, the pile shortened, the weighted door was gradually raised, and the cloth strata developed folds very much resembling those of the Berwickshire killas. Afterward, he showed the Society a machine in which the same process could be duplicated.

The horizontal thrust whose existence had now been demonstrated, Hall continued, arose as a natural consequence of Huttonian elevation through subterranean heat, the same agency we see manifested in volcanoes. Recalling some observations he had earlier made regarding Vesuvius (in his 1805 paper "On the Effects of Heat Modified by Compression") and Sir George Mackenzie's discoveries in Iceland, Hall related phenomena in both places with the strata now considered. Though immediately decisive field evidence was lacking, he remained convinced that some upward-moving plutonic intrusion of liquid granite had been the agent of distortion. Limestone strata of the Alps, he pointed out, were similarly convoluted. In reasoning on them, Saussure had considered (but rejected) the possibility of deformation by subterranean fire. The insufficiency of Saussure's alternative theory (crystallization), however, had been demonstrated by Playfair in his *Illustrations* (para. 207).

Hall next appeared before the Royal Society of Edinburgh on 16 March 1812 to read "On the Revolutions of the Earth's Surface," which continued but expanded his previous line of thought. "We are never more disposed to give credit to a philosophical system," Hall began, echoing Playfair, "than when we meet with a case of its successful application unknown to the author or containing circumstances which he had not taken into account when he formed that system" (1815b, 139). He now chose to argue for the "operation of immense torrents" in the geological past which, though not a part of the Huttonian theory as conceived by either Hutton or Playfair, was nonetheless compatible with it. Though long supportive of what he considered the essence of the Huttonian theory, Hall "could never help differing from Dr. Hutton as to the particular mode in which he conceived our continents to have risen from the bottom of the sea" (140).

Even while discussing Hutton's theory with him, Hall had retained an attachment to Saussure's Alpine observations (which Hall had verified for himself by personal inspection) and to the Russian ones of P. S. Pallas (of mammoths and other tropical animals found fossilized in Siberia), both of which seemed to prove the existence of torrential deluges. He had therefore been disposed to combine Hutton's insights with those of Saussure and Pallas.

Hall then reviewed the standard problem of erratic granite boulders, which had somehow moved fifty-four miles from Mont Blanc across the limestone Jura and Lake Geneva. Clearly, they had been transported by something. Saussure had ascribed their movement to a vast torrent or debacle of water, whereas Hutton and Playfair had thought ordinary river action sufficient. "The task," Hall declared flatly, "is beyond the power of any river that flows on the surface of the earth" (142). Alternatively, Deluc, while accepting Saussure's debacle, had supposed that the granite erratics had been ejected from below, from hypothetical caverns filled with hypothetical elastic fluids. Professor Karl Friedrich Wrede (1766–1826) of Berlin had even suggested that the boulders had been rafted into place atop floats of ice originating in the North Sea.

Hall also disagreed fundamentally with Hutton in two other respects. "The theory which he had advanced respecting the formation of valleys, by mere diurnal actions, appears to me liable to great objections," he confessed, endorsing arguments raised by Deluc against Playfair in that respect. He also concurred with Sir George Mackenzie in rejecting slow atmospheric erosion. Lengthy comparisons of Hutton, Playfair, Pallas, Saussure, and others followed.

In his previous paper, on the Berwickshire killas, Hall had pointed out that volcanic ducts opened during an eruption are then sealed by it as well, so that to begin the next eruptions new openings must always be made, occasioning a rather violent discharge. He now proposed that this same mechanism operated during Huttonian continental uplift: "However gradual and uniform the propelling force may have been previous to its accumulation," he declared, "the ultimate laceration must have been performed by a sudden and violent motion, producing an earthquake at the surface" (149). "It is well known," he continued in a note, "that Dr. Hutton conceived that the subterranean heat, as well as that of the volcanoes, was subject to short fits of activity, with ages of intermission" (150n). This concept of plutonic action, he believed, could fulfill all the conditions of Saussure's or Pallas's debacles. As part of his lengthy discussion of earthquakes, Hall described recent experiments in which he had exploded gunpowder under water in order to simulate what we now call tsunamis.

While uniting the ideas of Saussure and Hutton, Hall likewise accepted Wrede's suggestion that the granite erratics of Switzerland had been transported by floating ice. Though fully aware of glaciers and their moraines, however, Hall still required a "sudden diluvian wave" to connect the boulders with their source. He also attributed to his diluvian agency the large deposits of sand so evident in Holland and adjacent parts of Europe. As for lakes and valleys, they probably began as crustal splits occasioned by the violence of continental uplift. Both the coast of Berwickshire and Siccar Point were cited as further evidence. "I think myself authorized by the facts," he then concluded Part I by saying, "in deviating so far from the Huttonian hypothesis as to believe that the elevation of the land was performed by successive starts, similar to volcanic eruptions, though far more rare and more powerful" (166–167).

Hall read the second half of his paper on 8 June 1812, calling it "On the Revolutions of the Earth's Surface, Part II: Being an Account of the Diluvian Facts in the Neighbourhood of Edinburgh." "If such tremendous agents did in reality exert their influence in the Alps," he proposed without preliminaries, "it is not conceivable that other countries situated lower and composed of materials much more frail could have been spared" (1815b, 169). There was in fact evidence to that effect all over the globe, even locally. Hall then pointed out a variety of phenomena (almost all of them glacial, as we know but he did not) inexplicable by any ordinary cause. Mighty torrents, therefore, must have traversed these districts. Though the Huttonian doctrine of elevation, to which Hall heartily subscribed, suggested a series of inundations, however, all the local evidence seemed to derive from a single debacle of immense power but only a few minutes' duration. Western Scotland, he went on to say, seemed to have undergone two in succession, from nearly opposite directions. "In thus controverting some of the collateral opinions of Dr. Hutton and Mr. Playfair," Hall concluded, "I venture to hope that my arguments, which have been founded on their principles, and which have led me to acquiesce in their most general and important conclusions, may tend less to weaken than to confirm the results of their immortal labors" (210).

At the conclusion of his earlier paper on Berwickshire strata (researches in which his son, Basil Hall, had accompanied him), Hall acknowledged having just learned from "a young friend" currently at the Cape of Good Hope in Africa that at Table Mountain junctions of granite with killas and sandstone had been found. This letter and some further communications eventually became Capt. Basil Hall's "Account of the Structure of Table Mountain," as compiled by Playfair and read on 31 May 1813. The great mass of sandstone atop Table Mountain, Basil Hall proposed, had been forced into its present position by the intrusion of granite, a form of subterranean lava. This intrusion, Playfair then added, must have taken place beneath the sea, with quiet and regular uplift following. For him, Basil Hall's evidence established the plutonic origin of granite beyond question.

The fervor of the Huttonian-Wernerian controversy thus culminated between 1810 and 1813 in a series of publications favorable to some form of Huttonian belief; there had been almost no Wernerian replies of any consequence. By this time, moreover, the Huttonian faction had assumed virtually complete control of the Royal Society of Edinburgh. Sir James Hall was president; Lord Webb Seymour, one of its two vice-presidents; Playfair, secretary; Allan, keeper of the museum and library; Mackenzie, president of the physical class; and Hope, secretary of the latter. Despite this impressive predominance, however, European opinion as a whole and the public in general failed to support the new outlook. The next tasks would therefore be to preserve as many of Hutton's concepts as possible, to find further confirmations of them, and to convince the world of their validity.

[8]

The Triumph of Facts

In 1811 Humphry Davy gave another series of geological lectures; the first defined the Plutonian and Neptunian hypotheses as he understood them. Plutonism, in which our continents were simultaneously being decayed by elemental forces and renovated by central fire, had for Davy originated with Robert Hooke in the seventeenth century. But in his own time many persons found the existence of central fire unbelievable. Davy therefore proposed an alternative theory in which chemical and electrical forces within the earth might well accomplish the same effects. Among other geological writers, he recommended Jameson. His second lecture described granite as a primary rock, the highest and deepest. In lecture three, basalt was likewise sedimentary. The fourth lecture, on veins, seemed to affirm Werner. Lecture five acknowledged the decomposition of rocks but held that equilibrium was maintained through the effects of vegetation, the production of corals, and the operation of volcanic fires. In the sixth and final lecture, Davy considered the causes and effects of volcanoes, once more denying the existence of any global central fire but stressing (as Sir William Hamilton and other eighteenth-century writers had) their renovative function. The earth, he concluded, is a wonderful contrivance attesting the wisdom and power of the Deity.

These lectures reached the public in several forms, including summaries in *Philosophical Magazine* (Davy 1811) and in an Edinburgh pamphlet published by Thomas Allan, who objected privately to Davy's evident Neptunist leanings and consequent injustice to Hutton. "In my next course of lectures," Davy promised in reply, "I shall modify many of the doctrines, and certainly mould the whole somewhat more into a Plutonic form. I think you make me more unjust than I conceive I was (certainly more unjust than I intended to be) to Dr. Hutton and his enlightened and powerful philosophical defenders." Davy's own inclination, he claimed, "had always been to fire," but he also thought superheated water necessary to explain siliceous formations. Hutton's theory, considered as a theory, explained more geological phenomena with fewer inherent difficulties than any other. Even so, the remaining difficulties (which Davy did not specify) were very real and would have to be removed before the theory as a whole could be accepted. (J. Davy 1858, 134–135).

Encyclopedias

Transactions of both the Geological Society of London and the Wernerian Natural History Society of Edinburgh first appeared in 1811, but neither proved to be immediately effective in swaying public opinion. When ordinary citizens sought to learn something of the new science of geology, they more often consulted their encyclopedias—all of which supported Werner. *Nicholson's Encyclopedia* (London, 1809) considered Hutton significant but still slightly radical and devoted most of its article on geology to Werner, "to whom, in the opinion of his learned and zealous annotator [Jameson], we owe almost everything that is truly valuable in this important branch of knowledge." Its article "Rock" also followed Jameson on Werner, "on account of its so exactly corresponding with the appearances which masses of rocks everywhere present to our view." Concurrently, a second article on geology, this time in the fourth edition of the prestigious *Encyclopedia Britannica* (1810, 9:550–628), reviewed various theories of the earth at length, including Hutton's (596–599), objected to all of them, and referred its readers to John Murray's *Comparative View* (1802) and the works of Richard Kirwan (*Geological Essays*, 1799). This out-of-date appraisal was reprinted unchanged in 1815 and 1823. Charles König's "Geology" article in Abraham Rees's *New Cyclopedia* (London, 1802–20), published on 29 November 1810, derided the pretentions of Hutton's and all other theories of the earth while endorsing the more empirical system of Werner. König's "Mineralogy," for the same encyclopedia in 1813, was equally Wernerian. A more specialized work also intended for laymen, John Pinkerton's *Petralogy: A Treatise on Rocks* (London, 1811), offered yet another confident exposition of the superior Wernerian system. Almost the only defense immediately available to Huttonians was to criticize the frequently cumbersome and decidedly foreign Wernerian terminology, which they caustically dismissed as unintelligible jargon.[1]

Playfair's Reviews, 1811–12

Faced with such influential opposition, John Playfair attempted to gain a wider audience for orthodox Huttonianism (of which he was by now almost the only advocate) through a series of geological commentaries in *Edinburgh Review*.[2] The first of these definitely assignable to him, published in May 1811, was occasioned by the appearance of Charles Anderson's translation of Werner's *New Theory of the Formation of Veins* in 1809, the German original having appeared in 1791. As part of a general commentary on Werner, Playfair had to concede both the reality of his fame and the unquestionable superiority of his descriptive techniques, which had been of great service to mineralogy. But he then went on to summarize the Wernerian theory of veins (being filled from above by water), which through "certain and clear demonstration" he felt able to refute. Though Werner had unquestionably contributed to the advancement of both mineralogy and geology, his system was now outdated and had become an impediment to further progress.

That November (of 1811), Playfair—not Fitton, as earlier scholars supposed—reviewed the Geological Society of London's first volume of *Transactions*. Beginning with some clever mockery of the pretentions of early geological theorizers, he quickly dismissed Buffon, Moro, and even Werner ("of all others the most in vogue at the present moment") in order to advocate the dispassionate collection of facts, a program to which the Geological Society itself generally subscribed (Playfair 1811c). Henry Holland's paper on rock salt deposits in Cheshire, for example, seemed to be drawn up in the right style of natural history, uniting accurate detail with general views and describing phenomena without any contamination of hypothesis or theory. Yet Playfair defended Hutton's more theoretical description of the same deposit. The Pitch Lake of Trinidad, as described by Nicholas Nugent, "could not fail to be highly

[1] *Nicholson's Encyclopedia* was unpaginated; "Geology" is in vol. 3, "Rock," in vol. 5. For Rees's *New Cyclopedia*, see Pestana (1979).

[2] Playfair's reviews are listed individually in the Bibliography. My chief authority for attributions is again *The Wellesley Index to Victorian Periodicals*, vol. 1, which, for example, gives that of November 1811, reviewing *Transactions of the Geological Society of London*, vol. 1, specifically to Playfair. That it was not Fitton's is also clear from W. H. Fitton, "Articles Published in the *Edinburgh Review*, 1817–1849," a unique collection of offprints assembled and bound by Fitton, then presented by him to his son, Frederick Chambers Fitton, in October 1856; it is now in my possession.

gratifying to those who embrace the Huttonian theory of the earth" (Nugent himself remarked) because appearances there strongly supported the concept of subterranean fire. Observations on the geology of Devonshire and Cornwall by J. F. Berger of Geneva suffered from theoretical bias, overlooking the implications of phenomena apparently contrary to Werner—his erroneous theory of veins, in particular. Remarks by Arthur Aikin on the mineralogy of Shropshire affirmed the former fluidity of basaltic strata but stopped short of insisting that the fluidity had originally been igneous and did not clearly prefer either the Plutonic or Neptunian systems. Leonard Horner, describing the mineralogy of the Malvern Hills, remained similarly noncommittal. Finally, a sketch of the geology of Madeira by H. G. Bennett strikingly confirmed the volcanic origin of basalt, the experiments of Sir James Hall, and the Icelandic observations of Sir George Mackenzie. Slowly but surely, Playfair implied, the Huttonian position was coming to be established.

Playfair's next essay (1812a, appearing in February) dealt with Mackenzie's *Travels in the Island of Iceland* (1811). Like the book itself, most of Playfair's summarizing remarks were not specifically geological. In an attentive discussion of Mackenzie's mineralogical chapter, however, he proposed to review the leading facts "without any theory." Almost all the rocks of southwestern Iceland, Playfair noted, were either greenstone, basalt, or some kind of lava. They were often difficult to tell apart (indeed, we in the twentieth century know these terms to be virtually synonymous), but while calcite is often found in greenstone and basalt it is always missing from lava. Obsidian and pumice are also found in Iceland, and in circumstances which leave no doubt as to their volcanic origin (a point argued at some length). Mackenzie had then distinguished two distinct kinds of lava, common and cavernous. The latter gave no appearance of having ever flowed; rather, it had melted and bubbled up in place, often creating caverns when the huge bubbles burst. Sir George regarded this widespread phenomenon as proving the existence of subterraneous heat.

Playfair did what he could to promote geological observations favorable to Hutton's hypotheses; he also repeatedly emphasized perspectives unflattering to Werner's. In this regard, no geological advance seemed to him more opportune than the repeated discoveries of stratigraphical sequences not accountable by (and sometimes in obvious contradiction to) the supposedly invariable Wernerian one. His chief ally in this regard was the French comparative anatomist Georges Cuvier (1769–1832), who had once argued that the "great object of geology ought to be to determine the relative position of rocks" and whether that sequence was limited by any invariable law. Though Cuvier expressed his gratitude to the Wernerian school for their accomplishments, he also saw clearly that their dogmatic formational sequence was being contradicted at every turn.

Having alluded to Cuvier's significance in his review of Werner on veins, Playfair (May, 1811b) then publicized three of the Frenchman's epochal papers on fossil elephants. In confronting Cuvier's evidence for the extinction of vertebrate species, which he did not contest, Playfair had also to cope with the sudden floods or debacles such theorizing appeared to endorse. Steadfastly opposing any catastrophic explanation, Playfair contended that the accumulation of vertebrate fossils (in Siberia especially) was too great to have been formed in an instant; such prehistoric graveyards were rather the accumulation of ages. "This much at least may be considered as certain," he asserted, "that the explanation of these fossil bones is to be derived either from a submersion of the continents under water, quietly and without agitation, or from the accidents which occur in the ordinary course of nature. All other hypotheses seem to be excluded" (229). Thus, he implicitly rejected any geological corroboration of a unique global catastrophe.

In his review of the Geological Society of London's *Transactions* that November, Playfair returned more directly to the problem of stratigraphical sequences. As he advised pointedly in responding to Berger on Cornwall, "we cannot help thinking that the Wernerian geology is faulty in directing the attention of the mineral surveyor to some favorite points and withdrawing it from the rest. The order in which the strata succeeded seems to be the great object to which the mineralogists of that school are inclined to attend, and the order fixed on by Werner being very precise—and very different, we imagine, from that which nature has adopted—the person who would reconcile the one with the other has an abundance of work upon his hands" (1811c, 222). Similarly, in Aikin's paper on Shropshire, coal was found to be resting immediately on a Transition rock, "from which, on the Wernerian system, it is represented as extremely distant" (225). In closing, he urged the Geological Society as a whole to pursue delineations of strata.

In November the next year, Playfair noticed (1812c) Cuvier and Brongniart's innovative essay on the mineralogical geography of the environs of Paris (1811), a detailed examination of what soon came to be called Tertiary strata. In describing these layers, Cuvier and Brongniart had adapted the Wernerian term "formation" to designate beds of the same or different nature all formed at the same period. Most of the formations they discovered, however, had no place within the Wernerian scheme and were irreconcilable with it—the alternation of marine and freshwater deposits, for example. Cuvier and Brongniart had therefore proposed a totally new stratigraphic sequence that might be found elsewhere as well. Having established the vertical succession of that sequence, they traced the geographical extensions of various strata also, much as Playfair had only recently urged the Geological Society of London to do in England. Thus, for Playfair, Cuvier and Brongniart had now demonstrated the fallacy of Werner's supposed universal formations, overthrown a great geological idol (in the Baconian sense), exposed the absurdity of the worship paid to it, and removed a powerful obstacle to the further improvement of science.

On the whole, Playfair's attacks (or, more broadly, the considerable impact on British geology of Cuvier and Brongniart's discoveries) proved to be remarkably successful. For the next decade and more, such geologists as James Parkinson, William Buckland, Thomas Webster, John Macculloch, Gideon Mantell, Charles Lyell, W. H. Fitton, William Phillips, and W. D. Conybeare would often be concerned to establish correlations between French and British strata, with Cuvier and Brongniart the great authority and Werner sometimes almost totally forgotten. (The independent contributions of William Smith and his followers had already been acknowledged, but Smith himself was never a Huttonian and his story needs no retelling here.) After 1811, it is fair to say, the Wernerian sequence of universal formations was frequently ignored by English and many other geologists.

Bakewell, Townsend, and Cuvier (1813)

Robert Bakewell's often irascible *Introduction to Geology* (1813), considered the first modern textbook of the science in English, was primarily Huttonian. "I am inclined to think," its author wrote, "that the part of Dr. Hutton's theory which related to the igneous origin of basaltic rocks is as well established as the nature of the subject will admit of" (1813, 113), though other parts were far less satisfactory. "Mr. Werner and most of all of his disciples who deny the igneous origin of basalt," Bakewell continued, "have never visited active volcanoes and seem disposed to close their eyes upon their existence" (113–114). After pages of further discussion, Bakewell concluded that "the confident, not to say arrogant, manner" in which Werner's theory had been supported, "considering the preposterous claims which it makes on our credulity, is truly ridiculous and will form an amusing page in the future history of science." That theory, he prophesied uncharitably, "will be preserved from oblivion embalmed in its own absurdity" (229).

Bakewell's second edition (1815) took up the issue of stratigraphical disparity more directly. It was, for example, one of the first British tomes dealing in part with mineralogy to forego Wernerian terminology entirely. "The great mistake," Bakewell wrote in a new preface, "was in fixing these terms as the expression of invariable natural characters, and in asserting that the same succession of rocks that occurs in Saxony, which Werner first described, exists, or once existed, in every part of the world." "Since the universal evidence of facts in other countries made the Wernerian arrangement no longer tenable," he continued, "there has been such shuffling and shifting of rocks from one class to another that it threatened to upset the whole surface of the globe" (1815, xiv). Bakewell believed geological forces in the past to have been more powerful than present-day ones and thought the action of subterranean heat less ubiquitous or uniform than Hutton supposed (chap. 16) but otherwise adopted Huttonian positions. In 1819, Rees's *New Cyclopedia,* previously Wernerian, published new articles on geology and several related topics—including Werner himself—by Bakewell, thus becoming the first encyclopedic work to favor a Huttonian outlook.

In 1813—the same year as Bakewell's successful first edition—the Reverend Joseph Townsend affirmed *The Character of Moses Established for Veracity as an Historian, Recording Events from the Creation to the Deluge* (1813; vol. 2, 1815, considered events subsequent to the Deluge and was linguistic rather than geological). A competent geologist and no crackpot, Townsend ranked among the first British writers to recognize and promulgate the stratigraphical discoveries of William Smith, who also affirmed the Deluge. One section of his lengthy discourse comprised "General Observations on the Huttonian Theory of the Earth" (1:376–397). "He was not an Atheist," Townsend declared at the outset, "but . . . a firm believer in the wisdom and power of the Creator" (377). Yet Hutton had attributed to heat, rain, frost, and wind effects produced by submarine convulsions and the ravages of powerful currents before our continents emerged from the ocean. "The agents assigned by him," Townsend adjudged, "are not adequate to the effects produced; and, supposing them to have been adequate, would have required a longer space of time than other appearances of nature can admit of" (380–381). Not having followed subsequent controversies in detail, one supposes, Townsend based his dismissal of Hutton on Deluc's letters in *Monthly Review* for 1790 and 1791. That same periodical, moreover—in discussing Townsend—thought the old objections still valid and considered theories of the earth per se rather tedious. "At all events," it concluded, "the discussion of the two rival systems of fire and water has been too frequently agitated to afford either instruction or amusement to our philosophical readers" (235).[3]

As happens so often, there was a distinct correlation between the

[3]On Townsend: *Monthly Review* 74 (1814): 225–238. Townsend was probably responding to Deluc 1809b, in which the 1790s letters were reprinted.

geological beliefs that made sense to Europeans just emerging from the devastating Napoleonic era and the more recent history to which their lives had been at pawn. Whatever the rationalizations, endless time and the slow creep of Huttonian explanation had been almost universally abandoned. For those who had endured the world of Napoleon, history consisted of newspaper headlines filled with immense, sudden changes and widespread, inescapable destruction, even death. Not surprisingly, the public—a more important faction within the scientific community than formerly—preferred a metaphoric geology reflecting their own, more limited experiences.

The book that satisfied this need in exemplary fashion and became a best seller in consequence was *Essay on the Theory of the Earth*, a translation of Cuvier's preliminary discourse of 1812 by Robert Kerr, with additions by Robert Jameson. Published around October 1813 in Edinburgh, it achieved five editions. Metaphoric history aside, one further reason for this surprising popularity was that Cuvier, as represented in this version, returned geological speculation to the basis established for it in the earlier eighteenth century through religious and scientific consensus. In doing so, he effectively bypassed such problems as the origin of basalt, the importance of volcanic forces, and the age of the earth. Cuvier was, in fact, scarcely concerned with causation at all. On the other hand, he directed scientific attention to the fossil record, and particularly to vertebrate remains, which earlier researchers had generally regarded as incidental.

Agreeing with Werner, Cuvier believed that the sea had deposited a series of rock types in orderly procession. Older deposits tended to be widespread, if not universal, and uniform; younger ones were more limited and varied. As the fossil record abundantly confirmed, when the chemical composition of the ocean changed, its complement of life changed also. Fossils tell us that life-bearing continents have often been immersed by subsequent inundations and that portions of ocean bottom have become exposed. These calamitous changes, neither slow nor gradual, are invariably sudden (as the frozen mammoths of Siberia impressively attest). "Life, therefore, has been often disturbed on this earth by terrible events—calamities which, at their commencement, have perhaps moved and overturned to a great depth the entire outer crust of the globe, but which, since these first commotions, have uniformly acted at a less depth and less generally." Whole species have been extinguished by these catastrophes, leaving no memorial of themselves behind except some small fragments that the naturalist can scarcely recognize (16–17). Having established that the history of the earth consists in large part of numerous sudden revolutions, Cuvier next argued that such present-day phenomena as rain, rivers, the sea, and volcanoes could not have engendered them: "No cause acting slowly could possibly have produced sudden effects" (38), he affirmed. Cuvier regarded volcanoes as local phenomena (hence no central fire) totally incapable of elevating non-volcanic mountains, and he dismissed any theory involving the earth's axial tilt as well. Among the former theories he curtly summarized was one in which "the materials of the mountains are incessantly wasted and floated down by the rivers, and carried to the bottom of the ocean, to be there heated under an enormous pressure, and to form strata which shall be violently lifted up, at some future period, by the heat that now consolidates and hardens them" (46); a footnote identified this theory as Hutton's. Brilliant as each of these mistaken theorists may have been, according to Cuvier, all pursued chimeras. What progress had actually taken place in geology was owed to Werner and the numerous enlightened pupils of his school.

For Robert Jameson, Cuvier's original French edition (1812) could not have appeared more opportunely. It was almost certainly at his instigation that Robert Kerr (1755–1813), an experienced scientific translator, undertook the Cuvier project, which he did not live to see into print. Jameson added to Kerr's text an unjustified preface in which he emphasized Cuvier's affirmation of the historicity of Genesis, almost to the exclusion of his scientific arguments. In some appended notes, he also disagreed with Cuvier's assumption that all strata had originally been deposited in a horizontal position, criticized his characterization of the primitive rocks and crystallized marbles, cited Salisbury Crags as evidence of the recency of the Deluge, and rejected Cuvier's poor opinion of Werner's paleontological knowledge.

Regarding the last he may have been right, but Jameson's relative inflexibility—specifically, his unwillingness to see Werner as a contributor rather than a law giver—failed to achieve its end. Despite their superficial agreements, the Wernerian aspects of Cuvier's position were largely vestigial and would soon wither away. In any case, it was now Cuvier rather than Werner who became the authority. A signal advantage accruing from this change of figureheads was that, overall, Cuvier offered more a program of research than a dogmatic theory. As opposed to the heavy-handed Werner, Cuvier took the advancement of knowledge and his own contributorship for granted. With the by no means new but relatively neglected field of vertebrate paleontology (not yet so named) now central, moreover, amateur participation in the field became not only possible but necessary. As a result, Cuvier's *Essay*—with its few but vivid concepts—brought geology to a previously unattained level of popularity.

Fact, not Theory

During the next five years, a series of at least semipopular works attempted to satisfy the public appetite for a reasonable understanding of geological phenomena. As we have already seen, both Robert Bakewell's *Introduction to Geology* and Cuvier's *Essay on the Theory of the Earth,* representing primarily Huttonian and Wernerian viewpoints, respectively, were published throughout this period. Yet the predominant body of British opinion lay somewhere in between. Though based more on Cuvier's *Essay* than any other one source, this school (which also reflected the influence of the Geological Society of London) rejected most theorizing as premature, accepted some form of catastrophism (a recent deluge in particular), reconciled itself only vaguely with Genesis, increasingly emphasized the importance of stratigraphy and fossils, and gradually recognized the igneous origin of basalt. From this point of view, both Hutton and Werner as wholes were necessarily passé.

Some of the resulting warfare enlivened general-interest periodicals. In the *Edinburgh Review* of October 1813, for example, John Leslie—Playfair's successor in mathematics—while reviewing Leopold von Buch's *Travels through Norway and Lapland* (1813; notes by Jameson) at length, first described and then rejected (as "wildly hypothetical") the Wernerian system. W. H. Fitton also confronted Werner's system directly, for Nicholson's *Philosophical Journal*, much to the same effect. John Fleming, contrarily, accepted long-acting causes and rejected catastrophes but nonetheless appeared before the Wernerian Society as a loyalist. Serving as his complementary opposite, John Playfair reviewed Cuvier's *Essay* for *Edinburgh Review* in January 1814. Not all geological changes, he admitted, can be attributed to such slowly operating causes as we now see at work around us; local catastrophes, moreover, certainly exist and rivers have not necessarily formed the valleys in which they now run. Though he declared all geological systems presently available to us defective, Playfair clearly linked fossils with strata, accepted progression in the animate world, rejected biblical chronology, and thought the Deluge too short to be of geological significance. His remarks, like those of the other reviewers cited in this paragraph, were anonymous.

As periodical exposure made eclectic positions increasingly acceptable to the public, books with named authors soon followed. In his introduction to *A Descriptive Catalogue of the British Specimens Deposited in the Geological Collection of the Royal Institution* (1816), for example, William Thomas Brande summarized the Wernerian and Huttonian theories at some length before concluding that the latter, despite admitted flaws in detail, was "more bold and perfect than that of Werner, . . . involves fewer difficulties and contradictions, and is less at variance with nature"

(1816, xv-xvi). Some lectures Brande delivered at the Royal Institution that year were then published in the next as *Outlines of Geology* (1817). In these he identified the two currently prevailing theories as Werner's and Hutton's, then reiterated his partial endorsement of the latter. Brande unhesitatingly regarded granite and basalt as igneous and preferred the Huttonian term "stratified rocks" to Werner's "Transition series." Theorists always fly to extremes, he pointed out; and though he would not "be deemed assentient to every clause of the Huttonian doctrines," Brande nonetheless endorsed plutonian beliefs "as least at variance with facts, as least hypothetical, [and] as best entitled to the appellation of a theory of the earth" (1817, 142).

Facts, not theories, were now regarded by many as the basis of geological understanding. In 1818, William Phillips compiled and published *A Selection of Facts from the Best Authorities, Arranged so as to Form an Outline of the Geology of England and Wales* (1818). His arrangement, based on William Smith's, was stratigraphical. One of Phillips's sources, John Kidd's *Geological Essay on the Imperfect Evidence in Support of a Theory of the Earth* (1818), pleaded for tolerance on both sides, stretched the language of Scripture a bit farther, and raised points in opposition to both prevailing theories. Nevertheless, Kidd felt "inclined to admit the probability of Dr. Hutton's proposition that the materials of even the natural strata of this earth have been originally prepared by [rivers and] similar agents" (1818, 183). That same year William Knight presented *Facts and Observations towards Forming a New Theory of the Earth* (1818; some copies 1819). In it, he reviewed several earlier theories ("entirely hypothetical"), then those of Werner and Hutton at length. Among other points, he accepted the igneous origin of basalt and granite and the origin of coal from wood, but he denied that the present world derived its contours from causes now active, that central fire exists, and that general decay prevails. Though professedly devout, he also disparaged any attempt to reconcile geological evidence with Scripture. Hutton, Knight concluded, "can only be regarded as the framer of a system which explains but a part of the phenomena of the world" (1818, 17–18).

Greenough's *Critical Examination of the First Principles of Geology*, 1819

This factual, antitheoretical phase of Huttonian-Wernerian controversy extended primarily from 1815 to 1820 and was, in its way, reflective of the general disillusionment, even cynicism, which followed Waterloo and accompanied the last years of the Regency. That world-weary skepticism culminated in G. B. Greenough's *Critical Examination of the First Principles of Geology* (1819), a not very popular demonstration of scientific uncertainty.

Attacking the prevalent reliance on so-called facts and observations as naive, Greenough began by examining current assertions as to the alleged stratification of granite. Baldly juxtaposing quotations from prominent authorities, he was soon able to prove that experienced geologists discussing precisely the same exposures were entirely capable of interpreting them in diametrically opposite ways. He recalled, for example, that Playfair admitted to the stratification of granite in particular instances whereas Hutton had not. All Huttonian writers, moreover, denied the stratification of basalt.

Turning to the more general topic of stratification as such, Greenough also weighed conflicting opinions regarding the inclination of strata: were they deposited at such angles or subsequently distorted from an original horizontality? Citing Playfair's *Illustrations* (para. 121), Greenough thought the Huttonian position on this issue curious: "Assuming that all rocks have been forced up once from the bottom of the sea," he argued, "and primitive rocks twice, it supposes that the primitive rocks have acquired their vertical or inclined posture in consequence of this violent elevation, the secondary in spite of it" (1819, 57n). Having just explained the consolidation of primitive strata, Playfair saw them as being "broken, set on edge, and raised to the surface" (1802, 123). To account for unconformities like the one at Siccar Point, however, he then suggested a second submergence, during which the secondary strata were deposited, and a second elevation producing no further distortions.

"All masses and strata," Greenough next asserted, "are subject to curvature and angularity" (1819, 61). Hutton's position, as summarized by Playfair (1802, 45), was that horizontally deposited strata were uplifted while still flexible and ductile; gravity and the resistance of the mass itself then created a lateral and oblique thrust, which produced large-scale contortions that were among the most striking and instructive of all geological phenomena. Since Playfair himself believed Hutton's theory "nowhere stronger than in what relates to the elevation and inflexion of the strata" (1802, 234), Greenough was surprised to find Sir James Hall (in his paper of 1812) proposing the alternative of horizontal compression. Neither Playfair's explanation nor Hall's seemed acceptable to Greenough, but the Wernerian one (crystallization) failed to convince him also.

The points so far considered were among those raised by Greenough in "On Stratification," the first of his book's eight essays. Essay 2, "On the Figure of the Earth," dealt less with the Huttonians, who, recognizing in nature no trace either of a beginning or an end, "know of nothing original" (1819, 93). An important section of this essay, "On the Proximate Cause of the Inequalities Now Subsisting on the Surface of the Earth," strikingly affirmed the role of current geological changes, particularly with regard to the formation of valleys, but it nonetheless agreed specifically with Cuvier's preliminary discourse that the geological forces of the past had been considerably more efficacious. Citing the classic example of erratic granite boulders atop Jura limestone in Switzerland, Greenough quickly reviewed and dismissed the theories of Deluc (that they had been thrown up by gaseous explosions in place), Playfair (transported by the Arve), and Hall (hinting at glaciers). Neither rivers nor seas, Greenough thought, were sufficient either to move the boulders from their mountains or to shape mountains and valleys as they now appeared—no matter how much time one alloted such agents to do their work. He therefore endorsed the opinion of Sir James Hall and others that a forceful deluge or debacle must have taken place.

Greenough's essay 3, on topographical inequalities, reviewed crystallization, deposition, subsidence, volcanoes, earthquakes, and running water. Essay 4 demolished the Wernerian concept of universal formations, and essay 5, "On the Order of Succession in Rocks," similarly established that attempts like Werner's to confine any given rock to a set period of the world's history were easily defeated by field evidence; granite and other "primitive" rocks recurred throughout the geological succession. Essay 6 reaffirmed, among similar points, that calcareous rocks could be as old as siliceous rocks and questioned the rigid Huttonian distinction between stratified rocks and unstratified ones. Nor were highly inclined rocks of either type necessarily older than horizontal ones. Contrary to some Wernerian assertions, the most elevated rocks are not always the oldest; some Alpine peaks are granitic but others are calcareous. Similarly, Werner's supposed relation between the nature of fossils and the age of the rock containing them seemed more fanciful than true. Fossils from older rocks, Greenough affirmed, are not invariably less akin to living forms than are those from younger strata. The utility of fossils for stratigraphic correlation, moreover, had been exaggerated by some writers (a point he elaborated in essay 7). Greenough's final essay, on mineral veins, rejected both the Huttonian and Wernerian positions.

In part because Greenough was currently president of the Geological Society of London (he had been its first president as well), his skepticism regarding current procedures was widely noticed. For *Monthly Review*, his book served mostly to explode Wernerian pretensions. Yet the Huttonian theory was not obviously better, both being premature. Insofar as the science of geology had established reliable facts, they were those of stratification, fossils, extinction, and catastrophes—in other words, the Cuvierian position. Another critique, by the Reverend John Fleming of

Flisk in *Edinburgh Monthly Review*, stressed the distortions of fact accruing from theoretical bias and premature generalization. Though Cuvier was a reputable comparative anatomist, he could not be regarded as an authority in geology, for his theory of the earth abounded in errors, both in its facts and in its reasoning. Greenough himself, moreover, appeared to be "totally unacquainted with the laws of evidence." A third, particularly lengthy review by Thomas Thomson was meanwhile running in *Annals of Philosophy*. Thomson could not imagine anyone more capable than Greenough of doing justice to the merits of all parties, or more likely to remove "the magical influence attached to great names which have stamped upon certain opinions an artificial value, to which of themselves they are by no means entitled" (302). But the essence of the Huttonian position, for poorly informed Thomson, was that "this globe had no beginning, and will have no end" (366). Contrarily, "Werner, or his pupils (I know not which), attempted a great deal too much when they undertook to detail the order of the formation of all rocks constituting the crust of the earth" (462). Yet Thomson himself tentatively endorsed the old-fashioned idea that the earth was composed of concentric shells increasing in specific gravity as they approached the center.[4]

Though he attacked and undermined each theory, Greenough proved most effective in confirming an increasingly widespread perception that the credibility of Wernerianism had died with its originator (in 1817). Whereas most of his objections to Hutton were limited to differences of opinion among the Huttonians themselves, or between themselves and others, Greenough's critique of Werner derived straightforwardly from nature. With few exceptions, it seemed, all of Werner's assured pronouncements about universal formations were now proven wrong; as with Mara in Buddhist legend, the earth itself had spoken out against him.

Concomitant with facts were impartial observation and description. Throughout the Regency, many important geological writers (often trained in Wernerian techniques) represented themselves as being without theoretical bias. Lt. Col. Ninian Imrie, for example, a frequent contributor to the Wernerian Society's *Transactions*, in 1812 described the Campsie Hills of Stirlingshire and some adjacent strata in a resolutely empirical manner. Highly polished, often striated basalts he found there led him to affirm diluvial currents, but that was not unreasonable. Having seen Vesuvius and Etna at first hand, he firmly upheld the volcanic nature of *some* columnar basalt (more than one origin being possible), pumice, obsidian, and the Lipari Islands. In closing, Imrie warned the

[4]On Greenough: *Monthly Review* 90 (1819): 376–393; *Edinburgh Monthly Review*, 4 (1820): 557–571; *Annals of Philosophy* 14 (1819): 301–309, 365–373, 456–464. See also *Edinburgh Review* 33 (1820): 80–91.

Wernerian Society against "violent support of theory," for "where prejudice has taken root, we must bid adieu to all candid geological description" (1816, 48).

Similarly, in his "Observations of the Mineralogy of the Neighborhood of St. Andrew's in Fife" of 1813, John Fleming was only superficially Wernerian. As part of his remarks, Fleming criticized the "closet mineralogist" who indulged in hypothetical speculations regarding the formation of minerals, citing Hutton as his example. Though the latter regarded ironstone septaria as valuable evidence of consolidation by fusion, he had never (apparently) examined such specimens in situ, which is to say, embedded in a clay slate that presented not the slightest sign of ever having been exposed to heat. "If, moreover, the regular forms of basalt induced Dr. Hutton to conclude that they furnished proofs of the action of a central heat, he would have found considerable difficulty in applying his heat to these inclosed masses of basalt without fusing the bed of tuff which surrounds them." As Fleming then needlessly remarked, "he who has the boldness to build a theory of the earth without a knowledge of the natural history of rocks will daily meet with facts to puzzle and mortify him" (1816a, 154).

Fleming's anti-Huttonian stance appeared resolute enough, but he modified it considerably two years later when writing "On the Mineralogy of the Redhead [an Old Red Sandstone outcrop] in Angusshire." Brothick Burn, a nearby brook, had apparently formed its own valley and terraces. Playfair's assertion that strata were always deposited horizontally seemed to be contradicted by local evidence. On the other hand, seacoasts and valleys proved (through differential erosion) that "many of the inequalities of the earth's surface owe their existence to the long-continued action of air and water" (1816b, 352). Though Fleming affirmed the Wernerian theory of veins, neither the Huttonian nor the Wernerian theory of agate formation seemed adequate. Implicitly, then, he remained with those who believed that theories should be subordinate to evidence from nature.

Perhaps the most respected of all Scottish observers was John Macculloch (1773–1835), who published an impressive series of field reports and other papers in the *Transactions* of the Geological Society of London. Among his nine contributions to the second volume (1814), for example, were observations on the junction of trap and sandstone at Stirling Castle (affirming Hutton); some miscellaneous remarks on the Western Isles (rejecting Werner, contradicting Jameson; basalt and granite unstratified; origin of granite still problematical; strata bend, affirming Playfair; killas is not greywacke, with much on the inadequacies of geological nomenclature); others on quartz rock (quartzite as *metamorphosed* sandstone, emending Werner and Playfair); and new information on the basaltic island of Staffa (elevation, erosion). Consistently, throughout these papers, Macculloch likewise presented himself as a skilled but neutral observer "uninfluenced by systems" (*TGS* II: 44) who thought it "at all times desirable to keep clear of those terms which involve an hypothesis" (*TGS* II: 485).

In 1816 Macculloch published with the Geological Society of London, of which he was now president, his lengthy "Sketch of the Mineralogy of Skye." In it, he described syenite dykes and their alterations of adjacent limestone; basalts, moreover—clearly the product of fire—had altered adjacent clay. An earlier paper by him on Hutton's classic locality of Glen Tilt explored similar themes, the intrusion of granite into limestone and other rocks especially. This paper then stimulated Lord Webb Seymour and John Playfair to complete an essay on the same locality, which they had visited years before. Their "Account of Observations . . . upon Some Geological Appearances in Glen Tilt" (1815), drafted by Seymour, pointed out stratigraphic occurrences inconsistent with Wernerian geognosy and upheld the igneous origin of syenite (Hutton's "granite" of 1790). Quotations from Hutton's unpublished third volume were included.

Playfair's Travels, 1816–17

By this time Playfair had decided to publish a second edition of his *Illustrations of the Huttonian Theory*, not only in defense of Hutton but as a dignified response to the heavy-handed attacks on himself as geological observer endlessly promulgated by Deluc in travel books of 1810, 1811, and most recently 1813 (*Geological Travels in Some Parts of France, Switzerland, and Germany*).[5] Toward this end, Playfair began to collect additional examples favorable to Huttonian interpretation. Restricted by ongoing war, he temporized with extensive summer jaunts throughout Britain, then (probably in 1814) visited Ireland and the basaltic Giant's Causeway. After Napoleon's escape from Elba and the ensuing Hundred Days, Playfair resurrected some even more ambitious travel plans, and in the summer of 1816 (at age 68) undertook a remarkable geological odyssey on Hutton's and his own behalf.

He began with six weeks in Paris (meeting Cuvier, Brongniart, and Humboldt), then proceeded directly to Neuchâtel in Switzerland to study the famously puzzling erratic boulders of the Jura. According to extensive geological memoranda Playfair made at the time (but which we

[5]On the origin of Deluc's campaign against Playfair, see *Edinburgh Review* 6 (July 1805): 515, and Robinson and McKie 1970, 395–410.

know only from a relatively condensed summary of them by his nephew), he saw correctly that nothing other than glaciers could have transported the boulders; the Alps must then have been much higher, thereby spawning glaciers of the required magnitude. In Lucerne and Geneva, he was prevented from visiting the mountains himself because of the uncharacteristic torrential rains that year. Nonetheless, circumstances permitted him to inspect several of the classic sites first described by Saussure, an author Playfair knew exceptionally well.

After remaining a month at Geneva and learning that he had been granted a leave of absence by the University of Edinburgh, Playfair went over the Simplon into Italy. Some supposedly stratified granite adjoining Lake Maggiore gratifyingly turned out not to be. At Milan, Bologna, and Florence, he met prominent men of science, many of them geologists. En route to Rome, he saw remarkably contorted limestone strata between Spoleto and Terni, extensive travertine deposits, and deeply eroded ravines in volcanic tufa. While wintering at Rome, he explored nearby quarries as well as the usual antiquities, society, and art.

Proceeding the next spring to Naples, Playfair found Vesuvius in eruption, an opportunity of which he took full advantage. He also examined the older crater of Somma, the Solfatara, Monte Nuovo, the Phlegraean Fields, Pompeii, and even the volcanic island of Ischia, where he spent three days. At all these locales, Playfair investigated the formation, occurrence, and weathering of volcanic rocks. Returning to Rome, and later Florence, he again recognized the essentially volcanic origins of these areas, though volcanoes were no longer to be seen. An interest in the formation of marble led him to the famous quarries of Carrara. Passing through Milan toward Venice, he explored alluvial deposits covering the plain of Lombardy and some limestones and basalts near Vicenza. Intense June heat, however, prevented an excursion to the Euganean Hills, which (from specimens) he nevertheless recognized as volcanic.

From Venice, Playfair recrossed the Alps to Innsbruck, then revisited Lucerne to enjoy two weeks of industrious mountaineering. By this time, he was no longer making extensive notes, presumably because of mental and physical exhaustion. Yet even while resting at Geneva he began an essay on Swiss geology. In it, he defined three different types of mountains, described their constituent rocks (utilizing modified Wernerian terminology), and emphasized the geological efficacy of rivers. The text of this essay, whatever its degree of completion, has not survived.

Proceeding to and beyond Lyons, he found granite veins penetrating mica slate and some weathered granitic hills traversed by greenstone dykes. Approaching Clermont, Playfair then entered the classic Auvergne district, with its many extinct volcanic craters, basaltic flows obviously associated with them, and isolated domed hills known as puys. Calling on experience gained in Italy, he easily recognized the ancient lavas for what they were—but was not yet willing to associate them with some nearby columnar basalts (which would have entailed contradicting Hutton on one of his most insistent points). Following this last—and probably most disturbing—geological revelation, Playfair returned to Paris and eventually to Edinburgh, where he died less than two years later, on 19 July 1819. The projected second edition of *Illustrations* never appeared.

After his death, Playfair was ambiguously eulogized by Francis Jeffrey, the still acerbic writer who had disparaged his exposition of Hutton in 1802. Playfair, Jeffrey acknowledged, was certainly one of the best writers of his age and a charming man, but "all attempts to establish a theory of the earth must, for many years to come, be regarded as premature." It was, therefore, perhaps regrettable that the deceased should have devoted so much of his time and effort to questionable, inconclusive speculations on behalf of the Huttonian theory. Jeffrey's opinion notwithstanding, when Playfair's collected works (four volumes, but by no means complete) appeared early in 1822, *Illustrations* properly received the place of honor in volume I.[6]

Macculloch's *Description of the Western Islands of Scotland,* 1819

We do not know that Playfair lived to see the appearance of John Macculloch's comparative *Description of the Western Islands of Scotland* (1819), but he was surely familiar with at least some of the papers from which it derived. Taken together, as here, Macculloch's three volumes constituted one of the most competent regional studies available for any district, set a new standard for British descriptive geology, persuasively rejected Wernerian terminology (and, less obviously, the theory underlying it), affirmed the igneous origins of basalt and granite, and presented authoritative field evidence that, on the whole, favored Huttonian positions. Not yet the more thoroughgoing convert he would later become, however, Macculloch still had the independence of mind to believe both of the rival theories inadequate, as was increasingly the fashion. Thus, his preface deplored that "want of coincidence between the present facts and our systems" (1819, 1: xiv). By presenting only reliable information—without obtrusive theory—he hoped to "assist others in laying the foundation of a more accurate and extended knowledge" (1: xiii). His

[6] Playfair's travels are described from extensive manuscript evidence by his nephew in Playfair 1822, 1: xxvii-lxi. Jeffrey's remarks immediately follow (lxii-lxxvi; quoting lxvi).

meticulous, island-by-island descriptions have been consulted by geologists ever since.

Though Macculloch himself struggled to remain neutral, the utility of his observations to the Huttonians was demonstrated effectively by Leonard Horner, who lauded Macculloch's book in *Edinburgh Review*. Stratigraphical similarities between the islands and the mainland showed that they had once been contiguous, thereby affirming gradual but inexorable erosion over a vast period of time. As Macculloch repeatedly asserted, granite veins are posterior to the rocks in which they occur; thus, "his views entirely coincide with those of Hutton and Playfair" (459). The various plutonic rocks grade into each other; fossiliferous shale has occasionally been indurated by intrusive basalts, but the latter are never fossiliferous themselves. The supposed alternation of basaltic and sedimentary layers (so often cited to establish the aqueous origin of basalt) proves in all cases to be illusory; the basalts are invariably posterior to the sediments. Contrary to Hutton, however, basalts and lavas are identical. Finally, the Wernerian concepts of primary rocks and universal formations are demonstrably wrong. Though Macculloch was thoroughly entitled to the honors attending original discovery, Horner concluded, his reputation as a geologist already stood so high, and his views "so much beyond those who would give battle about the discovery of a pebble," that no further emphasis on his merits was required.

Boué's *Essay geologique sur l'Ecosse* and Later Papers, 1820–22

Ami Boué (1794–1881), born in Hamburg of Franco-Swiss parents, received his medical training at Edinburgh, during which he also studied geology under Jameson. After graduating in 1816, he traveled widely throughout Scotland, exploring its natural history. According to Boué, the appearance of Macculloch's *Description* necessitated some extensive alterations in the proofs of his own book, *Essay geologique sur l'Ecosse* (Geological essay on Scotland, 1820; dedicated to Jameson), which, despite its more restrictive title, dealt briefly with England, Ireland, France, and Germany as well. Boué distinguished ten Scottish formations: granite, gneiss, mica schist, porphyry, and chlorite schist, all primitive or primordial; greywacke, transitional; sandstones and limestones, secondary; basalts, volcanic; and alluvium, superficial. Regarding volcanic rocks, he straightforwardly affirmed the igneous origin of Hebridean basalts, while remaining less certain about some others. Explicitly rejecting Hutton's plutonic theory, however, Boué proposed that basaltic veins had been filled from above, by lava flows passing over them.

In a public letter to Jameson, "On the Geognosy of Germany, with

Observations on the Igneous Origin of Trap," read 26 January and 9 February 1822 before the Wernerian Society, Boué demolished a considerable amount of Neptunian orthodoxy. The Clay Slate, hitherto taken as Primitive, he pronounced to be transitional, and to pass undistinguishably into greywacke. The Old Red Sandstone and First Floetz limestone alternated with each other. Other formations were not necessarily universal, though the Jura Limestone of France and Germany appeared elsewhere (especially in England) under other names. Conversely, the Chalk (of England) and the Plastic Clay (of France) were both widely distributed in Germany. As for the unstratified rocks, "we are yet very far from knowing either their origin or their true geognostic position" (1822, 101). Syenite is but granite, a rock occurring at various places in the geological succession. Porphyries, in some countries, alternate with greywacke; they too appear at various times. Trap rocks (i.e., basalts) can be found in the Floetz period and in the Transition. Sediments adjacent to basaltic veins have clearly been indurated or altered. "How such a change can be produced," Boué demurred, "I do not know, but such is the fact and everybody may see it." Indeed, he continued, "I can assert that the Erzegebirge [Werner's own area] contains many interesting facts and distinct appearances which might be adduced in support of the Huttonian theory" (103–104). Boué then straightforwardly rejected the Wernerian theory of veins. In another paper published the same year, he bluntly proposed that Werner's death in 1817 had prepared the way for an epoch of geological advancement in Germany.

[9]

Specific Problems

After 1817, dogmatic Wernerianism rapidly disappeared from British geological literature. The dead German's lamentable attempts at theorizing, William Conybeare proposed in 1822, "must now appear to all but his devoted adherents [as being] among the most unsuccessful and unphilosophical ever made" (Conybeare and Phillips 1822, xlv). Yet the new creed, though destined to reign for a decade or more, was not pervasively Huttonian.[1] According to this more viable consensus, which derived primarily from Cuvier, neither the origins of the earth nor its age nor even the forces primarily responsible for shaping it were known. Nonetheless, what facts there were demonstrated unmistakably that life on earth had progressively improved throughout an eventful history punctuated by numerous revolutions or catastrophes, each of which changed the fossil record irreversibly. Two further conclusions, moreover, seemed no less certain: that humankind had had a comparatively recent origin; and that there had also been a very recent global devastation of some kind by water, similar to the debacle of Greenough and other writers. Religiously inclined persons still tended to equate this debacle with the Flood, which would have placed it within human history, but the lack of manlike fossils made any such equation rather problematic. Greenough to the contrary, however, paleontological evidence and stratigraphical correlation now became major tools with which geological theorists endeavored to reconstruct the past. Among the

[1]According to H. T. De la Beche around 1826, "the old theories of the earth, being unfounded on facts, have for some time disappeared before modern researches, which have equally shown those of Hutton and Werner to be untenable" (McCartney 1977, 8).

many contemporary statements affirming all or part of this consensus, one may cite William Buckland, *Vindiciae Geologicae* (1820), Gideon Mantell, *The Fossils of the South Downs* (1822), W. D. Conybeare and William Phillips, *Outlines of the Geology of England and Wales* (1822), and Buckland again, *Reliquiae Diluvianae* (1823).

The Deluge

Like Cuvier, Hall, and Greenough before them, these new theorists still endorsed some kind of deluge. Of the several geological issues following from that assumption, none was of more current interest than the origin of valleys. Though "not disposed to agree with Dr. Hutton . . . that the main or longitudinal valleys of an alpine district have been excavated by the rivers which flow through them," for example, John Kidd (1818, 192) thought it undeniable that most of the transverse ones were. In his second essay, however, Greenough (1819) explicitly opposed the Huttonian theory that valleys had been excavated by the rivers flowing through them. Shortly after Greenough's book appeared, Buckland outspokenly concurred: "The occurrence of quartzose pebbles in such high situations as the top of Henly Hill," he argued in one of several essays, "goes far to prove the recent origin of the valleys through which the Thames and Evenlode now flow, and compels us to refer the excavation of them (at least in certain parts) to the denuding agency of the subsiding waters of the most recent deluge that has affected the earth" (1821a, 520). His assumptions being widely shared, other theorists also thought rivers too inefficacious to have created their own valleys and preferred to explain the latter through earthquakes, uplift, marine erosion under the sea, or (with Buckland) a sudden rush of water inland or withdrawing.

In 1822 Conybeare proposed that mountain valleys, at least, owed their first outlines to orogenic forces, though they were greatly modified thereafter by mighty currents of water. Regarding the latter, he did not mean "the streams (often inconsiderable rills) which now flow through them." Despite some geologists who seriously thought such a cause adequate to explain the observable effects, if given "a long lapse of ages" and "incessant action," Conybeare insisted on diluvial currents. To believe valleys formed by the rivers now within them, he concluded—whatever the length of time involved—required "the most direct physical impossibilities" (Conybeare and Phillips 1822, xxiii).

Buckland's *Reliquiae Diluvianae* (Relics of the flood), though primarily concerned with vertebrate fossils and caves, included appendices (based on his previously given papers) affirming the excavation of valleys by a transient flood. While loudly and often acclaimed for its religious implications, Buckland's work also represented prevailing geological opinion, as one well-qualified reviewer attested: "The effects of water upon the solid strata of the globe have been the subject of much geological debate," W. H. Fitton observed, "but it is now almost universally admitted that valleys have been excavated by causes no longer in action—contrary to the opinion of Dr. Hutton and Mr. Playfair, who maintained that they were formed by the long-continued erosion of the streams which actually run through them" (1823, 227). Fitton too affirmed a deluge of some sort, while questioning our knowledge of its particulars (196–234).

Belief in a deluge, though general, was far from universal. Many who accepted the biblical story in detail, for example, found it almost impossible to attribute so much geological efficacy to a short-term inundation that had seemingly arisen and subsided *gradually*. Thus, in May 1819 one clerical reviewer of a theological work rejected all *geological* proofs of the Deluge. Though he was answered the next year by Buckland, even the latter had already convinced himself that the Noachian inundation was not a unique geological event but only the most recent of a series.[2]

Buckland's subsequently reiterated affirmation of a geologically efficacious deluge in *Reliquiae Diluvianae* was applauded by most reviewers. "That there has been a deluge, affecting universally all parts of the earth's surface and producing everywhere the same or similar effects," Fitton responded enthusiastically, "no person who has duly examined the evidence can deny" (1823, 229). The only real question was whether that deluge had been, as Buckland held, recent, transient, and simultaneous. Replying to Buckland in June 1824, however, both James Smithson (1824) and John Fleming again rejected any form of geological proof. Fleming, the more substantial defector, attacked Buckland's contentions via the *Edinburgh Philosophical Journal* edited by Jameson, who noted helpfully that Werner himself had never advocated the "geological diluvian hypothesis" (1824, 299n). Another of Jameson's notes then repeated the powerful antideluge argument of a Professor Link; supposedly diluvial strata contained the bones of both extinct and surviving creatures, so there could not have been a *universal* deluge, which species now living would not have survived (304n).

Feeling his case grown stronger, or more necessary, Fleming spoke out again in April 1826, once more through Jameson, to declare "The Geological Deluge, as Interpreted by Baron Cuvier and Professor Buckland, Inconsistent with the Testimony of Moses and the Phenomena of Nature." In this much longer essay, Fleming reviewed the Flood fancies

[2]*Quarterly Review* 21 (May 1819): 53; Buckland 1820, app.

of such earlier theorists as Burnet, Woodward, Whiston, and Cuvier before summarizing his controversy with Buckland in greater detail. As his first major argument, Fleming contrasted the hypothetical deluge of the geologists with that of Noah (an event so gentle as not even to disturb the soil). He then opposed the debacle theory of valley formation and other aspects of Buckland's attempted proof. Realistic consideration of the latter's deductions from vertebrate fossil evidence, moreover, provided "the death-blow of the diluvian hypothesis" (1826, 234); "no such geological deluge ever occurred" (233). The whole theory, therefore, must now be discarded, like those of Burnet, Woodward, and Whiston.

Fleming's arguments were not immediately accepted, his being very much a minority position. Nevertheless, they appear to have convinced one important Wernerian. The Kerr translation of Cuvier had from its appearance in 1813 a preface by Jameson specifically endorsing asserted similarities between the geological and Noachian deluges. As of 1817, however, that formerly confident assertion no longer appeared. One decade later, in the 1827 edition, a new note on the deluge by him (p. 334) fully accepted Fleming's arguments of the previous year. Even more specifically—and rather courageously—Jameson cited Fleming (1826) at the end of a further, more lengthy note "On the Universal Deluge" (pp. 417–437), which denied the reality of its subject. If, therefore, Jameson had once been unusually outspoken on behalf of deluge geology, he was also the first of its important advocates to recant in public.[3]

Basalt, Granite, and Lava

Whether other theorists were yet to follow Jameson in rejecting the Deluge (some, inevitably, failed to realize that he had), much of the new thinking dealt with aspects of subterranean fire or subterranean heat. In this area, the most immediate issue had long been the origin of basalt. From our point of view, the identity of basalt and lava was satisfactorily established by French geologists during the late eighteenth century, to be repeatedly confirmed by a number of Werner's most esteemed pupils in their various inspections of the Auvergne. But the issue nonetheless dragged on for another thirty years or so, largely because three of the most active geological schools—Scriptural Neptunians (Deluc, Kirwan), orthodox Wernerians (Jameson), and strict Huttonians (Playfair)—all found reason to dispute the facts. A great deal was at stake.

By 1813, as Robert Bakewell had noted, that part of Hutton's theory relating to the igneous origin of basalt was well—but not decisively—

[3]See also Page 1969, who originated this line of argument.

established. For most British observers, decisive proof appeared three years later when William Conybeare read to the Geological Society of London J. F. Berger's extensive "On the Geological Features of the Northeastern Counties of Ireland" (1816) together with a few additional comments of his own. Conscientiously Neptunian, Berger affirmed the stratification of basalt, the excavation of valleys by diluvial currents, and a modified sequence of Wernerian formations. But he also fairmindedly noted the alteration of limestone by basaltic dykes and pointed out that the presence of water in basalt established only permeability, not aqueous origin.

Conybeare's "Descriptive Note Referring to the Outlines of Sections Presented by a Part of the Coasts of Antrim and Derry" then followed. Accompanied by William Buckland, and utilizing field techniques resembling John Macculloch's, Conybeare had visited northern Ireland in 1813, when wartime still made it dangerous to do so. Near Portrush (adjacent to the Giant's Causeway, a famous example of columnar basalt), Conybeare identified as flinty slate a dark, grainy stratum containing pyritized ammonites. This remarkable rock had, as he knew, "been the subject of much discussion among the supporters of opposite theories" (1816, 213). Neptunists, in particular, considered it a variety of basalt; for them, the presence of marine fossils obviously requiring an aqueous origin seemed an irrefutable proof of their theory. But Conybeare explicitly agreed with Playfair's earlier contention (1802, para. 252) that the fossiliferous layer was distinct from basalt though indurated through contact with it.

Conybeare spoke out no less frankly regarding basalt: "Desiring to keep that description of facts which must serve as the groundwork of theory—and which seems in the present state of science the most useful employment of the geologist—distinct from conclusions merely speculative, I have hitherto studiously refrained from expressing the views which I have been led to form on the origin of basalt" (1816, 208n). Once confronted with the striking basaltic cliffs adjoining Ballycastle Bay, however, he could do nothing other than agree completely with the igneous theory.

The Kenbaan Cliffs, he thought, directly contradicted what was known to be true of aqueous rocks: normally, the youngest rocks are the least elevated, but these rivalled the Primitive mountains in height; normally, the youngest rocks are the least consolidated, but the basalt here is as hard an any; dykes obviously associated with this basalt cut through rocks of all kinds indifferently, whereas veins of other kinds are restricted to particular rocks. In addition to this negative evidence against the Neptunist hypothesis, Conybeare accepted a series of arguments favoring the Vulcanists: (1) basalt and lava are chemically identical; (2)

basaltic rocks regularly occur in volcanic districts; (3) Wernerians themselves confess that the basalts of Auvergne are of igneous origin; (4) those personally familiar with volcanic areas (including Dolomieu and Spallanzani) have regularly maintained the igneous theory; (5) the basalt at Kenbaan gives every sign of having been injected from beneath; and (6) Berger provides as least four examples of how basaltic dykes altered adjacent rock. Conybeare then daringly affirmed that the Irish (and, by implication, Scottish) basalts had been formed by submarine volcanoes active at some "very remote" period preceding that last great convulsion which had excavated the earth's valleys and left extensive alluvial deposits behind.

By 1816, an increasingly compelling accumulation of examples from both Wernerian and Huttonian geologists attested to induration accompanying basaltic veins or dykes. In that year's *Transactions* of the Geological Society of London, for example, no fewer than five essays did so explicitly.[4] Together with additional evidence, much of it from the Auvergne and Italy, these scarcely contestable instances eventually swayed reasoning persons throughout Europe. Soon, following the detailed analyses of Scottish basalts by Macculloch in 1819 and Boué in 1820, little room for argument was left; thus, Conybeare and Phillips observed in 1822 that the weight of geological authority decidedly preponderated in favor of the igneous origin of basalt (1822, xvii). By 1825 Boué could declare (in Jameson's journal) that all knowledgeable geologists accepted extinct volcanoes; that most accepted tertiary basalts as volcanic; and that many also accepted the volcanic origin of earlier basalts (1825, 130). In 1826, moreover, Charles Daubeny described indurative basalt dykes in the Siebengebirge of Germany, a region hitherto cited by Wernerians on behalf of the aqueous origin of basalt. "That all of the trap rocks are, of whatever period or periods, analogous or identical with volcanic rocks, not only in their constitution but in their origin and causes," Macculloch enjoined the same year, "has now at length been admitted by all geologists deserving of that name" (1826, 370).[5] Eventually, Jameson himself confirmed the "interesting displays of changed stratified rocks [in Scotland] . . . pointed out long ago by Hutton" (1833, 386), which was to say, examples of chemical change "induced in neptunian rocks by plutonic agency" (388). By 1833, then, Jameson had become something of a plutonist as well.

Once basaltic dykes and sills had been recognized as igneous, the acceptance of igneous granite soon followed, though never so com-

[4]Macculloch on Skye, Berger on Ireland, Berger on dykes, Aikin on trap, and Macculloch on Glen Tilt; all in *Transactions of the Geological Society of London* 3 (1816).

[5]For Macculloch and Boué especially, see the articles by V. A. Eyles and Arthur Birembaut, respectively, in Gillispie 1970–80.

pletely. As Macculloch noted in 1819, "It is not easy to admit the arguments derived from . . . appearances in favor of the igneous origin of trap and refuse them in the instance of granite" (1: 219n). Conybeare and Phillips (1822, xviii) concurred. In 1820, Ami Boué still had his reservations. Only three years later, however, he was himself proclaiming (in Jameson's journal) that "the true origin of granite" had "been discovered by Hutton" (1823, 131). During these years rocks once regarded by Werner as Primary were now explained as subterranean lavas—by G. P. Scrope in 1825, for example. Concurrently, Boué reported that an increasing number of geologists, including himself, regarded granite, syenite, and porphyry as igneous (1825, 130–131). "That granite is identical or analogous with the trap rocks," Macculloch wrote in 1826, "is another fact established by geologists of a class yet too logical and too philosophical to have made converts of the baser multitudes" (1826, 370). For him, both granite and basalt belonged to the volcanic rocks. "In mineral composition," he asserted the next year, "a gradual and insensible passage may be traced from the best characterized specimen of primitive granite to the newest basalt that has flowed over the surface of the chalk." All the unstratified rocks, moreover, "however differently modified in regard to the proportions of their constituent parts, and to the period of their protrusion among the strata, have had one common origin, the interior of the earth" (1827, 306). Other investigators, too, supported this position.

By 1826, few geologists of any prominence still accepted Hutton's assertion that basalt and lava were distinct. In 1813 Macculloch had explained the secondary formation of zeolites, agates, and spar in basalt. Three years later, in his notes on the geology of northeastern Ireland (1816), J. F. Berger described a basaltic cavern on the island of Rathlin in which calcite stalactites had formed. There being no adjacent limestone, Berger argued, the lime creating the calcite must have been leached by ground water from the basalt itself. Spar and zeolite nodules often found in basalt were probably formed similarly, making their presence or absence irrelevant to any supposed distinction between basalt and lava. That same year, Cordier and Conybeare separately affirmed the chemical identity of basalt and lava. Despite some ineffectual opposition from Playfair (again in 1816), Boué (1820) and Macculloch (1824) supported this swiftly strengthening position. G. P. Scrope (1825) similarly affirmed the identity of basalt and lava in a note to his first page. Daubeny (1826) thought them different only in their modes of formation. Macculloch then straightforwardly rejected *any* distinction between basalt and lava in his reviews of Scrope and Daubeny, leaving the issue pretty much resolved. Inevitably, the identification of basalt as lava made volcanoes per se less foreign to the major concerns of Huttonian geology.

Volcanoes, Central Fire, and Internal Heat

Throughout his *Considerations on Volcanos* (1825), George Poulett Scrope utilized extensive personal investigations of Vesuvius, Etna, Stromboli, the Auvergne, and other volcanic areas to make points of broad geological applicability. He readily defined lava as "any rock in a state of natural liquefaction by heat," irrespective of its disposition as a flow, vein, or dyke. On their consolidation, he added, "lavas become rocks of trachyte, basalt, etc., according to their mineral composition" (1825, 1n). Jointed columns of lava and basalt seemed to him fully identical. Among his detailed analyses of volcanic processes and products, Scrope described cones and domes (basaltic and trachytic, respectively), rejected previous analyses "rather too much in the spirit of the Wernerian school" (95n), and compared erupted fragments of limestone partially fused under volcanic pressure with similar ones produced by the experiments of Sir James Hall, which he repeatedly endorsed (107, 111, 148n, 159n). Concurring with Hutton and Macculloch, Scrope thought that active island volcanoes in the Pacific Ocean were slowly creating a new continent (185). In its notice of the book, *Monthly Review* thought Scrope's system "evidently borrowed from Hutton" and Scrope himself "an advocate for that theory" (29).

Slightly later but more immediately respected, Charles Daubeny's *Description of Active and Extinct Volcanos* (1826) augmented geological lectures given by its author at Oxford. Back in his own student days, during the winter of 1816–17, Daubeny had attended Jameson's class in mineralogy at Edinburgh. Jameson then thought the origin of basalt still debatable, so Daubeny hoped to investigate existing European volcanoes and provide the necessary Wernerian evidence.

The first two parts of Daubeny's book, both based on actual travels, described extinct volcanoes in France and Germany and the volcanic districts of Hungary, Italy, Sicily, and the Lipari Islands; the third part borrowed descriptions worldwide; and the fourth generalized from all these available data. Among his other conclusions, Daubeny decisively rejected the aqueous origin of basalt. "Some allowance . . . ought to be made for Werner," he declared sympathetically, "when we consider the advanced period of life to which he had attained before the evidence in favor of the igneous origin of trap rocks had arrived at that degree of conclusiveness which would have justified a decided opinion on the subject." It was Werner's misfortune to have outlived his own system, remaining stationary just when geologists elsewhere were speeding forward; consequently, "his services have been as much depreciated latterly as they had been overrated before" (1826, 427n). Like Scrope, Daubeny praised both the experiments of Hall and Watt and the lucidity of Playfair.

While the igneous origin of basalt now seemed established, renewed emphasis on plutonic and volcanic processes involved considerable disagreement as to the source of thermal energy. Was there, as had been repeatedly proposed since antiquity, actually a central fire within the earth? If so, how was it aerated and fueled? If not, was there a central heat of some other kind? Or were all plutonic and volcanic phenomena caused by local and relatively superficial conflagrations? "There has not been much more gross ignorance displayed in any department of geology, abounding with this as it does," Macculloch proclaimed in his best fashion, "than in that which relates to the seat and cause of volcanic heat and action" (1826, 366–367). The spirit of Werner, he continued, a "true and lasting obstacle to the progress of geology, . . . yet rides the science and its imaginary followers from the grave, like a nightmare" (367). To him geology was indebted for the (unoriginal) opinion that volcanic fire derived from the inflammation of coal strata. But there were no such strata beneath any volcano known; their shallow depth would be inadequate to account for known effects; they would have burned out beneath many still-active volcanoes long ago; volcanic ejecta gave no sign of such an origin; and coal mines accidentally set afire had no resemblance to volcanoes. Surely a modern science could do better.

Whatever the cause of volcanic heat, it would not be found within the volcano itself. "The really probable theory of volcanoes," Macculloch proposed, "is that they depend on a permanent and deep source of heat in the earth, or on what has been termed a central fire, called into temporary action in consequence of the accumulation of chemical changes, or possibly by the casual admission, in certain instances, of that water which so often takes an active part in the exertions of volcanoes" (1826, 369). The earth as a whole he believed to be a hot, mostly fluid body covered by an irregular crust. "There is beneath us a fire and a cause", Macculloch prophesied, "which at some future day will reproduce revolutions similar to the preceding, and, when the time shall come, subvert the world, again to give birth to a future earth and new inhabitants" (370). Macculloch would subsequently elaborate this opinion in his mostly written but not yet published *System of Geology* (1831).

The concept of central fire was by no means agreed on. When John Leslie, for example, proposed "the existence of a subaqueous bed of air . . . to feed the numerous fires which occasionally rage in the bowels of the earth, and occasionally burst forth on the surface in volcanic spiracles," one reviewer of 1823 thought his idea "the wildest conceit that has ever figured in a sober work on philosophy."[6] Scrope (1825) rejected central fire in favor of subterranean lava whereas Daubeny (1826) ac-

[6]Anonymous review of Leslie's "Treatise on Meteorology" (supplementing the fourth and fifth editions of the *Encyclopedia Britannica*), *Quarterly Journal of Science, Literature, and the Arts* 14 (1823): 177, 172–185.

cepted a chemical explanation (water acting on metallic bases at a great depth). Humphry Davy's theory preceded Daubeny's similar one but was retracted in 1830. W. T. Brande, however, Davy's successor at the Royal Institution, taught both the Huttonian theory of "submarine fire" and its theological justification. Meanwhile, certain older explanations of volcanic "fire" had by now been decisively rejected. Among these was the Wernerian one (also held in modified form by Breislak) that volcanoes were fueled by deposits of coal and petroleum; other traditional versions utilizing the action of water on pyrites, or the combustion of sulphur, were equally inadequate.

Even more fundamental than debate about the existence of central fire was the broader question of the earth's internal heat. Controversy in this area began effectively with John Murray's assertion in his *System of Chemistry* (1806–07) that the Huttonian concept of sustained internal heat was physically impossible because any residual warmth existing within the earth would soon diffuse itself equally throughout its mass (3: app., 23–64). Though answered emphatically by Playfair in 1812, who distinguished equilibrium from uniform diffusion, Murray defended his position three years later (1815). The empirical fact, however—already well established by studies in France and Cornwall—was that temperatures steadily increased as one descended into mines.

An important series of mathematical papers by Joseph Fourier and some others in France then significantly advanced current knowledge of the physics of heat distribution and specifically affirmed (in 1824) the existence of a residual heat within the earth, presumably the result of its cosmological origin as a gradually cooling blob of nebular matter.[7] Thus, Scrope could write in 1825: "The observations lately made as to the temperature of mines, which *increases with their depth*, lead to the conclusion that the interior of the globe, at no great vertical distance, is at an intense temperature. This internal accumulation of caloric must be continually endeavoring to put itself in equilibrio, by passing from the center towards the circumference, wherever the conducting powers of the substances enveloping the globe permit" (30). Among others, Sir Alexander Crichton defended a version of this position the same year, as did Daubeny in 1826, Macculloch in 1827, and both Jameson and Macculloch in 1831.

While the theoretical work of Fourier and others gave an additional (but sometimes ignored) dimension to the volcanic studies of European geologists, Sir James Hall independently pursued the question of heat and Huttonian theory in his more empirical British manner. Accordingly, his last important paper, "On the Consolidation of the Strata of the

[7]Fourier's contributions are discussed by Herivel (1975, 197–202); see also Jerome R. Ravetz and I. Grattan-Guinness in Gillispie 1970–80. Lawrence (1977) adds an important discussion.

Earth" (presented 4 April 1825), assumed the general agreement of most geologists that both stratified and unstratified rocks had, in the process of consolidation, undergone both chemical and mechanical changes. For Hall, only the "unabated activity" of the earth's internal heat could account for these phenomena. Noting the frequent violent exertions of that heat under our oceans, said Hall, Hutton had thought this one force sufficient basis for a rational theory of the earth. He had not, however, attempted to explain the origin of that heat. Stimulated by his conversations with Hutton about heat during the 1790s, Hall soon undertook his famous experiments, proving first the identity of basalt and lava, then the artificial production of marble from lime. His most recent experiments, now revealed, dealt with the consolidation of strata.

Critics of the Huttonian theory, Hall noted, had often insisted that no amount of heat would consolidate loose sand or gravel into solid rock. This point being well taken, Hall looked for some necessary flux to effect the fusion—and found it, as he believed, in the ocean's salt. This idea had come to him in the summer of 1812 when, returning from a geological excursion to Lammermuir, he noticed a gravel bank traversed vertically by a conglomerate-surrounded basalt dyke, with the conglomerate grading by degrees into unconsolidated gravel. The intrusive basalt, he reasoned, had through its heat and something else consolidated loose gravel into conglomerate. Not far away, Hall then found (near the sea) sandstone cliffs displaying an efflorescence of salt. It soon occurred to him that heat applied to the bottom of the sea (beds of sand and gravel, drenched with brine) could consolidate both the salt and the sand.

Having conceived such a theory, Hall quickly submitted it to the test of experiment. When placed at the bottom of a crucible and then heated,

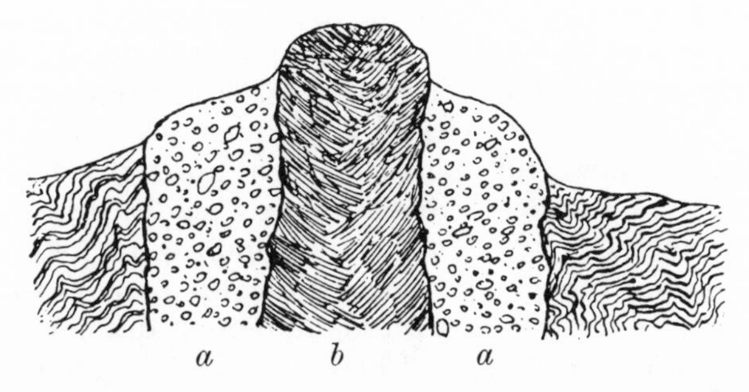

16. Section of agglomerate neck (a, a) with lava plug (b). "Some vents of agglomerate or tuff are pierced by a plug of lava, as may be instructively seen in many of the Carboniferous and Permian necks of the centre and south of Scotland" (Archibald Geikie, *The Ancient Volcanoes of Great Britain,* 1897, I, 64–65). Hall may inadvertently have been describing the same structure.

he found, dry salt and sand invariably produced solid stone—though some examples, to be sure, were more solid than others. Fumes of the salt (also used in glazing pottery, Hall pointed out) no doubt acted on the quartz sand as a flux, cementing adjacent particles together, as in sandstone. Additional experiments, "furnishing an unbounded variety of results," substituted brine for dry salt in order to approach the natural situation more closely. He again achieved consolidation, but only with high heat applied for many hours. Hall nonetheless believed that he had found the key to the agglutinated gravel seen by him in nature, to sandstone in general, and perhaps to stratified rocks of all kinds. He was even now extending these results to basalt, granite, and greywacke and soon hoped to lay before the Royal Society of Edinburgh, which he was addressing, these further confirmations of the Huttonian theory of the earth. Such confirmations did not, however, appear, and Hall's mistaken theory of saline flux proved not to be influential. Consolidation therefore remained an area of uncertainty.

Uplift and Erosion

Throughout the 1820s an increasing number of geological phenomena were attributed to elevation from beneath. As William Buckland, for example, continued to investigate the origins of British valleys, he came to see that any one theory or attempted explanation would be inadequate. Accordingly, in February 1825 he presented to the Geological Society of London a further paper proposing the formation of *some* valleys by elevation of the strata enclosing them. Valleys of a certain type he thought utterly impossible to explain by denudation alone without reference to "a force acting from below, and elevating the strata along their central line of fracture" (1829, 123). For him, the Weald of Kent and Sussex (in modern terms, a truncated anticline) was such a valley. Though water might well have cleaned up a large number of shattered fragments accruing from its elevation, no amount of water could have shaped it without powerful uplift from beneath. Many other similarly open valleys, Buckland held, "though largely modified by denudation, owe their origin to an antecedent elevation and fracture of their component strata" (125), a phenomenon he thought common to geological formations of all ages. Though Buckland himself did not associate the theory with Hutton's, his Valley of Elevation idea was a most significant extension of Huttonian concepts.

That same year, but in a rather different context, G. P. Scrope also emphasized the topographical significance of Huttonian uplift. Attributing a constant expansive force within the earth to subterranean lava, he regarded volcanoes as safety valves designed to prevent the earth from expanding too much. The transfer of heat from the earth's center to its circumference, causing both volcanic eruptions and earthquakes, had acted with undiminished energy throughout geological time. "How completely the appearances presented by the strata of our continents correspond with what would thus appear a priori to be the necessary results of the process in question," he remarked, "must be obvious to everyone." Having been ably and eloquently enumerated by advocates of the Huttonian theory, they "compose an irrefragable mass of evidence in favor of the progressive elevation of our continents by a force acting from below upwards" (1825, 191). Almost necessarily, Scrope agreed with Buckland that certain valleys called by him "anticlinal" (213), including the Weald, had been created tectonically—for Scrope, by the combined forces of elevation and subsidence. But he also accepted occasionally paroxysmal expansion, affirmed diluvial action, and rejected the Huttonian consolidation of sedimentary strata by heat (213n, 216, 222). In another book two years later, Scrope specifically attributed most of the earth's surface irregularities, including the basins of its seas, lakes, and rivers, to subterranean expansion rather than erosion. Less ambitiously, P. J. Martin also assigned seismic origins to the Weald.

As Scrope's example suggests, uplift became a popular geological explanation in many contexts. One of the most noticed scientific papers (and all the more so for having been written by a woman) described cataclysmic uplift along the coast of Chile (Graham 1824). Macculloch (1823) based part of his increasingly explicit support for the Huttonian theory on an analysis of uplifted volcanic islands in the South Pacific. The volcanic origin of Auvergne domes and the Euganean Hills had been recognized by Scrope, Macculloch, and others. Meanwhile, Leopold von Buch, one of Werner's most distinguished former students, observed unmistakable evidence of uplift adjacent to cratered volcanic cones and reasonably supposed that they too (like domes) had been pushed upwards from beneath rather than being constructed from accumulating ejecta (1821, 425). In so arguing for his Craters of Elevation theory, von Buch associated himself specifically with Hutton. While fully accepting the latter's plutonism, however, he ignored Huttonian emphasis on the erosion of surface relief.[8] His was therefore another of the more or less loyalist theories—like those of Hall, Scrope, Daubeny, and Macculloch—which accepted large parts of Hutton's own thinking but demurred at important points.

Anyone who accepted the igneous origin of either granite or basalt necessarily explained their widespread surface occurrences by pos-

[8]See the article on von Buch by W. Nieuwenkamp in Gillispie 1970–80.

tulating the removal of previously overlying sediments. As Macculloch argued in 1831, "The immense deposits of materials which now form the alluvial tracts of the globe, the enormous masses of secondary strata which have been produced by ancient materials of the same nature, all prove the magnitude of the destruction which mountains have formerly experienced, which they are now daily undergoing. Let imagination replace the plains of Hindustan on the Himalaya, or rebuild the mountains which furnished the secondary strata of England, and it needs not be asked what is the extent of ruin, modern or ancient" (1: 154). Even more obvious examples were available in Europe, where Saussure's observations and the fall of the Rossberg in 1806 had long ago made "the decay of the Alps" a scribbling traveler's cliché. For the great majority of theorists denudation was not at issue, but the means by which it had been accomplished were. As we have already seen in part, for example, fluvial erosion (which both Hutton and Playfair affirmed) was not widely believed in, largely because rivers seemed insufficiently powerful to create their own valleys. Similarly, subaerial erosion (in which wind, sand, rain, frost, and snow are agents) was accorded only minor significance by most theorists, though Hutton had stressed it also. Instead, geologists during the 1820s regularly preferred three other forms of erosion: coastal, diluvial, and (if necessary) marine.

Though evidence for coastal erosion could be found at almost any juncture of land and sea, it was nowhere more prominently displayed than along the southern coast of England, where Hutton himself had investigated it. "The encroachments of the sea along the coast of Sussex," Gideon Mantell wrote in 1822, "have continued from time immemorial. . . . By the incessant action of the waves the cliffs are undermined, and at length fall down, and cover the shore with their ruins. The softer parts of the strata . . . are rapidly disintegrated and washed away while the flints and more solid materials are broken and rounded by the continual agitation of the water and form those accumulations of sand and pebbles that constitute the beach" (293–294). Though Mantell was far more scientific, observers since Elizabethan times had preceded him in affirming the same destruction.

Like many others, however, Mantell also supported the concept of a much briefer and even more efficacious erosional agent, the one or more vast deluges affirmed by most Cuvierians. The chalk strata of Sussex, he believed, had (subsequent to their consolidation) suffered extensive destruction, several upper beds having been swept away; later deposits atop the remaining chalk were then "broken up, and in a great measure destroyed, by an irruption of water in a state of violent commotion—a catastrophe to whose powerful agency the present form of the surface of the earth, and the accumulations of beds of gravel, sand, etc. are to be attributed." The present effects of the ocean, he added, "appear to be wholly inadequate to produce changes like those which have formerly taken place" (1822, 304). As we have seen, belief in such a debacle was somewhat *de rigueur* in British geology at that time.

Energy and Time

Unlike Hutton and Playfair, most geological theorists of the 1820s took it for granted that geological forces in the past had been stronger than those in the present, a position firmly reinforced by Cuvier, whose influence continued to predominate. When we compare geologically past with present, Cuvier emphasized, "The thread of operations is broken; the march of Nature is changed, and none of the agents which she now employs would have been sufficient for the production of her ancient works" (1813, 14). As he then stressed, "It is in vain that we search among the powers which now act at the surface of the earth for causes sufficient to produce the revolutions and the catastrophes, the traces of which are exhibited by its crust" (20). Though he did not sufficiently elaborate these dramatic remarks, Cuvier obviously implied that the geological agencies of the past were not only stronger but of an entirely different sort.

William Buckland's argumentation on behalf of the Deluge was akin to Cuvier's. In 1820, for example, he cited "valleys and basins . . . drained by chasms and precipitous gorges of enormous depth, which could not have been produced by the most violent torrents that now flow through them, but must be referred to the disruption of mountain masses at the epoch of ancient revolutions that have overturned the globe" (1820, 16). Diluvial phenomena everywhere evident to us, moreover, were impossible to explain without reference to "the grand fact of a universal deluge" (23). Three years later, he thought it equally impossible that the excavation of valleys "could have been produced in any conceivable duration of years by rivers that now flow through them" (1823, 237) and therefore referred their formation to a transient inundation. As we have seen, many of Cuvier's followers reasoned similarly.

Several theorists accepted the reality of present-day geological forces while limiting their effectiveness. Buckland, for example, recognized the existence of modern geological catastrophes but considered them puny compared with those of the past. For him, the geologist's task was to decipher monuments of the "mighty revolutions and convulsions" our globe had suffered, "convulsions of which the most terrible catastrophes presented by the actual state of things (earthquakes, tempests, and volcanoes) afford only a faint image, the last expiring efforts of those mighty disturbing forces which once operated" (1820, 5). Similarly, Conybeare and Phillips assumed that volcanic forces had been more virulent in the past; "we now experience," they proposed, echoing Buckland, "only the expiring efforts, as it were, of those gigantic powers which have once ravaged the face of nature" (1822, xviiin). Karl von Hoff added (1824, 2: 51, 67–68, 367–371) that volcanic forces had not only been stronger in the past but more frequent and widespread as well. (He also argued for difference in kind; specifically, that only ancient volcanoes could produce basalts.) Writing his own book about volcanoes, G. P. Scrope identified three modes of rock formation (precipitation, subsidence, and elevation), all of them "still in operation," though "with diminished energy, it is true" (1825, 242). Constant Prevost (1825) and others agreed with this conception. Less scientific writers, too, fastened on the supposedly more powerful geological forces of the past as one of the period's most popular commonplaces.

Even so, the perhaps limited but nonetheless formidable efficacy of present-day forces was often acknowledged. Robert Jameson, for instance, affirmed the effectiveness of both rain and rivers in his Wernerian treatise of 1808. "Every long-continued rain," he argued, "convinces us of the powerful mechanical effects of water on the surface of the earth" (26), including the formation of valleys. For the obscure but significant John Carr in 1809, few things seemed more evident than that valleys had been excavated by the streams presently flowing through them. "The earth," William Smith proclaimed more generally in 1815, "is formed as well as governed, like the other works of its great Creator, according to regular and immutable laws." That such laws existed and were discoverable by human reason had of course been a recurring conviction throughout the Baconian-Newtonian scientific tradition of even the eighteenth century, of which Hutton himself was a legatee. Playfair had spoken out in 1816 to affirm that "the laws which are every day in action are those which it is most important for us to understand" (1822, 2: 83). In his *Observations on the Geology of the United States* (1817), the originally Scottish geologist William Maclure likewise proposed to limit geological speculation within "probable effects resulting from the regular operations of the great laws of nature which our experience and observation have brought within the sphere of our knowledge" (iv).

Contesting Thomas Jefferson's catastrophic interpretation of the Natural Bridge of Virginia, in 1818 Francis Gilmer followed Hutton (*Theory*, II, chap. x) in attributing its actual formation to "the very slow operation of causes which have always, and must ever continue, to act in the same manner" (1818, 188). That same year the Royal Scientific Society of Gottingen offered a prize for the best essay linking the great geological upheavals of the past with changes presently occurring. In his first volume (1822), the winner, Karl von Hoff, accepted—at least in part—the excavation of valleys by rivers, the accumulation of sediments at river mouths, and a variety of volcanic phenomena. For John Macculloch, the elevation of Italian strata was clearly attributable to "the same causes which are now, or have recently been, operating in producing smaller changes," perhaps catastrophically but more probably "through a long series of ages" (1823, 277, 281). About the same time, Adam Sedgwick of Cambridge defended the igneous origin of basalt dykes. "If we exclude volcanic agency," he argued, "what power in nature is there capable of producing such an effect? By supposing such phenomena the effects of volcanic action, we bring into operation no causes but those which are known to exist" (2: 291) and are adequate to effects even more extensive than those he had described. Theorists of all schools, then, freely called on present-day forces whenever their agency seemed plausible.

William Buckland's insistence on catastrophes and deluges in a series of much-noticed geological papers, the more important of which then reappeared in *Reliquiae Diluvianae,* convinced several less prestigious observers that Cuvierian emphasis on catastrophic explanations was fundamentally wrong. Thus, in 1825 and 1826 John Fleming publicly challenged Cuvier and Buckland's assumption that geological forces in the past differed absolutely from those in the present. Similarly, G. P. Scrope probed the question in his books of 1825 and 1827. Contesting Buckland's seismic interpretation of valley formation, Scrope thought the great majority of valleys "wholly and entirely excavated by the slow but constant and powerful action of the same causes which are still continually in force" (1825, 214). Two years later, he attributed erosion in the Auvergne to "rains, frosts, and other meteoric agents" rather than a violent deluge (1827, 160, 162). Daubeny conceded that aqueous forces may have been stronger and different in the past but thought that earthquakes and volcanoes—even if weaker—had remained basically the same (1826, 2). Though John Phillips, illustrating the geology of Yorkshire (1829), fully accepted the real existence of the Deluge, he likewise affirmed an at least limited acceptance of Huttonian subaerial erosion. That same year, Henry De la Beche again found "existing causes" undeniable and therefore attempted to distinguish valleys excavated catastrophically from those excavated gradually. But a well-known paper by Conybeare (also 1829) then denied the efficacy of riverine erosion altogether. Though a surprising number of writers during the 1820s acknowledged present-day geological forces, almost no one accepted the originally Huttonian position that those same forces had always prevailed.

As we have already seen, theorists who advocated the steadfastness of present-day geological forces through past time could explain their supposed efficacy only by assuming that their relatively undramatic effects had been allowed to accumulate over a vast span of years. What John Carr in 1809 called "sufficiency of duration" then became a recurring need. Throughout the Regency, however, it gained relatively few adherents, in part because extended geological time (which catastrophism did not require) had long been associated with antiscriptural intentions. But there were also geological objections, most of them based on the inability of theorists to document meaningful geological change in the present.

As we might expect, the topic interested G. B. Greenough, who after quoting both Hutton and Playfair rejected the latter's plea for vast periods of years: "Time graduating into eternity—nay, Eternity itself; what use could he make of it? What profit can a man expect from putting zeros out to interest?" (1819, 148). If seas and rivers could not produce visible effects within human history, he argued, they would obviously do no better in a million centuries. Similarly, Buckland thought it quite impossible that the necessary removal of sediments from river valleys "could have been produced in any conceivable duration of years by rivers that now flow through them" (1823, 237). Later, Conybeare argued in like manner regarding the Thames valley: "The drainage of the atmospherical waters has here produced no sensible effect for more than fifteen centuries," he declared recklessly; "it is inferred, therefore, that to assign to this cause the excavation of the adjoining valleys, 600 or 700 feet deep, is to ascribe to it an agency for which we have no evidence; the evidence, indeed, as far as it can be examined, being adverse" (1829, 146). This position made sense to many.

Given the popularity of catastrophism, those who advocated greatly extended geological time found it hard to come up with convincing demonstrations from nature on behalf of their beliefs. During the latter part of the eighteenth century, the most notorious geological evidence adduced on behalf of an elongated chronology (with its necessarily anti-Mosaic implications) had been the lava flows of Mount Etna, as studied by the Canon Recupero of Catania and subsequently publicized by the incautious Scottish traveler Patrick Brydone (William Robertson's son-in-law), whom Hutton may well have known. After almost fifty years, the "certain false conclusions" Macculloch cited in 1826 "as to the age of eruptions, and idly enough, as to that of the world itself" (366) were still Brydone's. Nonetheless, Macculloch's review of the next year called attention to Daubeny's "striking evidence" proving the high antiquity of Etna (1827, 311), an argument later repeated by Daubeny with regard to the Auvergne. In his own analysis of that volcanic region, Scrope then requested "an unlimited allowance of time" (1827, 162). As of 1831,

Macculloch himself was advocating the more daring chronology: "Let us contemplate Time as it relates to the Creator, not to ourselves," he proposed, "and we shall no longer be alarmed at that which the history of the earth demands" (1: 506). By then, however, that surly but precise investigator was no longer the most prominent Huttonian geologist.

[10]

Lyell

Destined eventually to be the most influential British geologist of the nineteenth century—as Hutton was of the eighteenth—Charles Lyell (1797–1875) was born the same year James Hutton died, on 14 November 1797 at Kinnordy, his family's estate, in Forfarshire, Scotland.[1] During his boyhood, however, the Lyells preferred to live at Bartley Lodge, Hampshire, in England, seven miles west of Southampton. After preliminary schooling at Ringwood, Salisbury, and Midhurst, Charles entered Exeter College, Oxford, in February 1816. During one of that year's vacations, he read his father's copy of the second edition of Robert Bakewell's *Introduction to Geology* (1815), with its outspokenly anti-Wernerian preface and insistence on the immensity of geological time. Before then, though an avid naturalist (mostly beetles and butterflies), Lyell had scarcely known that a science of geology existed.

Modern Causes

Returning to Oxford from spring vacation in May 1817, Lyell elected William Buckland's mineralogy course and soon became enthralled. Visits to London and Norwich enabled him to meet some prominent collectors of fossils and to compare his own geological analyses of landscapes

[1]Katherine M. Lyell, ed., *Life, Letters, and Journals of Sir Charles Lyell, Bart.*, (1881) is still valuable. A much more thorough life, by Leonard Wilson (1972), ends at 1841. The only complete modern biography, by Edward Bailey (1962), is superficial. Note that John Playfair was also a native of Forfarshire.

with those published earlier. At Yarmouth (decisive for Hutton's geological interests also) he stayed with Dawson Turner, a botanical correspondent of his father's; there he had time to examine, consider, and discuss (with Dr. Joseph Arnold, a knowledgeable local naturalist) the geological processes so evidently at work around him. Through detailed examinations of the river Yare, its delta, and its former channels, as well as the rapidly changing seacoast, Lyell derived an appreciation similar to Hutton's preliminary one of the efficacy of present-day geological forces.

After three weeks at Yarmouth, Lyell rejoined his father at Bartley Lodge, then accompanied him on an overland journey to Scotland. As the York coach carried them out of London, Charles noted that Highgate Tunnel was "dug through the London Clay almost entirely." In ascending some hills, they "passed up a very deep narrow glen, without a river (unfortunately for the Huttonians)" (K. Lyell 1881, 1: 45–46; see also 53, 56); the hills themselves were of magnesian limestone (i.e., dolomite). At Edinburgh, the Lyells called on Professor Jameson and explored basaltic Calton Hill. Following a week at Kinnordy, Charles alone undertook a geological expedition to the western island of Staffa, famous for its columnar basalt and spectacular Fingal's Cave. Buckland had asked Lyell to investigate the cave in order to refute a Huttonian explanation of its formation proposed by Leopold von Buch. Though he did so to Buckland's satisfaction, pupil and mentor were already beginning to disagree. On his return to London, Lyell studied mineral specimens in the British Museum and contemplated the rise and fall of Wernerian certitude regarding the origin of basalt. He was already a Huttonian in part.

In June 1818 the Lyells visited Paris, where Charles was unable to meet Cuvier (who was in England) but nonetheless explored the extensive collections of natural history there and read Cuvier's and Brongniart's major publication on the Paris Basin, which emphasized geological catastrophes and the alternation of freshwater and saltwater strata. From Paris, the Lyells went on to Geneva and Chamonix—"primitive country," as Lyell called these ruggedly scenic places (1881, 65). Here in the Alps he explored sites first described by Saussure and was particularly interested in glaciers. Before returning to London, the family geologized its way across Switzerland and over the Simplon Pass into Italy, then back via France. Throughout this itinerary, Charles collected fossils, rocks, and observations avidly. In March 1819 he was elected a fellow of the Geological Society of London (1: 57–111)

After taking a year off from serious endeavors (because of eye trouble), Lyell recovered his geological interests in 1821, when they soon competed for his time with the law studies he had rather diffidently undertaken. A meeting in Lewes, Sussex, that October with the accomplished provincial fossil collector Gideon Mantell proved to be an effective stimulus, in part because of the enthusiastic geological correspondence between them which followed. Lyell soon discovered that he found geology a good deal more interesting than law and coordinated his efforts with Mantell's in hopes of deciphering the geology of southeast England. To Mantell, Lyell confided his emerging preference for "the agency of known causes."[2]

While visiting France in July and August 1823, Lyell spent several days geologizing with Louis Constant Prévost, whose studies of French stratigraphy had also led him to prefer gradualistic explanations to the catastrophic ones of Cuvier and Brongniart. He therefore stressed analogies between present and past conditions, a methodology most congenial to Lyell. In all probability, it was Prévost who encouraged Lyell to conclude not only that present-day causes were important (of which he had already been persuaded) but, far more significantly, that no other causes had ever existed. On the other hand, Prévost, like many, also assumed that geological forces had been stronger in the past. He subsequently visited Lyell in England, and they spent much of June 1824 on a geological fieldtrip to southwest England. During this excursion, Lyell may have emerged as an even more convinced advocate of modern causes than Prévost himself.

In any case, Lyell seems to have been fully aware that he, like the geological community as a whole, was moving rapidly toward increasingly more comprehensive endorsements of Huttonian positions. In September 1824, only a year after his teacher's *Reliquiae Diluvianae* had been published, Lyell invited Buckland to Kinnordy. A joint visit to the Icelandic explorer Sir George Mackenzie followed. Buckland and Lyell then investigated sites of geological interest in the western and northern Highlands, including the parallel roads of Glen Roy and Hutton's own Glen Tilt, where the theorist's opinion regarding the origin of granite had been affirmed. From Blair Atholl, Buckland and Lyell continued south to Edinburgh; they geologized the neighborhood and inspected Jameson's mineral collections. At Dunglass, an elderly Sir James Hall took them to see Hutton's unconformity at Siccar Point. Though Lyell had earlier been commissioned to delineate Forfarshire on behalf of John Macculloch's forthcoming geological map of Scotland, his itinerary clearly indicated larger ambitions. A ferocious coastal storm off Sussex on 23 November then became for him a persuasive reminder of how much change could be effected by the forces of nature even in a day (he referred to it three times in his completed book). By the end of 1824 Lyell had rather definitely parted company with the school of Cuvier

[2]Lyell to Mantell, 6 June 1822 (Alexander Turnbull Library; Wellington, New Zealand).

and Buckland and was actively seeking evidence to support his augmenting case on behalf of modern causes (1881, 1:151–159).

Early Papers and Research

Lyell's first attempt at scientific publication, a paper read to the Geological Society of London on 17 December, established that some freshwater limestone in Forfarshire had been formed recently. This discovery was in direct contradiction to Cuvier and Brongniart, who had argued in their study of the Paris Basin that the freshwater limestones found there had no counterpart in modern lakes. Lyell's paper also expressed some mild skepticism regarding Buckland's concept of "diluvium"—that is, sedimentary deposits attributable to the Flood. In May 1825 Lyell analyzed what he took to be a dyke of serpentine, affirming Huttonian intrusion. That July he discussed Buckland's "valley of elevation" concept with Mantell and rejected any suggestion that the Weald of Sussex had been significantly altered by denudation. Yet by March 1826 Lyell was writing a paper on the Plastic Clay formation in Hampshire; his remarks included further skepticism regarding diluvium and an unexpectedly literal affirmation of Playfair. Thus, Lyell now advised, "The size of the valleys is in general in proportion to that of the streams flowing in them, and their excavation appears referable for the most part, if not entirely, to the long-continued agency of these streams" (1826a, 286). Buckland, contrarily, had argued that valleys, previously formed, determine the size of their occupying streams.

By this time, Lyell evidently felt his position strong enough to sustain a frontal attack on the opposition, which he attempted in a published essay reviewing papers appearing in the most recent volume of the Geological Society's *Transactions*. Though often concerned with vertebrate paleontology, a field in which Cuvier's authority was unquestionable, Lyell's essay also reflected his concern with matters Huttonian. Aware of Cuvier's likewise formidable reputation and following in geological theory, however, he initially emphasized catastrophic aspects of Huttonian thought. As Lyell recalled,

> It is now twenty years since Playfair observed that the land has been raised by "expansive forces acting from below," and there is reason to think that continents have alternately ascended and descended "within a period comparatively of no great extent." When the Huttonians first advanced these doctrines, no geologists disputed that there existed proofs of former changes in the relative level of land and sea, but Playfair's hypothesis appeared extravagant to many, and those were deemed "fearless of para-

> dox" who, as Mr. Greenough expressed it, "attributed to the waves constancy, mobility to the land" (1826b, 514).[3]

Though Playfair had examined various anomalies and disturbances within the rocks attentively, he did not then know of the great series of alternating freshwater and saltwater deposits established by Cuvier and Brongniart. His argument, therefore, lacked what would now be considered the most decisive evidence in its favor.

It would be incorrect, however, to explain the alternation of deposits by supposing them a record of fluctuating sea levels worldwide. Rather, Lyell supposed, it was the level of the land that had abruptly changed (as in the Chilean earthquake of 1822). No one, he continued, could read the several accounts of such recent upheavals without being tempted to inquire "whether the causes now in action are, as Dr. Buckland has supposed [1820, 5], 'the last expiring efforts of those mighty disturbing forces which once operated'; or whether, as Hutton thought, they would still be sufficient in a long succession of ages to reproduce analogous results" (1826b, 517).

Lyell then quoted Cuvier's opinion that geological forces in the past were entirely different from those operating in the present. Though many theorists agreed with such catastrophic assumptions, Lyell pointed out, a rapidly accumulating body of reliable facts and observations suggested another interpretation. "In the present state of our knowledge," he argued, "it appears premature to assume that existing agents could not, in the lapse of ages, produce such effects as fall principally under the examination of the geologist." Cuvier's assumption of discontinuing geological forces, moreover, seemed "directly calculated to repress the ardor of inquiry, by destroying all hope of interpreting what is obscure in the past by an accurate investigation of the present phenomena of nature" (518).

Carefully selected examples from the papers under review and other recent geological publications opposed suggestions that the geological past revealed either chaos or decline. The secondary strata show us proofs of occasional convulsions, it is true, but they document long intervening periods of order and tranquility as well. "The notion of a continually decreasing energy in nature's power to modify and disturb the earth's surface," Lyell supposed, "first originated in the observation that strata of the highest antiquity have suffered the greatest and most general derangement. But such must be the necessary effect of the uniform action of the same cause throughout a long succession of ages, and the frequent unconformability of strata clearly shows that disturbances have

[3] Lyell is here quoting Playfair (1802, Note xxi)—I have repositioned the quotation marks—and Greenough (1819, 191).

taken place at many and different periods" (518). Convulsions were not, therefore, to be assigned a particular period of their own.

Throughout later portions of his essay, Lyell continued to attack the unacceptable disorder implicit in catastrophist assumptions. As opposed to the outmoded Wernerians, he noted, a great majority of geologists now ascribe trap rocks to the eruptions of ancient submarine volcanoes. Previous theorists had often investigated regions in which stratigraphical derangement was unusually conspicuous. They "studied the exceptions before the rule," Lyell declared, with flagrant disregard of nature's present operations. It was not surprising that their theories, however much indebted to facts and observations, soon proved to be mutually contradictory. Thus, the generalizations of Werner and Hutton, "though bearing impressed upon them the decided mark of genius," have since required considerable modification (533). Nonetheless, one of the two theories, his readers would infer, was clearly superior to the other.

Lyell's remarkably persuasive closing, which did much to assuage nervous opposition from the religious establishment, emphasized the essential order so prominently attested by the geological record. Forces once thought capable only of defacing and ravaging the earth's surface, he argued, actually insure the perpetual renovation of the globe and are reflections of still more general laws "conceived by consummate wisdom and forethought." Animate creation, similarly, a sequence of successive plant and animal types, forms "one connected plan," and a progressive one at that (538). By the time this largely Huttonian essay appeared, Lyell had already decided, as he privately informed Mantell, to write a book "in confirmation of ancient causes having been the same as modern ones." He began working on it in March 1827.[4]

The first way station en route to Lyell's *Principles of Geology* was another essay review by him, this one based on G. P. Scrope's *Memoir on the Geology of Central France* (1827), an important study by the relatively neglected author of *Considerations on Volcanos* (1825). Scrope's more recent book dealt with the Auvergne, a region famous for its extinct volcanoes, which he had visited in 1821 and written up the following year, though publication had then been delayed. Meanwhile, the Auvergne had been described in a book of geological travels by Robert Bakewell (1823), by Scrope himself to some extent in 1825, and by Daubeny in 1826. Lyell regretted that Scrope's *Memoir* had not appeared earlier, for its detailed observations would have made some of the speculations in *Considerations on Volcanos* less problematical.

A principal question at issue between Scrope and the other English geologists who had visited the Auvergne concerned the formation of

[4]Lyell to Mantell, 2 March 1827; Alexander Turnbull Library. See also K. Lyell 1881, 1: 169, with comments emphasizing Huttonianism on pp. 160 and 163; and Wilson 1972, chap. 6.

valleys. "Whatever may be the merits of the rival theories," Lyell declared, "we consider this work the most able which has appeared since Playfair's *Illustrations of the Huttonian Theory* in support of the opinion that valleys, which decidedly owe their form to the agency of water, have not been shaped out by one sudden and violent inundation but progressively, by the action of rivers or of such floods as may occur in the ordinary course of nature" (1827, 477). That many other valleys were either wholly or partially the result of seismic forces, Lyell added, was now admitted by almost all geologists, though Buckland still required a transient deluge and loomed foremost among those refusing to acknowledge fluviatile explanations altogether. The Auvergne seemed ideally suited to decide this very question; Lyell, therefore, could not help lamenting Playfair's having died so soon after visiting it (in 1817), leaving his views unpublished. As for later travelers, Daubeny (1826) had divided the extinct volcanoes of Auvergne into those of "ante-diluvian" and "post-diluvian" origin, but Scrope (1827) specifically denied the validity of this. Lyell then cited John Fleming and other clerics of various persuasions who questioned the reality of a geological deluge; without attempting to resolve that issue more specifically, he ended his review by moderately rebutting Mosaic geology as such.

In May 1828, accompanied by Roderick and Charlotte Murchison, Lyell visited the Auvergne in person, spending no less than two months there to resolve a number of issues and particularly to confirm Scrope's suggestive evidence attesting to the fluviatile origin of valleys. On 5 and 19 December, with Lyell still abroad, a joint paper by Murchison and himself, "On the Excavation of Valleys, as Illustrated by the Volcanic Rocks of Central France," was read to the Geological Society of London. Largely descriptive, this major effort attempted primarily to establish two points: that no extraordinary inundation had ever disturbed the geology of the Auvergne, and that the excavation of its valleys by streams was demonstrable. Thus, fluviatile views previously advanced by Playfair and Scrope, among others, were now illustrated and confirmed by Lyell and Murchison, who willingly ascribed "almost unlimited power to ordinary rivers, when a sufficient lapse of time is assumed" (1829, 30). As no evidence whatever had emerged to substantiate a violent flood (but with strong evidence to the contrary), the result of this investigation—for Lyell, at least—was a striking confirmation of the persistence of present-day forces. In the debate that followed the conclusion of the paper, on 19 December, Scrope and Sedgwick supported the authors whereas Buckland and Greenough heatedly opposed.

Lyell and Murchison had written their paper at Nice in August 1828, while Murchison was attempting to recover his health after some imprudent overexertion. As Lyell wrote from there, "The whole tour has been rich, as I had anticipated (and in a manner which Murchison had not), in those analogies between existing nature and the effects of causes in remote eras which it will be the great object of my work to point out." For Lyell, no further doubts remained. "I scarcely despair now," he assured his father in the same letter, "so much do these evidences of modern action increase upon us as we go south (towards the more recent volcanic seat of action), of *proving* the positive identity of the causes now operating with those of former times." Murchison had advised Lyell to go on by himself through Italy to Sicily ("for your views, the great end") (K. Lyell 1881, 1: 199–200; see also 183–251; and Wilson 1972, chaps. 7–8).

Parting from the Murchisons at Padua, Lyell obediently worked his way south, on an itinerary he had not adequately prepared. "The effects of earthquakes on the regular strata, and the light thrown on the excavation of valleys by lavas," he informed Sir John Herschel, "are subjects to which I have directed a large share of attention" (K. Lyell 1881, 1: 201). He was therefore anxious to examine such parts of the Sicilian or Calabrian coasts as would provide evidence of elevation or subsidence, either through archeological remains (like the Temple of Serapis near Naples) or paleontological ones (living species of shells in uplifted strata). He also requested advice about Sicily from several of his geological friends who had already been there, including Scrope, Daubeny, Buckland, and Herschel.

While en route to Sicily, Lyell geologized at Monte Bolca and other parts of the Vicentin ("The volcanic phenomena are just Auvergne over again," 1: 203), the Euganean Hills, the Val d'Arno, Florence, Siena, Viterbo ("Almost the whole way from Siena to Viterbo is a theatre of extinct volcanic action," 208), and Rome. The latter's famous hills, he found, were all cut by the Tiber and its tributaries through volcanic deposits that in some cases rested on strata containing very recent shells—yet all of this dated from eons long prior to the foundation of Rome. After spending a week in the Eternal City, Lyell proceeded to Naples. While awaiting a delayed steamboat to Sicily, he explored volcanic formations and archeological evidence on the island of Ischia (recent uplift, coastal erosion) and at Vesuvius, Pompeii, Herculaneum, Paestum, and Puzzuoli. At the latter, Lyell was especially gratified by the so-called Temple of Serapis, of which only three columns still remained upright. Each of the three had been bored into at a high level by marine bivalves, proving that the columns had been submerged and then reelevated at least twelve feet within historic times. Understandably, Lyell soon came to regard the Temple of Serapis as a symbol of the theory he was attempting to establish.

On 18 November, after a two-day voyage from Naples, Lyell arrived at Messina, from which he rode south on muleback to Taormina and

Catania, with magnificent views of domineering Mount Etna, the highest active volcano in Europe, along the way. Assisted by local geologists, Lyell investigated both Etna and its Valle del Bove zealously. (Though the latter is perhaps a caldera—the issue being still in doubt—Lyell considered it a valley explainable only through subsidence.) Because the Valle del Bove exposed a magnificent cross section of Etna itself, Lyell could see how infinitely complex the whole mountain was; as the product of innumerable eruptions, and built up by them in relatively thin layers, the cone must have taken an immense period of time to construct. Lyell remained in Sicily until 9 January 1829, when, full of rapidly maturing ideas and heightened authorial enthusiasm, he returned to Naples. "The results of my Sicilian expedition," he wrote Murchison from there, "exceed my warmest expectations in the way of modern analogies" (1: 234). His book, planned throughout and in part written, would attempt to establish sound reasoning in geology, its principal assertion being "that *no causes whatever* have from the earliest time . . . ever acted, but those now acting; and that they never acted with different degrees of energy from that which they now exert" (224). Further researches at Rome, Genoa, Turin, Geneva, and Paris delayed his homecoming to London until 24 February.

Though Lyell found his new beliefs thoroughly convincing, other members of the Geological Society did not. At the meetings of 15 May and 5 June, therefore, W. D. Conybeare read his long paper analyzing the origins of the Thames valley, in which he distinguished fluvialists like Hutton, Playfair, Scrope, Murchison, and Lyell from diluvialists like Greenough, Buckland, and himself. To account for the geological phenomena he had discovered, however, Conybeare had necessarily to postulate four recent episodes of marine erosion. The evident fragility of this hypothesis pleased Lyell a great deal, as was apparent from the exulting description of the first session with which he favored Mantell the next day: "A splendid meeting last night. . . . Conybeare's paper on Valley of Thames, directed against Messrs. Lyell and Murchison's former paper, was read in part. Buckland present to defend the 'Diluvialists,' as Conybeare styles his sect, and us he terms 'Fluvialists.' Greenough assisted us by making an ultra speech on the impotence of modern causes. 'No river,' he said, 'within times of history, has deepened its channel one foot!'" As Lyell then went on to assure Mantell, "Conybeare's memoir is not strong by any means. He admits three deluges before the Noachian! And Buckland adds God knows how many *catastrophes* besides, so we have driven them out of the Mosaic record fairly."[5] The discussion of 5 June went particularly well; Adam Sedgwick, the Society's president, appeared to be leaning in Lyell's direction.

[5]Lyell to Mantell, 16 May 1829 (Alexander Turnbull Library); K. Lyell (1881, 1: 252–253) misdates this letter and misreads "impotence" as "importance"; Wilson, 1972, 264–265.

That summer Sedgwick and Murchison undertook a major geological expedition to the northern flanks of the Alps, including parts of Germany, Bohemia, and Saxony. Their itinerary took them through Rotterdam, Bonn, Rhineland volcanics, Mainz, Frankfort, Gottingen, the Harz, Berlin, Prague, Vienna, Trieste, Salzburg, Munich, and Strasburg, then home via Paris. For Sedgwick particularly, the trip was a revelation. Though Saxony, for example, had been "the focus of Wernerian geology," he found it "the most decidedly volcanic Secondary country" he ever saw. "The granite bursts through on one side," he explained, "then sends out veins, and along the whole eastern flank the secondaries are highly inclined and often absolutely vertical. Near Goslar they are absolutely heels over head." Sedgwick was equally astonished to find rocks considered Secondary (transitional) in age by Werner intermingled with deposits that were clearly Tertiary.[6] As Lyell then wrote delightedly to John Fleming on 31 October, "Sedgwick and Murchison are just returned, the former full of magnificent views. Throws overboard all the diluvian hypothesis; is vexed he ever lost time about such a complete humbug; says he lost two years by having also started a Wernerian. He says Primary rocks are not primary but, as Hutton supposed, some igneous, some altered Secondary. Mica schist in Alps lies *over* organic remains. *No* rock in the Alps older than *Lias*! Much of Buckland's dashing paper on Alps wrong" (K. Lyell 1881, 1: 256–257). Still, Sedgwick's defection from Wernerian ranks did not necessarily ally him with Lyell.

Principles of Geology, vol. I, 1830

After several years of gestation, augmentation, and revision, Lyell's first volume (of an announced two) of his *Principles of Geology, Being an Attempt to Explain the Former Changes of the Earth's Surface, By Reference to Causes Now in Operation* (1830–33) appeared on 24 July 1830. It consisted of twenty-six chapters, the first four of which—written in the spring of 1829—traced the development of geology. Chapters VI to IX proposed a theory of climate through which Lyell hoped to explain the earth's changing complement of life. As part of that discussion, Lyell attempted to undermine Fourier's deduction (from experiments on the cooling of incandescent bodies) that the earth's central heat was diminishing. Having almost no facts to assist him, however, Lyell could only hope that further experiments might reveal more definitely whether there *was* internal heat and, if so, what laws might govern its distribution (1: 141–143). Though clearly unable to support this central Hutto-

[6]John W. Clark and T. M. Hughes, *The Life and Letters of the Reverend Adam Sedgwick* (1890), is still the only extensive treatment of Sedgwick's life and work. For his 1829 tour, see 1: 349–361; I quote p. 356.

nian proposition effectively, Lyell nonetheless cast his lot with a number of the others.

By 11 January 1830, as he then informed Mantell, Lyell had purchased a copy of Playfair's *Works* (1822) and had it bound. Volume 1 of that set, we know, included a reprinting of Playfair's *Illustrations of the Huttonian Theory*. Though Lyell had previously been familiar with the *Illustrations*, his commitment to ownership of it soon influenced his own book. Thus, his title, "Principles of Geology," endorsed a phrase originally coined by Playfair. Similarly, the volume's epigraph quoted Playfair's paragraph 374, asserting the uniformity of natural laws throughout geological time. Chapter V, written around January 1830, argued more lengthily on behalf of the same point. As a further quotation (demonstrably from the 1822 version of Playfair) makes evident, significant portions of this chapter reflected a just-completed—but nonetheless erroneous—perusal of the earlier work.[7]

The more strictly geological portion of Lyell's volume—its original beginning—was delayed until chapter X, which became the first of several reviewing basic changes in the organic and inorganic realms. These changes, Lyell pointed out, are of two kinds, aqueous and igneous. "The *aqueous* agents are incessantly laboring to reduce the inequalities of the earth's surface to a level while the *igneous*, on the other hand, are equally active in restoring the unevenness of the external crust" (167). A first group of aqueous causes included all those connected with the circulation of water from land to sea; a second, with tides and currents.

Lyell's analysis of running water, accompanied, as throughout the volume, by interesting examples, endorsed both the mechanical and solvent power of water, by freezing and dissolving, respectively. He also described the powerful attrition of which streams transporting sand and pebbles are capable. Stimulated in part by a recent paper of Scrope's on the excavation of valleys, Lyell argued at some length for the fluviatile position, citing "decisive" evidence from the Auvergne as proof that "neither the sea nor any denuding waves . . . have passed over the spot since the melted lava was consolidated" (177). In this chapter and those following, Lyell marshalled an impressive array of examples (only some of which he knew at first hand), including rivers adjacent to Mount Etna, the falls of Niagara, the Po, the Mississippi, and various local floods. The flood of Tivoli, in 1826, with which his eleventh chapter concluded, allowed Lyell (in an argument borrowed from Playfair) to equate slanders once directed against Aristarchus with similar "accusations founded on religious prejudices" raised against Hutton. "We might ap-

[7]Lyell to Mantell, 11 January 1830 (Alexander Turnbull Library). Playfair uses the phrase "principles of geology" in his *Illustrations*, at the bottom of p. 124. For the composition of the *Principles*, vol. 1, see Wilson 1972, chap. 9.

peal to the excavating power of the Anio [a river at Tivoli] as corroborative of one of the most controverted parts of the Huttonian theory," Lyell concluded (197).

Chapters XIII and XIV, describing deltas, emphasized the constructive activity of running water. Rivers both destroy and create land, Lyell affirmed, and have evidently done so in precisely the same manner throughout geological time. Though Hutton was himself unnamed (the topic being one he had not considered), Lyell's endorsements of general laws, balanced creation and destruction, and overall orderliness were fully in accord with his predecessor's. Regarding tides and currents (chapters XV–XVII), Lyell also endorsed "modern observations"—in fact, Hutton's—showing that the reduction of continuous tracts to offshore islands and then isolated rocks was still going on. The south coast of England and other sites provided striking examples of coastal erosion. But tides and currents were capable of constructive deposition as well. Thus, geologists willing to presume that "the course of nature has been uniform from the earliest ages, and that causes now in action have produced the former changes of the earth's surface" (311), would easily perceive effects analogous to present ones in the ancient rocks.

The remaining chapters (XVIII–XXVI) dealt with volcanoes and earthquakes. To begin, Lyell emphasized that geological forces, though undiminished overall, fluctuate, frequently abandoning one theater of operations for another:

> That we should find, therefore, cliffs where the sea once committed ravages and from which it has now retired, estuaries where high tides once rose but which are now dried up, valleys hollowed out by water where no streams now flow—all these and similar phenomena are the necessary consequences of physical causes now in operation; and we may affirm that, if there be no instability in the laws of nature, similar fluctuations must recur again and again in time to come. (313)

The geological quiescence of any formerly active region, therefore, tells us nothing about the magnitude of geological forces as a whole.

Utilizing von Hoff's second volume and other sources available to him, Lyell surveyed volcanic activity around the world. Chapters XIX and XX, however, dealt specifically with Vesuvius and other Neapolitan phenomena with which Lyell was personally familiar. In discussing the lavas of Vesuvius and their characteristics, he acknowledged "the Huttonian assumption . . . that lava cools down more slowly under the pressure of a deep sea than in the open air," which he supposed a corollary derived from Sir James Hall's experiments on compression, "whence it was inferred that vast pressure prevented water from expanding into steam" (348). A jibe at Werner followed three pages later.

Chapter XXI, while surprisingly brief on Etna, endorsed Hutton's position on the igneous origin of porphyritic dykes. It was then necessary to controvert opinions by Humboldt and von Buch that attributed the formation of volcanic cones to inflation and uplift rather than accumulation. This "craters of elevation" theory (XXII), though seemingly Huttonian in part, entailed strongly catastrophic implications and was therefore opposed by Lyell (as well as Scrope) for more than twenty years (see Dean 1980). Concluding his discussion of volcanoes, Lyell abruptly diminished their impressive energies to near insignificance as compared with those fires "in the nether regions" undoubtedly responsible for both fusing and consolidating great masses within the earth. Though they cannot be stratified, he theorized, subterranean volcanic rocks probably "separate into natural divisions" (397), as happens with many lava currents. Such "deep-seated igneous formations," moreover, "must underlie all the strata containing organic remains, because the heat proceeds from below upwards, and the intensity required to reduce the mineral ingredients to a fluid state must destroy all organic bodies in rocks either subjacent or included in the midst of them" (398). By a continued series of elevations, these igneous masses could be brought to the surface and upheaved into mountains—as indeed happened, to the extreme puzzlement of earlier geologists who supposed them representative of the first state of the earth.

Amid many vivid tales and consequential proofs, Lyell's four chapters on earthquakes (XXIII–XXVI) emphasized the importance of seismic movements in the formation of valleys and other fluvial effects. Assuming the gradual elevation and subsidence of mountain chains, he suggested, it is easy to explain how rivers have enlarged ravines into valleys and deepened existing valleys so much. To reconcile the slow action of ordinary rivers with the great width and depth of their valleys, we have only to posit a long succession of routine earthquakes occurring at the usual intervals of time. Chapter XXV included a long analysis of the Temple of Serapis, which Lyell had also adopted as his frontispiece.

Volume I, and Lyell's consideration of earthquakes and other igneous phenomena, ended with chapter XXVI, in which some additional facts and a series of conclusions were set forth. Important as the visible effects of active volcanoes certainly are, Lyell reiterated, they indicate far greater processes at work within the earth. "That both the chemical and mechanical changes in the subterranean regions must often be of a kind to which no counterpart can possibly be found in progress within the reach of our observation," he affirmed, "may be confidently inferred" (460–461). Far beneath presently active volcanic regions, enormous masses of matter in constant fusion ceaselessly discharge heat, lava, and gas; to this consistent release, Lyell thought, "we probably owe the general tranquility of our globe" (470). In all exertions of geological force, he emphasized, renovating as well as destroying causes are unceasingly at work, "the repair of land being as constant as its decay, and the deepening of seas keeping pace with the formation of shoals" (473). The effects of aqueous and igneous agents "are continually counteracted by each other, and a perfect adjustment takes place before any appreciable disturbance is occasioned" (474). More specifically, Lyell concluded, "the constant repair of the dry land, and the subserviency of our planet to the support of terrestrial as well as aquatic species are secured by the elevating and depressing power of earthquakes." This cause, "a conservative principle in the highest degree," was, above all others, "essential to the stability of the system" (479), a conclusion possibly deriving not from Playfair but (in certain verbal respects, at least) from Hutton's original paper of 1788.

Lyell's first volume, then, endeavored to elevate several of Hutton's observations and theories into axiomatic principles of geology. Among these, we may note:

1. The uniformity of natural laws throughout geological time.
2. The destructive power of water.
3. The constructive power of subterranean heat.
4. The excavation of valleys by streams.
5. The igneous origin of dykes.
6. The importance of compression.
7. The former fluidity and common origin of igneous/plutonic rocks.
8. The uniqueness of subterranean processes.
9. The essential balance of creation and destruction.
10. The utility of geological processes for sustaining life.

Besides reaffirming points already established by Hutton, however, Lyell also enhanced the theory in several respects—most notably by completing symmetries of creation and destruction not fully developed by his predecessor (but sometimes proposed by his opponents). The major additions, some of them influenced by Playfair, emphasized the creative power of water, the fluctuations of geological forces, the importance of earthquakes, and the uniqueness of past life and climates. But it is not always easy to distinguish between substantive and merely verbal differences. Though Lyell clearly dichotomized igneous and aqueous forces, for example, Hutton had done so less distinctly. Overall, Lyell preferred to emphasize the identity and uniformity of past and present geological causes rather than the balanced cycle of destruction and creation postulated by Hutton.

Lyellian Controversy, 1830–31

However skillfully Lyell may have argued on behalf of his modified Huttonian positions, he did not escape determined and prolonged opposition. After Conybeare's generally ineffectual attempt in 1829—that paper on the valley of the Thames—the next opponent to throw down his gage was Henry De la Beche, who in March 1830 published "Notes on the Formation of Extensive Conglomerate and Gravel Deposits," which had much broader implications than its title suggested. "At the present time," he began, "when actual causes are by some geologists considered adequate to the production of nearly all the phenomena which we observe in the structure of the earth's crust, it becomes important to ascertain, as far as our knowledge will permit, the value of such causes." (1830a, 161). It had been imagined by those same mistaken geologists that extensive deposits of gravel and conglomerate had been formed by causes similar to those now existing. Attention to the present actions of seas and rivers, he believed, would disprove their hypothesis.

Though De la Beche quoted Playfair with approval, he plainly opposed the Huttonian notion that gravels from the land are transported far into the ocean's depths to be solidified into conglomerate. On the contrary, the ocean throws such gravels back upon the land. Having just returned from a voyage to Jamaica, De la Beche could not accept Playfair's description of wave action. Rivers, similarly, probably do not carry gravels for great distances. The Huttonian explanation for the origin of conglomerate was therefore unnatural. We can explain these vast deposits only by recognizing "some greater and more general force than the action of seas on their coasts or rivers on their beds," which could only mean "debacles, . . . frequent and great" (171). De la Beche's subsequent *Sections and Views Illustrative of Geological Phenomena* (1830b) disavowed all theories and was critical of many, especially the Huttonian view that valleys have been excavated by rain water and the rivers flowing through them.

Once Lyell's first volume had actually appeared, it was favorably reviewed, through prior arrangement with the author, by G. P. Scrope, who emphasized the theological acceptability of uniformitarian geology. After summarizing his friend's arguments on behalf of both aqueous and igneous agencies, Scrope specifically endorsed Lyell's preference for observable causes but questioned his assumption that they have always acted in precisely the same way. "The general invariability of these laws," he proposed, "is no way called into question by our supposing the condition of the whole globe to have varied gradually with respect to temperature or subterranean forces of expansion, or to have been once under wholly different circumstances—in a fluid or nebulous stage, for instance—and to have passed through several progressive stages of existence previous to its acquiring the precise character in which we at present view it" (1830b, 465). Like Playfair, Scrope believed that uniformitarian geology ought to consider the changing fossil record and our planet's cosmological origins more fully. For him also, but not Lyell, the history of the earth was progressive. Certain rocks, he noted, "do not appear to be anywhere produced in the present circumstances of the globe" (468), and the nonappearance of fossils in the oldest sediments was a problem that neither Hutton nor Lyell could solve. "To account for the greater consolidation, more crystalline structure, and absence of animal impressions in the earlier sedimentary rocks," Scrope noted, "Dr. Hutton was driven to suppose them altered by central heat. Mr. Lyell, who rejects this as untenable (many of these formations being interstratified with loose beds and soft shales clearly unaffected by heat), refers these general characters to the effects of time, infiltrations, and mysterious agencies, such as chemical affinities and electricity" (469), a seeming inconsistency that did not prevent Scrope from regarding Lyell's book as the beginning of a new era in the history of geology.

Despite Scrope's initial argumentation, the theological implications of uniformitarian geology continued to attract notice, most of it strongly hostile. In August 1830, for example (only a month after volume one of the *Principles* had come out), William Conybeare offered a general and public rebuttal to Lyell's advocacy on behalf of existing causes and the uniformity of nature. He regarded the whole book as nothing more than an extended commentary on "the celebrated Huttonian axiom that 'in the economy of the world no traces of a beginning or prospect of an end can be discerned.'" Leaving moral objections aside, Conybeare regarded Hutton's infamous concept as "one of the most gratuitous and unsupported assertions ever hazarded" (1830, 218).

Writing for a theological journal, William Whewell characterized Lyell's volume as an elaboration of "the well-known Huttonian doctrine that the strata at the surface of the earth have been formed by the agency of causes which still continue to act in the usual course of the world" (1831, 185). Though Hutton's insistence on unlimited drafts of time alienated many from his theory, Lyell remained undeterred by the same requirement. Even if his book were to be consulted "merely as a new exposition of the Huttonian theory," Whewell adjudged, "there is no comparison in the fullness and variety of his facts, in the clear intelligence of their true nature and bearing, between this and any previous work of the kind" (186). Within it, however, Lyell had failed to explain the succession of animate life, which Whewell took to be "a distinct manifestation of creative power, transcending the operation of known laws of nature." Geology, then—but not Lyell's—"lighted a new lamp along the path of natural theology" (194).

That November, aiming clearly at Lyell, Conybeare began his multipart "Examination of Those Phenomena Which Seem to Bear Most Directly on Theoretical Speculations" (1830–31). By proceeding chronologically through a series of selected geological phenomena, he hoped to demolish the uniformitarian theory while affirming his own catastrophic one, especially the violent excavation of river valleys by diluvian currents. Among the phenomena cited, he included the general form of the terraqueous globe (a spheroid of rotation); the marine origin of stratified rocks (deposited gradually and slowly over a long period of time), igneous intrusives, and inclined strata. Beyond these mostly Huttonian points, Conybeare also affirmed the decreasing violence of geological convulsions in successive periods, analogous gradations in the texture and consolidation of rocks, and a steady diminution of world temperature, as demonstrated by the fossil record. The distribution of volcanic rocks, moreover, also suggested greater energies in the past.

The second major portion of Conybeare's "Examination," appearing in March and April 1831, discussed aqueous causes. "Mr. Lyell," he reiterated, "believes that the forces which act on our planet have been, and are, ever constant and invariable. . . . I, *e contra*, have endeavored by a tolerably detailed examination of [geological] phenomena to show that the only fair inference from them is the direct contradictory" (1830–31, 188–189). Conybeare opposed Lyell's fluvial theory of valley formation with his own diluvial one, which, he insisted, did not necessarily entail the Mosaic deluge. Beds of conglomerate and gravel, he argued, indicate several periods of violent diluvial action, the valley of the Thames being a further example. He attributed most valleys of excavation to marine currents, both when the sediments were still being accumulated beneath the ocean and when the ocean was gradually retreating from the land. The existence of waterfalls also seemed to him inconsistent with the fluvial hypothesis. Assuming the Niagara River and the Thames to be of approximately the same age, Conybeare proposed, the former should have eroded far more effectively than it had.

Ironically, by the time Conybeare's extended rebuttal had been concluded, he was no longer Lyell's chief antagonist. Adam Sedgwick, then president of the Geological Society, assumed that role through an address given by him in February 1831. "The study of the great physical mutations on the surface of the earth is the business of geology," he affirmed, "but who can define the limits of these mutations?" (1831, 302). Since any form of a priori reasoning seemed unacceptable to Sedgwick, he thought it necessary to criticize Lyell's *Principles*, a volume for which he otherwise had high praise. Unfortunately, Lyell had chosen to appear not only as a historian of the natural world but as "the champion of a great leading doctrine of the Huttonian hypothesis," a thesis "framed by its distinguished author without any knowledge of the most important facts of Secondary geology" (303). In unwisely presenting himself as the defender of a theory, Lyell had allowed his thinking to be warped by the hypothesis he sought to defend; in adopting the language of an advocate, he had sometimes forgotten the character of a historian. Clearly disturbed by the powerful arguments of both Hutton and Lyell affirming the ubiquity of denudation, Sedgwick (rather like Hutton himself) sought desperately for some countervailing principle.

"The forces of degradation," Sedgwick declared, "very often of themselves produce their own limitation." Enlarging on a theme originally promulgated by Deluc, he recognized several inherent conservative checks on the "mighty ravages" of erosion: "The mountain torrent may tear up the solid rock and bear its fragments to the plain below, but there its power is at an end, and the rolled fragments are left behind to a new action of material elements" (303). And what was true of a single rock applied also to a mountain chain or other vast region: once some episode of spoilage had been completed, landscapes were refurbished by vegetation and other processes, nature's "vast counterpoise to all the agents of destruction" (304).

Despite Sedgwick's objection on the same grounds to Lyell, a modern-day observer cannot help but notice how his own thinking—both geological and theological—accorded with Hutton's. "For we all allow," Sedgwick continued, "that the primary laws of nature are immutable, that all we now see is subordinate to those immutable laws, and that we can only judge of effects which are past by the effects we behold in progress. Whether there be, or be not, any physical traces of a state of things anterior to the commencement of our geological series of deposits is a question of no importance" (304). To this extent, then, Sedgwick was a Huttonian himself. But there was one major principle—implicit in Hutton, explicit in Lyell—with which Sedgwick could not agree. To assume, as Lyell did, that "the secondary combinations arising out of the primary laws of matter have been the same in all periods" of the earth's history seemed to him "an unwarrantable hypothesis with no *a priori* probability" (304). On the contrary, Sedgwick argued—and this was the core of his position—"the earth's surface presents a definite succession of dissimilar phenomena" (304). Such Lyellian phrases as "the undeviating uniformity of secondary causes," "the uniform order of physical events," "invariable constancy in the order of nature," and so forth (all of them accurately quoted from the *Principles*) were therefore essentially meaningless. To be sure, each geological formation may have required a great deal of time, but we are not then entitled to invent long, imaginary cycles. "In the very first step of our progress," he insisted, "we are surrounded by animal and vegetable forms of which there are now no living types" (305). One either affirmed the doctrines of spontaneous generation and transmutation of species ("with all their train of monstrous consequences") or assumed that conditions in the past had changed. "These subjects, indeed, are not yet touched upon by Mr. Lyell," Sedgwick noted, "and I throw out these remarks only to show by what difficulties the Huttonian hypothesis is encountered—of a kind, too, never present to the mind of its inventor" (305). The appearance of man, Sedgwick continued, was a geological phenomenon of vast importance and "absolutely subversive of the first principles of the Huttonian hypothesis" (306).

The progression of life aside, Sedgwick also found obvious discontinuity in the stratigraphic record. Alluvium and diluvium, for instance, were of differing origins and rarely continuous; examples from the Secondary strata seemed even more impressive. It was almost universally admitted, he adjudged, that highly inclined strata had been deformed through elevation. But this (unacknowledgedly Huttonian) position had been greatly developed in recent years by the orogenic theories of Leoncé Elie de Beaumont, a French theorist who had proved to Sedgwick's satisfaction that "whole mountain chains have been elevated at one geological period" (308), though different ones at different times. Accordingly, there had been several successive periods of extraordinary volcanic energy—a finding directly and irreconcilably opposed to the principle of uniform causes that Lyell was seeking to establish.

Before closing, Sedgwick further stirred his audience by reviewing the concept of a universal flood. "The vast masses of diluvial gravel scattered almost over the surface of the earth," he now believed, "do not belong to one violent and transitory period" (313). To equate such evidence with the Noachian deluge was natural at one time but no longer defensible. Sedgwick therefore ended his remarks with a striking formal recantation, specifically disavowing the biblical flood in which he had once believed.

Lyell, then, had achieved at least that much. But it was apparent from the strength and breadth of his opposition that prevailing geological thought had not agreed to be revolutionized. (One should emphasize, however, that much of Hutton's thinking had become established by this time and many of his errors discarded.) Yet, in part because of Lyell's strident advocacy of his two major principles—the uniformity of causes *and* the uniformity of effects—other geologists deliberately emphasized contrary positions. Among them, Lyell declared privately to his sister, "Sedgwick's attack is the severest, and I shall put forth my strength against him in volume II" (K. Lyell 1881, 1: 318).

For Lyell, 1831 was a critical year, his tasks being to defend arguments he had already proposed in his *Principles* and to extend them by writing the remaining volume. Among his various activities, he read John Macculloch's Huttonian *System of Geology* (1831), geologized the Rhineland, prepared further arguments against the Mosaic deluge, and followed with great interest the brief appearance and rapid destruction of a new volcanic island in the Mediterranean. Visiting Scotland in September, he saw Macvey Napier, the editor of *Edinburgh Review*, who had many interesting reminiscences of John Playfair and his manner of composition. Henry Cockburn, of the same generation, thought Playfair's finest demonstration of style was one of the pamphlets he wrote on behalf of John Leslie's professorship at Edinburgh in 1805. Lyell then read Playfair on the progress of the physical sciences (1816) and planned to recast his definition of geology as a result. Nor did he neglect several controversial articles by Elie de Beaumont, who (inspired, perhaps, by Sedgwick's 1830 address) was specifying his differences from Lyell. "As I am contending at once seven or eight distinct points with no less persons than Sedgwick, Conybeare, E. de Beaumont, and three or four others," Lyell advised his fiancee, "you may well suppose how very carefully I must weigh every word and opinion."[8] Under the pressure of time, he decided to issue the three hundred pages he had already written as volume 2 (now, of three); it appeared in January 1832.

Principles of Geology, vol. II, 1832

Much of Lyell's second volume comprised his extended reaction to arguments presented by J. B. Lamarck (*Philosophie Zoologique*, 1809) on behalf of organic transmutation ("evolution"). Insofar as they were a response to Sedgwick, Lyell's arguments supported a generally Huttonian position. Unlike Playfair later on, however, Hutton had never endorsed the concept of extinction. Excepting the creation of man, life for him remained essentially unchanged—within certain limits of variability. To reconcile this desired Huttonian stability with the now unquestionable facts of extinction and new species, Lyell supposed that types were solely the product of environmental conditions and could therefore recur in the course of geological time (though none, apparently, ever had). Except for his volume's epigraph, therefore, we do not find Lyell citing Playfair until chapter XII, in which one of the latter's minor arguments (regarding vegetable mold) is disputed. Lyell then moved to confront Sedgwick's 1831 address directly, by undermining his claims for vegetation and other conservative causes as effective counterforces to erosion.

As part of his advocacy of Elie de Beaumont, Sedgwick had accepted a

[8] K. Lyell, 1881, 1: 330–348, quoting p. 347. On p. 338, "Dr. Reid's" is Lyell's error of recollection for "Leslie's."

view of geological history in which long periods of repose alternated with outbursts of "feverish, spasmodic energy," during which whole mountain chains were elevated suddenly, creating immense ocean waves that then inundated the continents and annihilated large numbers of species. But, as in 1829, Lyell cited recent, loosely consolidated volcanic cones personally examined by him in France, Spain, and elsewhere to affirm that no flood had ever passed over them. Some remarks on the formation of peat in chapter XIII derived from John Macculloch's *Western Islands* (1819) and *System of Geology* (1831). In chapter XVIII Lyell discussed the formation of coral reefs, further controverted von Buch's theory of elevation craters, and rejected Hutton's theory of the wholly animate origin of limestone, which Macculloch had endorsed.

In the one major review (solicited beforehand by Lyell) that volume II attracted, William Whewell first used the terms "Uniformitarians" and "Catastrophists" to designate two apparently opposing schools of geological theorists. Catastrophism, he affirmed, "has undoubtedly been of late the prevalent doctrine, and we conceive that Mr. Lyell will find it a harder task than he appears to contemplate to overturn this established belief" (1832, 126). Writing Lyell soon afterwards, Scrope challenged Whewell's dichotomy: "I do not see any but an imaginary line of separation between you," he concluded. For him, it was only a dispute about degree, a "plus or minus affair" easily resolved in perfect cordiality with minor concessions on both sides. But in discussing Scrope's remarks with Whewell, Lyell himself rejected any such compromises and promised to address the subject further in his forthcoming third volume.[9]

Principles of Geology, vol. III, 1833

Volume III, completing the *Principles*, appeared in April 1833, together with a long preface detailing Lyell's geological travels on the volume's behalf. Chapter I, printed during the spring of 1832, sharply contrasted the uniformitarian and catastrophic modes of theorizing. The second chapter, on the origin of rocks, noted sedimentary and volcanic ones briefly before discussing the primitive, or primary, rocks, and especially granite, in greater detail. "Nothing strictly analogous to these ancient formations can now be seen in the progress of formation on the habitable surface of the earth," Lyell conceded surprisingly, "nothing, at least, within the range of human observation" (11). He then

[9]Scrope to Lyell, 20 March 1832 (American Philosophical Society, Philadelphia; Wilson 1972, 351); Lyell to Whewell, ca. 22 March 1832 (Trinity College, Cambridge; Wilson 1972, 351). During the interim, second editions of his first and second columes appeared. Intended primarily as a stopgap to satisfy continuing demand, they embodied only minor changes and did not attract the interest of reviewers.

reviewed the (unnamed) Wernerian theory—since almost entirely abandoned—regarding the formation of Primitive rocks and the Huttonian alternative that succeeded it. Despite Werner, granite was soon found to be unstratified, intrusive, of no set age, and ultimately plutonic. Geologists, moreover, had noticed how the intrusions of volcanic and granitic veins and dykes altered adjacent rocks, often forcing them to recrystallize. It then became highly probable that larger effects of the same kind might be taking place, as "immense masses of fused rock, intensely heated for ages, came in contact at great depths from the surface with sedimentary formations. The slow action of heat in such cases, it was thought, might occasion a state of semi-fusion, so that, on the cooling down of the masses, the different materials might be rearranged in new forms, according to their chemical affinities, and all traces of organic remains might disappear, while the stratiform and lamellar texture remained" (13). Thus, Lyell argued, many so-called primitive rocks may actually have been formed at different times from sediments considerably older. This seemed to him a much more satisfactory explanation for the existence of stratified primary rocks than Werner's.

If the German theorist's term "Primary" could no longer be retained, his further category of "Transition" rocks was no better. Though originally created to account for rocks of similar composition (part chemical and part mechanical), it was now also dependent on a defined group of included fossils. At present, the whole concept was extremely confused; "Transition" ought wholly to be abandoned, Lyell argued, as any kind of historical designation. (Actually, the term "Secondary" was already becoming commoner, and significant attempts toward more acceptable periodization had been progressing for some years.)

Lyell then began a long discussion of the Tertiary formations, with which this volume was primarily concerned. After citing major recent studies of Tertiary strata and fossils, Lyell contrasted aspects of the Secondary and Tertiary as wholes, his major problem being to account for "the superposition of successive formations having distinct mineral and organic characters" (26) without resorting to the literalistic catastrophism of previous investigators. Chapters IV and V, including Lyell's innovative subdivisions of the Tertiary, then discussed the dating of rocks, with argumentation opposing both Werner (37) and the catastrophists. Chapters VI to IX drew on Lyell's fieldwork (November 1828–January 1829) in Sicily, with a few later additions. Throughout these pages, he emphasized the intermittancy of volcanic processes, the immensity of time evidently involved, the unlikelihood of diluvian inundations, the ability of subterranean lava to uplift sediments, the consolidation of various kinds of igneous (plutonic) rocks from that lava, and the probable alteration of earlier sediments by it.

In these chapters, Lyell demurred from attributing the formation of most Sicilian valleys solely to the excavating power of running water. Instead, he preferred to stress the fracturing brought about by ongoing subterranean uplift. Citing sea cliffs and other evidence, Lyell proposed that "the principal features in the physical geography of Sicily are by no means inconsistent with the hypothesis of the successive elevation of the country by the intermittent action of ordinary earthquakes" (113). But the magnitude of those valleys, and their similarity to others elsewhere, proved as well that slow, gradual, and pervasive uplift could not be disregarded. Lyell also endorsed the efficacy of submarine erosion (in valleys not yet raised above the level of the sea) but, rather like Deluc and Sedgwick, foresaw a time when those forces would become relatively unimportant. An enumerative recapitulation then restated the argument of these chapters more concisely. All of those so far reviewed had been written and printed by the summer of 1832.

Chapter X (completed November–December 1832, as were the remaining ones) endeavored to substantiate Lyell's Sicilian conclusions with evidence from elsewhere. His analysis of the dykes of Mount Somma, for example, uncovered phenomena "in perfect harmony with the results of the experiments of Sir James Hall and Mr. Gregory Watt" (124). Further evidence from the Neapolitan area again discredited the supposition of diluvial waves while supporting that of repeated earthquakes. Chapter XI then stressed that there had been no "alluvial epoch"; rather, such formations had originated in every geological period. As for the parallel grooves found in rocks by Sir James Hall and others, they may have been caused by large boulders rolling along the ocean floor while the later-elevated topography was still under water.

In later portions of his argument, Lyell devoted chapter XIX to the Auvergne, recapitulating his fieldwork in 1828 with Murchison, then disputing catastrophic explanations of alluvium found there; he again rejected Daubeny's attempt to divide the region's extinct volcanoes into antediluvian and postdiluvian epochs and came as close as he dared to labeling the deluge of Moses a fable. As a geological hypothesis, at least, Lyell believed the Flood to be entirely irrelevant. Chapters XXI and XXII, accordingly, interpreted the Weald of Sussex as a gradually uplifted landform excavated primarily by marine currents. Its famous transverse valleys, however, had been the result of fracture.

Chapter XXIV reviewed and attempted to refute Elie de Beaumont's orogenic theories regarding the successive, and catastrophic, uplifting of mountain chains. These theories assumed the gradual cooling of the earth's fiery nucleus, a hypothesis unacceptable to Lyell, who preferred to emphasize "the intermittent action of subterranean volcanic heat" as "a known cause capable of giving rise to the elevation and subsidence of the earth's crust without interruption to the *general* repose of the habitable surface" (339). Elie de Beaumont also held that his sudden mountain uplifts had generated worldwide deluges and attributed to each uplift specific geological dates that fieldwork could then dispute. Several reputable geologists had already published opinions contrary to Elie de Beaumont's, as Lyell was pleased to note.

Lyell's concluding chapters, XXV and XXVI, were also his most Huttonian. Elaborating a brief discussion of the primary rocks from chapter II, Lyell now referred to unstratified ones as "Plutonic." "Granite, porphyry, and other rocks of the same family," he affirmed, "often occur in large amorphous masses, from which small veins and dykes are sent off, which traverse the stratified rocks called 'primary' precisely in the manner in which lava is seen in some places to penetrate the secondary strata" (353). The discussion of granitic veins that followed utilized contributions by John Macculloch, Captain Basil Hall (Table Mountain, South Africa), and several investigators of Cornwall. Lyell next reproduced Macculloch's sketch depicting the junction of granite with stratified limestone and schist in Glen Tilt, citing details from the latter's paper of 1816. He then continued to rely on Macculloch's observations (*System of Geology*, 1831) while affirming that granite, once thought the oldest rock, "has been produced again and again, at successive eras" (357, cf. 359); that syenite and granite are not distinct; that submarine eruptions of basalt are perfectly volcanic; and that no valid distinction between "trappean" and volcanic rocks exists. The line between plutonic and volcanic was similarly arbitrary at best, "there being an insensible passage from the most common forms of granite into trap or lava" (362).

Chapter XXVI, on stratified primary rocks, considered the enigma of gneiss, mica schist, and clay slate, all of which occurred in layers. Gneiss, moreover, often passed into granite. How could one resolve the origin of a rock that appeared on one hand to be sedimentary and on the other to be igneous? Only the Huttonian hypothesis, Lyell thought, offered a satisfactory resolution of the problem. According to that theory, he continued, the materials constituting gneiss were originally deposited from water as a sediment, but these strata were then altered into a granitelike form by the effects of nearby granite (or other plutonic rocks) in a state of fusion. In his review of the evidence on behalf of such transforming indurations, Lyell cited the well-known example of Salisbury Crags, as reported by Playfair and later Macculloch. He then derived further instances from Berger's 1816 paper on Antrim, with that important addition regarding the origin of basalt by Conybeare. Experiments by Sir James Hall and Gregory Watt, and more passages by Playfair and Macculloch, were then invoked. Because "Primary" could no longer be considered a chronological designation, Lyell proposed instead to substitute the ahistorical term "Hypogene" (subterranean-formed); stratified primary rocks he designated "Metamorphic" (changed form).

It was now possible for Lyell to encapsulate his modified Huttonian position into a summary statement resembling the sweeping visions of his predecessor: "As the progress of decay and reproduction by aqueous agency is incessant on the surface of the continents, and in the bed of the ocean," he concluded, "while the hypogene rocks are generated below, or are rising gradually from the volcanic foci, . . . there must ever be a remodelling of the earth's surface in the time intermediate between the origin of each set of plutonic and metamorphic rocks and the protrusion of the same into the atmosphere or the ocean" (380). Still, a final problem remained; it had once been assumed that the unfossiliferous primitive rocks attested to a period in the earth's history before the creation of life. Now, however, the oldest strata indicated that animate life was then already in existence. As Playfair had earlier remarked (1802, 119), it was one thing to say that we do not find traces of a beginning and quite another to say that there never was a beginning. Defending himself against certain theological objections, therefore, Lyell emphasized reaches of time beyond human comprehension; nonetheless, he held that geology attests within its bounds to "a perfect harmony of design and unity of purpose" in nature, which we know to be the creation of "an Infinite and Eternal Being" (385).

Principles of Geology, third edition, 1834

Though Lyell's *Principles of Geology* had reached its expected length, it remained conceptually unfulfilled in some respects. Given the comprehensive nature of the enterprise, his book always would (despite twelve editions in all), but having the full work now before him Lyell revised it carefully into his more coherent third edition (1834). Of the important changes indicated in his preface, two were particularly Huttonian:

1. Lyell's initial discussion of central heat (1830, I: 141–143; unchanged in 1832–33) was recast in a more assertive form. "There are undoubtedly some grounds for inferring from recent observations and experiment that the temperature of the earth increases as we descend from the surface," Lyell now boldly averred, "but there are no proofs of a secular decrease of heat accompanied by contraction" (1834, I: 203–204). Though he emphasized our ignorance of the sources of volcanic heat and conceded that both light and heat are constantly radiating from the earth, Lyell expected that some form of compensating cause would eventually be discovered.

2. Some brief remarks by Lyell on the causes of volcanic heat and earthquakes (1830, I: 463) now became two new chapters (IX and X). Because volcanoes and earthquakes tended to occur in the same regions,

Lyell assumed that they were related. Active volcanoes necessarily overlaid enormous masses of intensely heated matter, but their activity, like that of earthquakes and other related phenomena, generally consisted of paroxysmal convulsions followed by long periods of tranquility. Such appearances were evidently connected to some extent with the passage of heat from the earth's interior to its surface. What, then, was the source of this heat?

Though it was popular to assume that the earth's heat derived from its origin as a cooling body originally fluid, Lyell explained our planet's spheroidal shape as the result of gradual and still-existing causes. Recent experiments, moreover, affirmed immense compression within the earth. It was more probable, Lyell thought, that "after a certain degree of condensation, the compressibility of bodies may be governed by laws altogether different from those which we can put to the test of experiment" (1834, II: 277), but the whole subject was still so obscure that one could hardly wonder at the variety of conjectures regarding the earth's core, whether it be fluid, solid, or cavernous. Researchers had established that temperatures in mines increased with depth; according to responsible calculations, a descent of only twenty-four miles would bring one to the melting point of iron, "a heat sufficient to fuse almost every known substance" (279). Cordier and others to the contrary, however, recent experiments supported the conclusion that this maximum temperature would not then continue to increase. Inequitable heat distribution throughout a fluid mass, moreover, would tend to be rectified by the ascent of hotter—and the descent of colder—currents, suggesting relative stability. Thus, he concluded, one may safely assume that "no part of the liquid beneath the hardened surface is much above the temperature sufficient to retain it in a state of fluidity" (283). He also proposed that a freely convecting molten interior might well melt its fragile crust.

The outcome of this reasoning was that Lyell rejected his previously held concept of central heat altogether. "Instead of an original central heat," he tentatively proposed, "we may, perhaps, refer the heat of the interior to chemical changes constantly going on in the earth's crust, for the general effect of chemical combination is the evolution of heat and electricity—which, in their turn, become sources of new chemical changes." Subterranean electric currents, he supposed, might "exert a slow decomposing power . . . and thus become a constant source of chemical action, and consequently of volcanic heat" (294). The oxydation of metals and alkalis (when in contact with water) is capable of producing intense heat; this reaction, then, might well be one of the principal means by which internal heat and the stability of volcanic energy are preserved.

In chapter X, a continuation of the same subjects, Lyell contrasted the immense bulk of the globe with the relatively insignificant mountains protruding from it, emphasizing the comparative repose and inertness of the earth. As before, he regarded the periodic discharges of volcanic energy in one form or another as a stabilizing device. Earthquakes, he surmised, might well be caused by the sudden expansion or escape of subterranean volcanic gases previously condensed (possibly liquefied) by intense pressure. The same process might also explain Huttonian elevation and subsidence, and even the eruptions of volcanoes. These forces, and earthquakes in particular, therefore seemed to Lyell the great restorative force in nature both Hutton and he required.

Lyell's changed position with regard to central heat particularly interested G. P. Scrope, who commented at length on this new feature as part of his *Quarterly Review* evaluation of the entire work. Though there were many theories respecting central heat, Scrope noted, none of them disputed its existence or the importance of its effects.[10] The chemical theory (oxydation of the metallic core) largely derived from Humphry Davy, with additions by Daubeny and Ampere, but Davy himself subsequently repudiated it. Cordier and other French geologists preferred a theory that derived central heat from the still-molten interior of a gradually cooling planet. De la Beche, in his *Researches in Theoretical Geology* (1834), had supported both views. Now Lyell was suggesting yet another mechanism, one based on electrical currents.

Though the nature and causes of heat, as well as the nature of the interior of the globe beyond a few miles, were quite unknown, Scrope nonetheless felt very unconvinced by either the electrical or the chemical theory. He continued to believe instead that "the globe is gradually cooling down, and still retains an intense temperature below its surface" (1835, 415). This temperature, however, in no way implied a state of fusion, for the incumbent pressure of the crust would probably prevent it, except locally. Though the earth's *surface* might originally have been fluid, he pointed out, this did not by any means imply that it had been fluid throughout; beyond a certain depth, he believed, gravity would maintain the globe's essential solidarity.

In his essay Scrope devoted particular attention to Lyell's original third volume, which had not previously been reviewed. When he and Lyell disagreed, Scrope was sometimes the more Huttonian. He argued, for example, that "There does not appear any reason for introducing the sea into the Weald valley, as Mr. Lyell does, for the purpose of effecting its denudation and the removal of the materials of its ancient strata. The agency of rain and rivers acting through an indefinite time

[10]In his 1834 address as president of the Geological Society of London, G. B. Greenough (1834) questioned both the idea of elevation and the existence of a central heat. There is no evidence that either position was taken seriously.

may have alone accomplished this" (440). Similarly, when Lyell considered the stratified primary rocks sedimentary deposits altered by subterranean heat, Scrope affirmed that "this is the theory of Hutton" (443); regarding metamorphism, however, Scrope himself was not yet a convert.

In one essential matter, Lyell had even mistaken the real nature of his own argument. "It is not the constancy of the laws of nature which he is contending for," Scrope declared; "this, no-one disputes. His real theory is that there has been no progressive variation in the intensity of the forces which modify the earth's crust, but that a cyclical succession of such changes (of equal amounts in equal periods) has been going on throughout all time" (448). Lyell to the contrary, Scrope held, all the evidence suggested that the earth has undergone, and continues to undergo, a series of progressive changes; subject like everything else in creation to the law of integration and disintegration, it had a beginning and will have an end.

By 1835, Lyell had replaced Hutton as the center of uniformitarian geological controversy. Though issues originated or emphasized by Hutton would continue to be debated, the debate would focus on Lyell's writings, with the names of Hutton and Playfair becoming increasingly scarce. On many issues, the Huttonians had won. Henry S. Boase's *Treatise on Primary Geology* (1834), for example, was the last book of importance to oppose Huttonian plutonism (as modified by Lyell in his volume III). Similarly, the Edinburgh Geological Society, founded in 1834, was so obviously Huttonian that it effectively superseded the almost moribund Wernerian Society. Finally, there were numerous indications at the British Association for the Advancement of Science meeting at Edinburgh (also 1834) that real opposition to many Huttonian issues had dissipated. From now on, Hutton and Playfair belonged primarily to the historians.

[11]

Hutton and the Historians

By 1835 or so it was commonly understood throughout Europe that speculations about geological phenomena had been proposed off and on since antiquity by well-qualified (but not necessarily prescient) thinkers. Yet, however brilliant these earlier theorists may have been, they generally approached their task as philosophers or theologians rather than as scientists (it is revealing that the word "scientist" probably originated in 1834). Despite their awareness of these precursors, writers of Lyell's generation frequently reiterated their conviction that the *science* of geology, like its name, scarcely existed before about 1780. As the new endeavor developed, clarified, and organized, it became more obviously self-conscious and collective. In this case as in others, the emergence of assured identity necessitated historical explanation: how, then, had the science of geology come into being?[1]

Even in classical antiquity, Aristotle and other philosophers had been aware of their predecessors. But it was not yet popular to see one's own thought as an outgrowth, so Aristotelian intellectual history consisted in large part of a review of errors. The earliest major British speculators, including Burnet, Woodward, and Whiston—all seventeenth-century figures—treated their progenitors in this same ungenerous manner and were similarly ridiculed by later theorists whose work had been made possible by their own. In general, the idea that human knowledge *evolves* did not become prominent until sometime after the battle of the An-

[1]For a history of the word "geology," see Dean 1979. The present chapter is not so much concerned with the history of geology per se as with the development of Hutton's reputation.

cients and the Moderns (around 1700) had been won by the latter—thanks largely to the example of Isaac Newton—and Baconian "progress" established. Even then, however, it was by no means apparent that a given theorist had contributed understanding rather than error. For many nineteenth-century geologists, some of whom became historians of their science, this was the fundamental problem to be solved in assessing the significance of James Hutton.[2]

Fitton's Review of William Smith, 1818

The first British chronicler of modern geology was William Henry Fitton (1780–1861), whose influential advocacy on behalf of six publications by the stratigrapher William Smith in *Edinburgh Review* (February 1818) also included historical remarks. Continental writers having preceded British ones in sensing the importance of such understanding, Fitton was largely indebted to a survey of earlier publications by Nicolas Desmarest (1794). On the whole, Fitton thought, this "chaos of philosophers" with their "medley of opinions" did not warrant much respect. Though we in the twentieth recognize the late seventeenth-century debate over the nature of fossils as an episode fundamental to the history of geology, for example, Fitton simply found it incredible that anyone could ever have doubted their biological origin. Necessarily, he recorded Martin Lister's advocacy of geological maps (in 1684) but noted the latter's opposition to animate fossils also. Though he had been a far better fossilist, wrote Fitton, John Woodward's views were "warped by the then prevailing taste for antediluvian history," since, like most others in his time, he accepted the authority of Genesis (1818, 316). Buffon (1749; 1778) did much to popularize the study of organic remains, which he specifically associated with more general problems regarding the structure and history of the globe. In retrospect, however, the lesser known Rouelle has proven to be of more significance, for he was the first theorist to suggest (as William Smith would later) that particular fossils were invariably associated with particular strata. Together with Lehmann and others, moreover, Rouelle distinguished primary from secondary mountains and illuminated intermediate strata that Werner would later call his Transition series. Lehmann (1756), meanwhile, preceded Werner in asserting that a universal sequence of strata existed.

The most important stratigraphical observations of all were undoubt-

[2]This chapter is limited to British figures. For American recognition of Hutton, see George W. White, "The History of Geology and Mineralogy as Seen by American Writers" (1973): Samuel Miller, p. 201; Jeremiah Van Rensselaer, p. 205; G. W. Featherstonhaugh, pp. 206–207; William W. Mather, pp. 208, 209; and the Gibbonses, p. 210.

edly those of John Michell, whose paper "On the Cause and Phenomena of Earthquakes" appeared in the *Philosophical Transactions* of the Royal Society of London for 1760; according to Fitton, Michell was the first writer to understand the stratification of England. He was then followed by John Whitehurst (*Inquiry into the Original State and Formation of the Earth*, 1778), a writer too often misled by his unfortunate taste for cosmogony who nonetheless ably supported Michell's contentions regarding strata and fossils. Both writers affirmed a regular sequence of layers.

The last theorist Fitton considered was Werner, who had died only the previous year, regrettably without any biographical memorial as yet. In one of his last acts, Werner sold his distinguished collection of minerals to the School of Mines at Freiberg for an extremely minimal sum, the larger part of which then reverted to the school at his death. Werner's contributions included his opinion as to the aqueous origin of basalt (1787), his doctrine of universal formations (1790–91), and his emendation of the latter to include the Transition class (1795–96). Otherwise, Fitton thought, Werner's doctrines were little more than a selection from those of preceding writers. Jameson, however, had attested in 1811 that Werner preceded Smith in devising a plan to create geological maps in which successive strata would be represented by different colors. Following this admission, Fitton concluded his review of pioneering stratigraphers—in which, astonishingly, James Hutton was never mentioned.[3]

Conybeare and Phillips's *Outlines of the Geology of England and Wales,* 1822

William Conybeare contributed a fuller and altogether more adequate history of geology to his and Phillips's *Outlines* (1822). Deriving many references and quotations from James C. Prichard's *Egyptian Mythology* (1819), Conybeare reviewed classical observations on geological phenomena by Aristotle, Pliny, Lucretius, and other writers. "In some of their general physical notions," he proposed, "we may almost fancy we see the germ of more modern theories" (1822, xxxix). Aristotle, for example, preceded Buffon in emphasizing the displacement of the sea. Similarly, the "wild but splendid conception" of frequent destructions and renovations affirmed by many ancient schools strongly reminded Conybeare of the Huttonian theory—but he supposed the Stoic version

[3]It is certainly true that Hutton was never a stratigrapher in the manner of William Smith. On the other hand, his predictions of unconformities, his clear recognition of their significance once found, and even his more general views about distortions of strata from the horizontal surely qualify him as a pioneering contributor to the science of stratigraphy.

to have derived from abstract reasoning rather than observations and induction.

Further portions of Conybeare's succinct narrative alluded to Avicenna and other medieval Arabs; Italian writers on fossil shells, including Fracastoro; Palissy in France; and Agricola in Germany. A late sixteenth-century British writer, George Owen, made "undoubtedly the earliest attempt to establish the important and fundamental geological fact that the same series of rocks succeed each other in a regular order throughout extensive tracts of country, and to elucidate the geological structure thus indicated" (xl-xli), but his work had remained in manuscript until only recently. Though Burnet, Woodward, and Llwydd could be dismissed quickly, Lister ("the first proposer of regular geological maps") required more extensive notice. Meanwhile, papers contributed to the *Philosophical Transactions* (1661ff.) began to place geology on a more empirical basis, to which the splendid genius of Buffon—wasted on theoretical speculations—added little or nothing.

Guettard, in 1746, created the first actual geological maps, and Lehmann, in 1756, established the fundamental distinction between primary and secondary rocks that had been obscurely indicated almost a century earlier by the Florentine writer Steno. (Though Desmarest had emphasized Rouelle's contributions, Conybeare noted, the latter had promulgated his views only through unpublished lectures having little influence.) Unfortunately, Lehmann preceded Werner in supposing erroneously that the few strata he had actually observed were universal. Michell, in 1760, then became the first theorist to place the regular succession of strata on a modern basis. Though infected with cosmogony, Whitehurst (1778) masterfully explicated the limestones and coal strata of his district.

In 1788, Conybeare then noted, "Hutton published his 'Theory of the Earth,' a work which has exercised a lasting influence over the writings of a large class of English geologists" (xliv). Because of Playfair, the theory had become so well known that no summary of it seemed required. In Conybeare's opinion,

> Hutton had the merit of first directing the attention of geologists to the important phenomena of the veins issuing from granitic rocks and traversing the incumbent strata, and of bringing forward in a striking point of view the circumstances which seem to corroborate the igneous origin of trap rocks. The wildness of many of his theoretical views, however, went far to counterbalance the utility of the additional facts which he collected from observation. He who could perceive in the phenomena of geology nothing but the *ordinary* operation of actual causes, carried on in the same manner through infinite ages, without the trace of a beginning or the prospect of an end, must have surveyed them through the medium of a preconceived hypothesis alone. (xliv)

This was, then, the first relatively impartial attempt to ascertain Hutton's place within the history of his science.

Later figures in Conybeare's history included Werner (as quoted earlier, "his attempts at theorizing must now appear to all but his most devoted adherents [as being] among the most unsuccessful and unphilosophical ever made"), Saussure, Pallas, Smith ("a name which can never, in tracing the history of English geology, be mentioned without the respect due to a great original discoverer," xlv), Webster, Greenough, Macculloch, and Sedgwick. He also noticed the "Edinburgh school" as a whole, contrasting Playfair ("a genius of a very high philosophical order") with the meritorious but misguided Jameson. Even so, Conybeare could not help regretting "the injurious effects . . . produced by that excessive addition to theoretical speculations which has converted the members of that [Edinburgh] school into the zealous partisans of rival hypotheses and led them to contribute far less then they might otherwise have done to the real progress of inductive geology" (xlviii). For him, Hutton and Playfair had been insufficiently Baconian.

Fitton's Presidential Addresses, 1828–31

On 15 February 1828, Fitton addressed the Geological Society of London as its president. "In the speculative department of geology," he believed, "nothing has been of late more remarkable, with reference to its history in this country, than the universal adoption of a modified Volcanic theory and the complete subsidence, or almost oblivion, of the Wernerian and Neptunian hypotheses." It was no longer denied, he continued, that volcanoes have been active and important throughout the earth's history; that strata originally horizontal have been raised, shattered, and bent by subterranean forces; and that the same strata have been intruded, perhaps repeatedly, by veins of liquid rock. "Whatever, therefore, be the fate of the Huttonian theory in general", he proclaimed, "it must be admitted that many of its leading propositions have been confirmed in a manner which the inventor could not have foreseen" (1828, 55). Fitton also eulogized the geological writings of Playfair; their literary excellence, he believed, had accelerated the progress of the science and probably outweighed in value "the partial success of the speculations for which he so strenuously contended" (56).

In his second presidential address, on 20 February 1829, Fitton predicted that "as the doctrine of Werner, which ascribed to volcanic power an almost accidental origin and an unimportant office, has long since expired, so the more recent views, which regard a certain class of causes as having ceased from acting, will probably give place to an opinion that the forces from whence the present appearances have resulted are in geology, as in astronomy and general physics, permanently connected with the constitution and structure of the globe" (1829, 133). The geologist, like the astronomer, he continued, is called on to trace the effects of powerful forces over long periods of time. Chemical and mechanical changes going on around us, he thought, are forever varying in their energies and effects. Yet, when viewed in relation to the vastness of geological time, even these seemingly irregular phenomena acquire a kind of uniformity. Thus, that part of the Huttonian theory in which geological changes are compared to planetary movements (Playfair 1802, para. 387)—after a long interval, the same configuration is certain to recur—seemed to him increasingly probable, philosophically just, and imaginatively captivating.

Recognized as a foremost historian of his field, Fitton expanded his endeavors into the significantly augmented "Notes on the History of English Geology" that appeared serially in *Philosophical Magazine* between August 1832 and January 1833. As before, he stressed the importance of William Smith, to whom the first Wollaston Medal, for accomplishment in geology, had been awarded by the Geological Society of London in 1831. As before, he depended heavily on the earlier account of Desmarest. It now seemed, however, that the latter's advocacy of Rouelle as prescient stratigrapher had been overzealous, Michell once again emerging as a more important figure. Werner seemed equally deserving of attention, but Fitton had nothing new to say of him except to deplore the Huttonian-Wernerian controversy that had diverted geological attention from the actual structure of the globe. George Owen (manuscript, 1595) seemed to him "the Patriarch of English Geologists" (1832–33, 443), and he ignored Hutton altogether, presumably for being Scottish rather than English, though numerous foreign authors had been admitted.

Lyell's *Principles of Geology*, 1830

In 1826, Charles Lyell undertook a review of the most recent volume of the *Transactions* of the Geological Society of London for *Quarterly Review*. As a preface to his more specific remarks, Lyell alluded to "Werner and his scholars in Upper Saxony" and the "powerful minds of Hutton and Playfair" (1826b, 508) before surveying recent paleontological discoveries. He contrasted Hutton's uniformitarian emphasis on causes presently in action with Buckland's catastrophism and noted vaguely that "the generalizations of Werner and Hutton, though bearing impressed upon them the decided marks of genius, have required considerable modifications" (533).

The first volume of Charles Lyell's *Principles of Geology* (1830) included three chapters and some additional remarks adumbrating the history of the science. Although Lyell's purpose throughout (as he freely admitted in private letters) was polemic rather than disinterested, his conceptualization of earlier developments proved to be no less influential than his theory as a whole and was even more dominant in that few serious rivals to it appeared for more than half a century.[4]

Ancient Hindus and Egyptians, Lyell began impressively (chapter II), agreed in ascribing the creation of the world to a divine being who has since destroyed and refurbished it in cycles. Such myths, he remarked archly, evidently derived from exaggerated traditions of normal geological events; nonetheless, they still influenced the thinking of some modern philosophers. Basic processes of geological change were generally well understood by the time of Pythagoras (Ovid's *Metamorphoses*, actually), Aristotle, and especially Strabo, all of whom Lyell interpreted as modern-day uniformitarians. They had not, however, recognized changes in the world of life.

After the fall of the Roman Empire (chapter III), Lyell observed, geological speculations were not much pursued until the tenth century, when two Arab writers, Avicenna and Omar the Learned, contributed important treatises. Lyell found no Christian equivalents before the sixteenth century; Fracastoro (1517) then declared fossil shells to be the petrified remains of once-living animals but not a product of the Deluge. Unfortunately, these clear-sighted views were disregarded, so debate concerning both issues—the nature of fossils and the reality of the Deluge—continued for almost three centuries. On the other hand, sufficient freedom of expression prevailed among Italian clerics to allow for genuine debate. Lyell therefore reviewed further contributions by Mattioli, Falloppio, Mercati, Olivi, Cardano, Cesalpino, and Majoli. For Lyell, the latter's suggestion in 1597 that marine sediments and fossil shells might have been deposited on the land by volcanic upheavals represented "the first imperfect attempt to connect the position of fossil shells with the agency of volcanoes, a system afterwards more fully developed by Hooke, Lazzaro Moro, Hutton, and other writers" (1830, 1: 27). Unlike Fitton and Conybeare, Lyell particularly stressed the contributions of Nicholas Steno (1669), whose work attested not only to the superiority of Italian geological research but to the obstacles it had yet to overcome.

The next century and a half was for Lyell a sad period in the history of

[4]For Lyell and his "anti-Mosaical" version of the history of geology, see Lyell to Murchison, 11 August 1829 (Geological Society of London); Lyell to Scrope, 14 June 1830 (K. Lyell 1881, 1: 268–271). Lyell's objectivity as a historian was seriously questioned in 1975 by several speakers at the Charles Lyell Centenary Symposium (*British Journal for the History of Science* 9, July 1976). See also Lyell to John Hawkins, 25 January 1830 (West Sussex Record Office, Chichester).

geology, as misguided theologians throughout Europe attempted to maintain the Deluge theory of fossils. "Never," he remarked, "did a theoretical fallacy, in any branch of science, interfere more seriously with accurate observation and the systematic classification of facts" (29–30). Those who pursued more enlightened views, contrarily, were unfairly accused of disbelieving the whole of Scripture. The history of geology between the late seventeenth century and the end of the eighteenth, characterized as it was by the frequent revival of exploded errors and relapses from sound to thoroughly absurd opinions, demonstrated retardations as well as advances.

During this time, several British figures of particular interest emerged. Martin Lister (1678), for example, recognized the real nature of fossils, proposed the concept of extinction, and utilized his awareness of large-scale stratigraphic continuity to suggest the construction of geological maps. Robert Hooke's *Discourse of Earthquakes* (1705; written 1668) was for Lyell the outstanding geological treatise of his time; its revival and development of classical insights (by Strabo and others) became an important step in the progress of modern science. Hooke, for example, mentioned that the Italian coast near Naples had been elevated significantly during the eruption of Monte Nuovo in 1538. Despite his recognition of the continuing efficacy of both earthquakes and volcanoes, however, Hooke did not find the basic key of gradualistic uniformitarianism, as his theories on the origin of fossils and the elevation of mountains reveal. Ray, Woodward, Burnet, Whiston, and Hutchinson then contributed a variety of insights and errors.

Dismissing these more extreme theorists, Lyell returned with pleasure to the geologists of Italy, particularly Lazzaro Moro (1740), who, like Hooke, had attempted to verify and extend the classical theory of earthquakes. The uplift of a marine shoal in the Grecian archipelago in 1707 led him to appreciate the elevating power of subterranean forces, an insight he then applied to whole continents, citing faults and other dislocations as evidence. Despite Moro's unnecessary attempt to accommodate the Mosaic account of Creation, Lyell regarded him as an important predecessor whose insights sometimes anticipated Hutton's. A defense of Moro's theory by Generelli (1749), moreover, emphasized the perpetual waste of mountains and continents while suggesting extended geological time and some kind of restorative principle. Since God would surely not have created a helplessly diminishing world, Generelli necessarily supposed that the same forces which had originally raised earth's mountains from the abyss continued to produce others.

Buffon, also publishing in 1749, attributed no influence whatsoever to earthquakes and volcanoes but instead reverted to the old idea of a universal ocean that successively withdrew (perhaps into subterranean caverns), leaving behind it increasing exposures of landforms, fossils, and erosional features. When this simple assertion proved unacceptable to ecclesiastical authorities, Buffon was forced to retract his (for Lyell) generally accepted beliefs that the mountains and valleys of the earth were of secondary origin (i.e., post-Creational) and that the same forces which had created them would in time destroy them and create others in their places.

The Lisbon earthquake of 1 November 1755 prompted Michell to compose an important essay on the cause and phenomena of earthquakes (1760) which included significant stratigraphical insights. Arduino and Lehmann, publishing just before, had established the fundamental distinctions of primary, secondary, and tertiary rocks and mountains. Raspe (1763) then specifically revived Hooke's theory of earthquakes, calling attention also to the formerly tropical climate of Europe and changing plants and animals throughout the fossil record. Though later theorists debated relations between past and present life, Boscovich (1772) once more advocated the efficacy of normal earthquakes, stressing their cumulative effects over time. Like Hooke, Moro, and other theorists, however, he believed geological forces to have acted more powerfully in the past than in the present.

Werner (chapter IV) was, for Lyell, the first modern geological theorist to emphasize the practical utility of geological knowledge—as applied to mining, particularly. The chief advantage of his influential system of instruction at Freiberg was that it directed the attention of field workers to the constant relations of certain mineral groups, including their order of superposition. But Werner's own dogmas had led him to misinterpret even the geological phenomena most familiar to him, in the immediate vicinity of Freiberg. Thus, "the supreme authority usurped by him over the opinions of his contemporaries was eventually prejudicial to the progress of the science, so much so as greatly to counterbalance the advantages which it derived from his exertions" (56). With regard to basalt and other igneous rocks, he had been entirely wrong and very much a retrograde force in the history of geology (Desmarest, Dolomieu, and Montlosier were sounder). His doctrines concerning geological forces, moreover, were nonuniformitarian and often fanciful. Despite Werner's failure ever to visit England, certain theorists there had enthusiastically adopted his most erroneous pronouncements.

Werner's contemporary Hutton, Lyell pointed out, held some very different views. His *Theory of the Earth* "was the first in which geology was declared to be in no way concerned about questions as to the origin of things; the first in which an attempt was made to dispense entirely with all hypothetical causes, and to explain the former changes of the earth's crust by reference exclusively to natural agents" (61). Hutton had attempted to give fixed principles to geology—as Newton had for astronomy—but the relatively undeveloped state of the science had prevented him from doing so.

Lyell's further remarks on Hutton, including supposed quotations, were not gleaned from Hutton himself nor even from Playfair's *Illustrations*. Instead, Lyell freely paraphrased Playfair's "Biographical Notice" of Hutton (1805, 52–55). Thus, his "questions as to the origin of things" (1830, I: 4, 61) derived from Playfair's assertion that "the object of Dr. Hutton was not, like that of most other theorists, to explain the first origin of things" (1805, 52); and his supposed quotation (61), from Playfair's pages 52–53. The phrase "volcanic heat," disingenuously quoted by Lyell, was neither Hutton's nor Playfair's. Later portions of the sketch concerning Hutton, granite, and Glen Tilt included numerous unacknowledged borrowings from Playfair's pages 68–69, which Lyell then belatedly cited from the 1822 reprinting. Thus, Lyell relied almost entirely on Playfair's biographical notice of Hutton, as he had earlier on Cuvier's of Werner.

Among his predecessor's accomplishments, Lyell acknowledged, Hutton had convinced himself of the igneous origin of basalt and similar rocks, and of their frequent injection while molten into older strata. "The compactness of these rocks, and their different aspect from that of ordinary lava, he attributed to their having cooled down under the pressure of the sea" (1830, I: 61–62, an erroneous reading). Through his investigation of granitic veins and the subsequent realization that granites had been formed throughout geological time, moreover, Hutton "prepared the way" for what would become the most important innovations associated with his theory: his (misquoted) inability to find traces of a beginning or prospects of an end in the economy of the world, and his evident belief that all past changes on it had been brought about by the slow agency of existing causes.

The theory's greatest defect, Lyell believed, was its undue emphasis on subterranean heat, to which both consolidation and uplift were attributed. "Hutton made no step beyond Hooke, Moro, and Raspe," Lyell adjudged, in pointing out how ordinary earthquakes might, over a long period of time, bring about significant geological changes (63–64). On the contrary, Hutton seems to have fallen well short of their views, imagining that the continents were first gradually destroyed; when their ruins had furnished materials for new continents, they were subsequently upheaved by "violent and paroxymal convulsions" (64). Hutton therefore required alternate periods of disturbance and repose. Lyell would repeat this egregious misreading in other portions of his *Principles*.[5]

[5]Lyell subsequently attributed the supposedly Huttonian idea to himself. "In my comments on the Huttonian theory," he advised G. P. Scrope on 25 June 1830, "I throw out that there have been, in regard to separate countries, *alternate* periods of disturbance and repose, yet earthquakes may have been always uniform, and I show or hint how the interval of time done may

Though Hutton's knowledge of chemistry and mineralogy was considerable, Lyell added, he knew little about fossils: "They merely served him as they did Werner, to characterize certain strata [another misreading] and to prove their marine origin. The theory of former revolutions in organic life was not yet fully recognized, and without this class of proofs in support of the antiquity of the globe, the indefinite periods demanded by the Huttonian hypothesis appeared visionary to many" (64). As his errors demonstrate, Lyell actually knew relatively little about Hutton's theory.

Quoting a lengthy passage from Playfair's life of Hutton (also used previously by its author in *Illustrations*), Lyell then discussed religious opposition to Hutton's theory during the French Revolution, an almost hysterical period throughout which any sort of heterodox speculation might too easily be associated with political and religious horrors. In his otherwise meritorious *Natural History of the Mineral Kingdom* (1789), for example, Williams had misrepresented Hutton's theory altogether while railing at its religious implications. Kirwan (1799) and Deluc (1809) were little better, but their pernicious influence succeeded in delaying for a time the triumph that would otherwise have followed Playfair's forceful and elegant *Illustrations*. Returning to Hutton and Playfair in chapter V, Lyell defended the modernity of his own strict uniformitarian assumptions (citing Playfair in support) but also foresaw a time when convinced gradualists would oppose the allegedly Huttonian doctrine of sudden continental elevations.

This, then, was Lyell's not altogether reliable conceptualization of his own predecessors, including Hutton; it is one he would repeat with only minor emendations (subheads and paragraphing) in every edition of his *Principles*, through the twelfth and last (1875). In all that time, the Hutton portion changed substantively only in three respects: the phrase "violent and paroxysmal convulsions" (64) was abridged to "violent convulsions" only; the disparagements of Hutton's doctrines in chapter V were dropped (both of these in the third edition); and chapter IV of the fourth and following editions added these remarks:

> The explanation proposed by Hutton and by Playfair, the illustrator of his theory, respecting the origin of valleys, and of alluvial accumulations, was also very imperfect. They ascribed none of the inequalities of the earth's surface to movements which accompanied the upheaving of the land, imagining that valleys in general were formed in the course of ages by the rivers now flowing in them while they seem not to have reflected on the excavating and transporting power which the waves of the ocean might exert on land during its emergence.

---

make the passage appear abrupt, violent, conclusive, and revolutionary" (K. Lyell 1881, 1: 273–274).

At no time thereafter did Lyell ever accept modern fluvialistic doctrine, which Hutton and Playfair had originated.

Finally, in his third edition (1834) and all others to 1875, Lyell somewhat disingenuously endorsed an earlier suggestion by William Whewell that "the advancement of three of the main divisions of geological inquiry [had], during the last half century, been promoted successively by three different nations of Europe—the Germans, the English, and the French" (I: 103). Werner, in Germany, had first described the different types of rocks systematically and precisely. Smith, of England, next identified the Secondary formations and their uniquely identifying fossils. For their parts, Cuvier and Brongniart had then traced the Tertiary formations of the Paris Basin, effectively systematizing that epoch as Smith had the Secondary. For both Lyell and Whewell, therefore, Werner, Smith, and Cuvier were the founders of modern geology.

By establishing close analogies between ancient and modern species, Lyell continued, comparative anatomists such as Cuvier and Lamarck soon accustomed geologists to a more coherent history of the earth. Through their discoveries, Lyell argued, "the mind was slowly and insensibly withdrawn from imaginary pictures of catastrophes and chaotic confusion, such as haunted the imagination of the early cosmogonists." Thus, numerous modern proofs attested to "the tranquil deposition of sedimentary matter and the slow development of organic life" (105). Though Cuvier himself had insisted on a geological past absolutely distinct from that of the present, his own researches tended to undercut this outdated supposition. "The growing importance, then, of the natural history of organic remains," Lyell concluded, "may be pointed out as the characteristic feature of the progress of the science during the present century" (106). Huttonian tradition, unmentioned in his assessment, was presumably irrelevant to that progress.

Paton, Brewster, and Whewell

Though the resulting book is scarcely considerable as serious history, some clever caricatures of late eighteenth-century Edinburgh personalities by John Kay (in 1787) were published for the first time in 1837 with a casual, lighthearted text by H. Paton. Most of the latter's often erroneous biographical remarks concerning Hutton derived from misreadings of Playfair's "Life" together with two colorful but not entirely trustworthy anecdotes capitalizing on the theorist's well known reputation for peculiarity. Besides emphasizing Hutton's friendship with Black, Paton also affirmed incidentally that the great geologist normally spoke Scots.

David Brewster's responses, in the *Edinburgh Review* for April 1837, to

William Buckland's Bridgewater treatise (1836) abjured frivolity of any kind. A former clergyman himself, Brewster emphasized the dangers of lingering traditions, meaning no longer tenable beliefs in a literal six-day creation and a universal deluge. "The highest demands of truth, and the best interests of mankind," he stressed, "are invariably sacrificed when religion is intruded into questions of science and civil policy" (1837, 4). The mistaken persecution of Galileo, though "fraught with deep instruction to every friend of religion," seemed less impressive than the recent triumphs of geological discovery, through which a science once branded as anti-Christian, even atheistic, had not only established its most deplored assertions beyond reasonable doubt but had demonstrated that these same revelations could be powerful auxiliaries of both natural and revealed religion.

"Our countryman, Dr. Hutton," he continued, "had the honor of sustaining the first assault from the enemies of reason" (4). In his "Theory of the Earth" (1788), a work never yet sufficiently appreciated for its soundness and ingenuity, Hutton "renounced at once the attempt to reconcile the phenomena of geology with the recent creation of the world; and, from the present state of our globe, he endeavored to trace the causes which have operated in the past and which are likely to continue in the future" (4). Brewster then quoted at length from Hutton's part 1 (217–219) and famous conclusion. The theory's "grand and exciting views," Brewster next pointed out, were in no way opposed to religion; on the contrary, Hutton repeatedly cited them as the strongest possible evidence on behalf of divine benevolence and design. Nonetheless, his theory had been attacked for its supposedly skeptical implications by such narrow-minded rival hypothesizers as Kirwan, Deluc, and Williams, who erected the outmoded doctrines of six-day creation and universal deluge into articles of faith. Playfair and Hall, on the other hand, were for Brewster noble and heroic rationalists motivated only by love of truth. The English school, unfortunately, though in a position to establish Huttonian geology in all its evident superiority, had chosen instead to sacrifice high principle on the altar of expediency; as a result, leadership in the field shifted to Cuvier in France, whose enlightened views (no less subject to religious slander) made their way in England only slowly.

Those same Cuvierian views, Brewster concluded, are "now the universally received doctrine of the English school, as geology has "renounced her ecclesiastical tenure, and demanded a lease of millions of years" (13). In an adjacent footnote, he commended the delicacy observed by Hutton, Playfair, and Cuvier when speaking of the length of time required by their theoretical remarks: some modern theorists, he noted, "employ language so specific and exaggerated as to alarm the timid and rouse the very prejudices which it had been their boast to disarm. . . . We must say [he admonished] that if geologists conceive that they add dignity to their science by the rash expression of *millions of millions* of years, they mistake the feelings as well as the judgment of the public" (13n).

William Whewell's ambitious *History of the Inductive Sciences, from the Earliest to the Present Time* (1837) included a long section on geology. For our purposes, it is unnecessary to review earlier portions of Whewell's exposition, except to remark that his history is both indebted to Lyell's and often superior to it. One cannot maintain, however, that Hutton's name is equally prominent. For Whewell, Werner attempted to defend several poorly substantiated doctrines, including the aqueous origin of granite, on which he was corrected by Hutton. But von Hoff and Lyell, rather then Hutton and Playfair, first established the efficacy of present-day forces. The whole Wernerian-Huttonian struggle, moreover, was premature. In a second edition (1847) Whewell elaborated on this last remark, pointing out that Playfair's often-lauded *Illustrations*—on which Whewell's own knowledge of Hutton apparently depended—contained such erroneous beliefs as the igneous origin of flint and the excavation of river beds by the rivers themselves. With regard to the latter point, he continued, anyone who had read Deluc's *Elementary Treatise on Geology* would "deem his refutation of Playfair complete" (1847, 3: 505). As for what Whewell called the Doctrine of Geological Uniformity, Hutton (in Lyell's characterization) "did not assert that the elevatory forces which raised continents from the bottom of the ocean were of the same order, as well as of the same kind, with the volcanoes and earthquakes which now shake the surface" (509). Nonetheless, Whewell gave Hutton, Hall, and Macculloch full credit for having established the metamorphosis of sedimentary rocks through heat and pressure (470). Though Hutton's theory, therefore, was, like Werner's, certainly premature, it contained far more anticipations of truth. "Many of its boldest hypotheses and generalizations," Whewell conceded, "have become a part of the general creed of geologists, and its publication is perhaps the greatest event which has yet occurred in the progress of Physical Geology" (505).

Lyell and Fitton Dispute, 1838–39

At a meeting of the Geological Society of London in March 1838, Charles Darwin read a paper (which Lyell thought favorable to himself) connecting volcanic phenomena with the elevation of mountain chains, particularly the Andes. In it, he associated the *slow* upthrusting of cordilleras with gradual, intermittent injections of molten rock from below. Responding with examples from the Alps, Henry De la Beche disputed that strata could remain flexible enough to be thus reshaped over a long period of time or that small, gradualistic forces could significantly reshape strata at all. When John Phillips rose to the Lyellians' defense, Fitton eagerly took him on, then turned to Lyell, whom he charged with not having sufficiently acknowledged Hutton's advocacy of gradual elevation. Lyell attempted to defend himself by replying that most critics had attacked him for *over*rating Hutton, and that Playfair had also understood Hutton to be a catastrophist. Whewell then intervened to suggest the futility of supposing what Hutton might have thought if he had known facts discovered since his death.[6]

The appearance of Lyell's *Elements of Geology* (fall 1838), a textbook of physical processes, enabled Fitton to pursue the relation between Hutton and later geology a bit farther, this time in *Edinburgh Review* for July 1839. Among their leading doctrines, Fitton observed, modern-day geologists accept the sedimentary origin of stratified rocks, the existence of stratigraphical *formations* in a certain order of superposition, and the igneous origin of unstratified rocks. The first two ideas had been developed over a period of time by Fuchsel, Werner, Smith, Cuvier, and Brongniart; in general, their claims as innovators had not been disputed. The third idea, however, originated with Hutton ("a geologist of the highest order," 1839, 441), who was too far in advance of his contemporaries to have been justly appreciated by them. Fitton's purpose in the present essay, therefore, was to establish that Hutton's views, once vehemently opposed by zealots, were the same as those now almost universally accepted. "No other name in the history of geological theory," Fitton declared, "can be placed beside his own" (442).

Fitton next proposed three questions regarding the Huttonian theory. First, to what extent was it true? Citing Lyell by name, Fitton observed that his friend's famous "principles" were nothing other than "the leading propositions of Hutton's theory confirmed and extended by modern observations"; the basics of Hutton's theory, moreover, would "probably be the foundation of every future geological system" (445). Specifically, Hutton had proved that granite and similar rocks are the product of fusion; that all, or nearly all, of the stratified rocks are sedimentary; that they have been indurated, elevated, and invaded by liquid plutonic masses; and that the force responsible for all the latter is some form of internal heat. But he had probably gone too far in ascribing the consolidation of all strata to heat alone. His assertions that flints in chalk were injected in a state of fusion and that pyrite was exclusively igneous, moreover, had since been entirely discredited.

[6]K. Lyell 1881, 2: 39–41, 281–282 (two versions of the same letter, Lyell to Horner, 12 March 1838); Wilson 1972, 453–456, 505.

Second, to what extent was Hutton's theory new? Most importantly, he had succeeded in proving that granite is igneous and comparatively recent; he also refuted the doctrine of primitive rocks and originated the concept of metamorphic ones. Eschewing, moreover, any attempt to explain the origins of the earth (unlike so many preceding theorists), Hutton insisted that the proper business of geology was to account for existing phenomena through existing laws of nature. His methodology—inferring generalizations from facts carefully ascertained—significantly distinguished Hutton's theory from all preceding ones, ensuring a tardy but ultimately complete acceptance of his major doctrines.

Elaborating Hutton's opinions on igneous fusion, Fitton quoted at length from "Observations on Granite," noted further contributions by later researchers, and emphasized those of Sir James Hall (1815). "It must be known and remembered," he cautioned, "that the attempt to account for the phenomena by the Neptunian hypothesis had loaded geology with the most extravagant suppositions, and retarded the progress of sound reasoning for nearly half a century" (448).

Hutton established that no genuinely primitive rocks exist (the term having previously been applied to nonfossiliferous crystalline ones "supposed to form the aboriginal and undisturbed nucleus over which all sedimentary strata were deposited," 449n). With occasional ambiguity he regarded as primary some of the most ancient of the sedimentary strata, which were frequently indurated and highly inclined, newer deposits being more commonly horizontal. Fitton thought both terms outdated (his primary rocks—Werner's Transition series—were now largely Cambrian and Silurian) but praised Hutton's chapter IV in *Theory* as having been written with great force and more than usual clarity. Those chapters on primary strata immediately following (V and VI), Fitton considered among his most interesting, repeatedly illustrative of "what have been called the *new* theoretic doctrines of the present day" (449). Fitton then quoted extensively, with much italicizing, to demonstrate Hutton's intuitive brilliance, "the anticipations of a genius" (450). Lacking space enough to extract all of his predecessor's innovations, he recommended Hutton's original *Theory* (which he was first to utilize for historical purposes) to all interested readers able to endure its frequent repetitions and occasional obscurity.

Third, what were the reception and progress of the theory? Fitton accepted Playfair's misleading observation that Hutton's views, initially obscure and imperfectly presented, had been lost among a welter of competing geological theories. More important, however, despite numerous passages within them attesting to divine benevolence, was the widely held conviction that Hutton's works encouraged religious infidelity. Even at the present day, Fitton could not help noting, "the supposed discordance or agreement of geological results with the small portion of the Mosaic history which relates to the Creation and the Deluge is a frequent subject of public discussion" (453). Such zealotry, Fitton suggested, is best ignored.

At the time of Hutton's death in 1797, he continued, few if any British geologists of comparable ability remained. Five years later, Playfair published his much-admired *Illustrations,* which usefully improved the quality of geological speculations in Britain but plunged Hutton's own writings into almost complete oblivion. As no copy of the latter's *Theory* was presently available in the libraries of the Royal, the Linnaean, or even the Geological Society of London, few present-day geologists could have seen it. Necessarily, then, most of Hutton's current critics derived their knowledge of him entirely from Playfair.

Huttonian-Wernerian controversy, for Fitton, began with the publication of Jameson's *System of Mineralogy,* III (1808) and was ultimately useful in that it diffused a taste for geological pursuits throughout Britain. After about 1810, he believed, plutonism was widely accepted in Britain, being furthered then by Macculloch and other members of the Geological Society. Adopting Hutton's theory almost entirely during the 1820s, Boué had attempted to claim some of his predecessor's insights for himself. Eventually, the Wernerian opposition collapsed, for even Jameson "has long since expressed his conversion to the Huttonian theory with the utmost frankness, and on several occasions given the weight of his testimony to the genius of its author" (455n). Thus, Hutton's views have come to be generally accepted. As Fitton then observed, "the connection of the preceding history with Mr. Lyell's works is by no means slight." Given so many later writers willing to appropriate accomplishments not their own—now that the long, lonely battle for acceptance had been won—he wished in all fairness to establish beyond question that "Dr. Hutton was really the founder of the TRUE THEORY OF THE EARTH" (466).

Having read this seemingly gratuitous attack on his new book (which scarcely mentioned historical background), and, indeed, on his integrity, Lyell replied promptly by letter to the hot-tempered Irishman's "elaborate disquisition on Hutton," which appeared to be excessive. "It has been useful to me," Lyell admitted, "as I found it difficult to read and remember Hutton, and though I tried, I doubt whether I ever fairly read more than half his writings, and skimmed the rest." Considering how late Hutton, as opposed to Steno, Hooke, Leibnitz, and Moro, came into the field, and consequently how much greater his opportunities had been, Lyell thought his geological perspective unduly narrow; he was therefore not entitled to "such marked and almost exclusive preeminence" as Fitton had proposed. Regarding the concept of metamorphic rocks, Lyell had not previously found a highly salient passage

(*Theory*, I: 439) quoted by Fitton. The latter might well have given more independent credit to Macculloch, "who so ably filled up Hutton's rough sketch of the metamorphic theory, but with scanty reference as usual to the merits of a great predecessor." Regarding Macculloch, Boué had once apologized to Lyell for having pretended to undue originality in his own views. Similarly, Lyell himself had never claimed to be the originator of his theory regarding the adequacy of modern causes. The quotations from Playfair affixed to his *Principles*, volumes I and II, had been intended to acknowledge how Lyell had been brought round slowly, against some of his early prejudices, "to adopt Playfair's doctrines to the full extent," being anxious to acknowledge his and Hutton's priority. All in all, then, Lyell felt that he had recognized Hutton and his accomplishments adequately, particularly when one compared his remarks with those of Cuvier, von Buch, Humboldt, Boué, Brongniart, and others, all of whom neglected Hutton completely. "I trust," he concluded, "that no book has made the claims of Hutton better known on the Continent of late years than mine."[7] The history of geology appended to Lyell's *Principles* therefore remained essentially unchanged throughout the next thirty-five years, influencing many derivative accounts.

[7]Lyell to Fitton, 1 August 1839 (K. Lyell 1881, 2: 47–50; quoting pp. 47–48, 48, 48, 49, 50). The whole letter should be consulted.

[12]

Toward Modernity

Throughout his twelve editions (1830–75), Lyell exuded a persuasive complacency that, however much applications and details might change, his establishment of uniformitarian geology as an empirical, inductive science would endure forever. That his own work climaxed decades of Huttonian initiative is certainly true, but not all of Lyell's impressive arguments would stand up to prolonged examination. As the science of geology continued to augment its rapidly increasing legacy of theories and observations, some further perspectives important to later evaluations of Hutton developed.[1]

Glacial Theory

Records dating back to the Middle Ages had long ago convinced observers of Alpine glaciers that these great currents of ice significantly waxed and waned, carrying boulders and other debris with them for some distance. Their first really systematic investigator was Horace-Bénédict de Saussure (1740–99), whose *Voyages dans les Alpes* (1779–96) detailed the formation, movement, and carrying power of glaciers. He also recognized the erosive potential of streams issuing from them and believed that the necessary melting had been caused by some kind of subterranean heat. Nonetheless, Saussure failed to appreciate the full

[1]Throughout this chapter, I am indebted to anthologies and surveys of nineteenth-century geological and geomorphological sources, including Mather and Mason 1970 [1939]; Chorley, Dunn, and Beckinsale 1964; and Davies 1969. Tinkler 1985 includes useful insights.

extent of former glaciation, attributing high-level deposits, erratic boulders, and enlarged valleys to a violent debacle or flood. He did not, in any way, propose an age of ice.

Hutton's knowledge of glaciers derived almost entirely from the first and second volumes (1779 and 1786) of Saussure's *Voyages*. In chapters iii and iv of the second volume of *Theory of the Earth,* he quoted Saussure and others on glaciers at length, accepting observations according with his own theory but criticizing Saussure for his reliance on an imaginary deluge. Even so, Hutton did not himself appreciate the fuller implications of glaciation, attributing to ordinary running water the same effects Saussure had invented his debacle to explain. After reading Saussure on his own, Playfair (1802, 388–389) followed Hutton's lead in emphasizing the ability of glaciers to move boulders of considerable size. Jameson (1808) and Deluc (1809) then argued similarly, the latter citing original observations. In his volume of Scandinavian travels (1810), von Buch tentatively recognized the widespread dispersion of erratics, some northern ones having traveled as far south as Germany. Hall's two-part paper of 1812, "On the Revolutions of the Earth's Surface," affirmed immense torrents similar to Saussure's debacle, calling attention to their effects throughout Europe while emphasizing phenomena immediately adjacent to Edinburgh. Four years later, Playfair examined the erratic Jura boulders at first hand, concluding afterwards that nothing other than glaciers could have transported them. Geologizing at Chamonix in 1819, De la Beche found both its glaciers and supposed debacles impressive. In 1826 Jameson published a paper (of 1824) by Jens Esmark citing relevant geological evidence to prove recent and widespread Scandinavian glaciation; in his university lectures the next year, he suggested that similar phenomena in Scotland were best explained by assuming the former presence of glaciers. His *Edinburgh New Philosophical Journal* soon became an important source of current information regarding them.

Meanwhile, in 1815, J. P. Perraudin, a local guide in the Swiss canton of Valais, proposed more extended glaciation in the past to account for certain boulders left hanging on the sides of valleys. This otherwise ignored conjecture was then revived and extended in 1821 and 1829 by Ignace Venetz, with partial support from De la Beche in 1832. That same year, Reinhard Bernhardi of Germany enlarged Esmark's views to include the idea of a polar ice cap that had once covered Europe as far south as central Germany. In 1833 Lyell (*Principles*, vol. III) discounted glacial transport of boulders in favor of an alternative theory emphasizing the carrying power of icebergs. Jean de Charpentier's reaffirmation of glacial agency in a paper of 1834 limited itself to Switzerland; Jameson reprinted his essay two years later. By then, Charpentier was exploring the glaciers of Savoy with Jean Louis Agassiz (1807–1873), who abandoned his former adherence to Lyell's iceberg theory as a result. Another, and more catastrophic paper by Charpentier, also published by Jameson, extended Swiss glaciers of the past as far as the tops of the Jura.

On 24 July 1837, at Neuchâtel in Switzerland, Agassiz presented a now-famous argument in which he unequivocally attributed all supposedly diluvial phenomena in Europe to glaciers. Believing that glaciers had at one time covered virtually the whole of northern Europe, Agassiz endorsed and extended suppositions previously advanced by Venetz. (The first unmistakable proposal that there had been an ice age, however, appeared earlier in 1837 in a poem by Karl Schimper.) For Agassiz, catastrophic uplift of not yet emergent Alps under a continental ice sheet had sent the rapacious, boulder-strewing glaciers on their way. During the next two years, Jameson published three papers on glaciers by Agassiz, including his Neuchâtel address.

At the Glasgow meeting of the British Association for the Advancement of Science (22 September 1840), Agassiz once more endorsed continental glaciation, extending his theory to include Asia and North America as well. Afterwards, he and Buckland (who had visited Agassiz in Switzerland two years before) toured parts of Scotland and England together. As a result, Britain's staunchest remaining advocate of deluge geology not only accepted the glacial theory himself but temporarily convinced Lyell to do so as well. It is significant, however, that both were persuaded only to interpret moraines and other local features as glacial; the idea of continental glaciation—of an ice age—remained far less acceptable. In London on 4 November, Agassiz read a paper summarizing the geological effects of glacial action in Switzerland and presenting evidence of similar effects in Scotland, Ireland, and England. He was then followed immediately by Buckland and Lyell, who for the rest of the year tendered further examples to the Geological Society.

The appearance of Agassiz's *Etudes sur les glaciers* (Glacial studies, 1840), a masterful work that dismissed both the deluge and iceberg theories of boulder transport, seemingly resolved immediate controversies. But most British geologists prudently distrusted Agassiz's conceptual flamboyance. As his particular claims became more and more extravagant (glaciers in the Amazon river valley, eventually), the larger insight—that there *had* been an ice age—gained adherents rather slowly. Lyell, for example, continued to defend his iceberg hypothesis in later editions of his books—until 1858, when he finally capitulated. Other major geologists, like Murchison, never conceded. Despite them both, augmenting observations in Britain, continental Europe, North America, and elsewhere eventually demonstrated how pervasive glaciation had been. After an initial suggestion by Agassiz in 1840, moreover, it was seen later on that there had been several episodes of Pleistocene glaciation rather than one. Between 1852 and the end of the century, finally, detailed proofs emerged that glaciations had taken place throughout the history of the earth.[2]

All in all, the glacial theory significantly modified Huttonian tradition:

1. It demonstrated the inadequacy of Lyell's theory of past climates (dropped from the *Principles* in 1847) and left his overall conception of the geological past in doubt.
2. It explained erratic boulders, a conspicuous anomaly that had previously given rise to a number of bizarre theories.
3. It reinforced the centrality of erosion and deposition as geological processes, in an important new context.
4. Creating a new class of sedimentary rocks, it explained massive "diluvial" deposits, striae, and numerous landforms, all of which had previously been used to substantiate either the Deluge or debacles.
5. More generally still, it discredited unique geological catastrophes, largely by emphasizing their recurring nature.
6. Finally, the glacial theory also discredited both catastrophism and uniformitarianism as then understood (recurring ice ages being somehow uniformitarian yet not very different from the recurring deluges of Buckland and Conybeare). The geological clock became a good deal more complicated thereafter.

Fluvialism

Glacial theory modified traditional uniformitarianism by requiring its adherents to accommodate spasmodic geological forces more readily. Thus, explanations combining slow, ongoing agents with violent occasional ones were more often proposed. In 1849, for example, J. D. Dana suggested that valleys in Pacific islands had originated through the interactions of four causes: convulsions (primarily volcanic), marine erosion, fluvial erosion, and gradual decomposition. Alexandre Surell (1841), T. J. Taylor (1851), and Alfred Tylor (1853) successively emphasized the surprising power of rapidly running water, as Robert Everest (1832) had earlier.[3] By 1857 Col. George Greenwood had become such a convinced

[2]In addition to the sources in note 1, see Imbrie and Imbrie 1986 [1979]. Lyell acknowleged "the glacial epoch, or the epoch immediately antededent to that in which all the species now contemporary with man were in being" at the end of chapter 6 in the *Principles*, 7th ed. (1847, 92), 8th ed. (1850, 92), and 9th ed. (1853, 91), but not until his tenth edition of 1867–68 was the subject taken up in detail.

[3]The anthologies and surveys cited in note 1 document the revival of fluvialism fully. Sources in this section not in my own bibliography are excerpted there.

fluvialist that he defiantly published a prescient but frequently outrageous book entitled *Rain and Rivers; or Hutton and Playfair against Lyell and All Comers*. Having learned geology solely through personal observation, Greenwood later insisted, he had become a fluvialist before ever hearing of Hutton. Despite his amateur standing, Greenwood forthrightly attacked Lyell's by then creaky theory of marine erosion and hastened its demise.

The second edition (1858) of G. P. Scrope's *Geology and Extinct Volcanos of Central France* reprinted most of the original 1827 text (parts of which were by then out of date), together with some later comments modestly celebrating his own foresight. After reaffirming the gradual excavation of valleys, Scrope conceded that "vast irregularities of level, elevations and depressions of the earth's surface, on every level of magnitude," have been occasioned by causes other than rain, frost, and similar phenomena—"chiefly submarine expansion" (207n). Mountains, valleys, oceans, lakes, and even rivers, he thought, owed their primary forms to such causes; sudden upheavals of large masses of the earth's crust, moreover, might even have sent repeated floods over the land. But this passage, accepted from 1827, was immediately undercut by a further remark that neither Lyell nor most other prominent geologists were "even yet sufficiently impressed with the immense amount of excavation or denudation effected on supra-marine land by the erosive force of the pluvial and fluvial waters—in other words, by 'Rain and Rivers.'" That story, he continued, was yet to be written, "but not by the eccentric author of the recent work under that title" (208n). By this time, Scrope had seemingly given up catastrophic explanations of geological phenomena almost entirely. The leading idea present in all researches, and accompanying every fresh observation, he noted, "the sound which to the ear of the student of Nature seems continually echoed from every part of her works, is—Time!—Time!—Time!" (208–209).

Despite Lyell's continuing opposition to fluvialist analyses, Scrope and those agreeing with him had begun to prevail. Thus, Thomas Oldham, writing for the Geological Survey of India (1859), thought it impossible to see how any littoral or normal marine action could have produced the long, deep, and sinuous canyons that he described. On the contrary, Oldham proposed, "these river gorges appear to me to have been excavated almost entirely by the force of the streams which have flowed and still continue to flow through them" (174). In 1861 two American engineers published an impressive paper documenting the amount of sediment carried annually by the Mississippi River into the Gulf of Mexico. Throughout the 1860s there was intense debate regarding fluvialism among such British geologists as A. C. Ramsay, J. B. Jukes, Archibald Geikie, G. P. Scrope, C. Le Neve Porter, W. Topley, and W. Whitaker. By the end of that decade, the fluvialist revival had overcome most of its Lyellian resisters.

Of several Scottish geologists associated with the fluvialist revival, Andrew C. Ramsay (1814–1891) most effectively linked Hutton and Playfair with his own generation. Like them, Ramsay saw materials eroded from the continents as forming new land beneath the sea. In his first major statement of theoretical allegiance, an essay of 1846 on the denudation of south Wales, Ramsay displayed a Huttonian ability to visualize changing landscapes over immense periods of time. On the other hand, he affirmed a theory of marine planation similar to Lyell's while disparaging the efficacy of rivers. Even so, he convinced some forward-looking contemporary minds that it was time to go beyond Lyell.

"Stuck at Hutton's *Theory of the Earth* and Playfair's *Illustrations* all day (Sunday) and before night read all, and made a complete abstract of the latter," Ramsay noted in 1847, after becoming professor of geology at University College, London (Geikie 1895, 116). His first series of lectures, adumbrating the history of geological thought since Strabo, would then stress the centrality of uniformitarianism. "Wrote a good bit of my lecture at night," Ramsay's diary entry continued: "Hutton every day strikes me with astonishment. Lyell does not do him half justice" (117). These convictions were reflected in *Passages in the History of Geology* (1848), the published version of his lecture, which became a landmark in the development of Hutton's reputation.[4]

Interested particularly in the erosive capabilities of glaciers, Ramsay published important papers on them over the next two decades, attributing to glacial action several present-day landforms. His first effort (1850) acknowledged former glaciation in north Wales. Two years later, after fieldwork in Switzerland, he found British evidence attesting to an earlier glaciation during Permian times. By 1855, Ramsay had decisively rejected any kind of precipitous marine debacle. "I have no faith," he wrote, "in violent currents of sea water (such as have been sometimes assumed to result from imagined sudden upheavals of land) washing across hundreds or thousands of square miles, and bearing along and scattering vast accumulations of debris far from the parent rocks" (1855, 198). Over the next decade, Ramsay gradually relinquished his previous adherence to Lyell's iceberg theory and marine denudation of the Weald. "Given sufficient time," he argued in 1862, "any amount of degradation may be produced by rain and running water" (1862, 378). Through defections by Ramsay and others, the Lyellian concept of corrosive marine inundations rapidly lost favor.

[4]In Ramsay's *Passages*, pp. 26–38 deal exclusively with Hutton, who is emphasized throughout. Though his remarks often derive from Playfair's, Ramsay also demonstrates his firsthand knowledge of Hutton's *Theory*.

In a parallel development, J. B. Jukes (1811–1869), a former student of Sedgwick's, read a paper of 1862 in Dublin and London analyzing river valleys in southern Ireland. Firmly restricting the agency of marine erosion, he insisted that the rivers had created their own valleys. Denudation, Jukes proposed, was indeed of two kinds, marine and atmospheric. But, as a basically horizontal force, marine denudation affected only sea coasts, at sea level; it could not normally produce ravines or narrow, winding valleys. Atmospheric denudation, on the other hand, worked vertically and ubiquitously, attacking land masses everywhere. Jukes then emphasized the ability of atmospheric denudation to remove sediment layers even hundreds of feet thick—given sufficient time. Uplift, folding, and limited marine denudation had all been at work in southern Ireland as well. Ramsay immediately endorsed this Huttonian scenario of Jukes's in a book on the physical geology and geography of Great Britain (1863) and a paper the following year on the erosion of valleys and lakes. Opposing Murchison, he quoted extensively in the latter from Hutton (*Theory*, II, chaps. ix, x) and Playfair (1802, 104, 111–113) to demonstrate that rivers and glaciers alone, if given sufficient time, were fully adequate to account for valleys, lake basins, and other surface features previously attributed to fractures, dislocations, and subsidence. Another major Scottish geologist had also been convinced by the Irish demonstration: "Although I have long held the belief of Hutton," Archibald Geikie (1835–1924) wrote in 1865, "that our valleys are mainly the work of atmospheric waste, the history of the process of their excavation was but dimly understood by me until the appearance of the admirable paper by my colleague, Mr. J. B. Jukes, on the river valleys of the south of Ireland" (1865, 141n). For many historians since then, Jukes's paper has also appeared to be decisive.

Geikie's *Scenery of Scotland Viewed in Connexion with Its Physical Geology* (1865) celebrated both the land and its geologists. As part of a general review of erosive processes, for example, Geikie's defense of fluvialism (the book's major theme) strongly endorsed earlier work by Hutton, Playfair, and Ramsay, as did his more general position on the development of landforms; he recognized rain, rivers, glaciers, and the sea as all having contributed to the shaping of Scotland throughout a long geological history of ceaseless change, that "gradual waste to which the surface of the earth is everywhere subjected" (39). Three years later, in a subsequently famous paper "On Modern Denudation" (1868), Geikie cited earlier calculations by others on the carrying power of rivers to establish that marine erosion was far less effective than subaerial. Applying figures derived from the Mississippi to Europe, he found that the whole continent might well disappear in about four million years. Such results, he thought, demonstrated that even reputedly slow geological processes (for which earlier theorists had required vast eons) might well have proceeded more rapidly. The Huttonian concept of almost unlimited geological time was therefore open to challenge.

Geological Time

Geikie had been stimulated to speculate on geological time and its perhaps exaggerated duration by William Thomson (later Lord Kelvin, 1824–1907), a physicist who had in turn been much impressed by earlier assertions regarding the possible diminution and eventual exhaustion of the sun's heat. In publications of 1862–68, Kelvin had argued that any possible age attributed to the earth was subject to physical laws, especially the second law of thermodynamics. Solar heat, for example, could not have been in existence for more than one hundred million years, so the earth itself must be younger. Charles Darwin's awesome proposal in the *Origin of Species* (1859) that the denudation of the Weald must of itself have required three hundred million years was therefore ridiculous. Kelvin also endorsed a progressively cooling earth, which meant that geological processes must have been more energetic in the past; Lyellian uniformitarianism, on the contrary, denied any difference in efficacy through time. On 27 February 1868, Kelvin addressed the Geological Society of Glasgow, reading "On Geological Time." "A great reform in geological speculation seems now to have become necessary," he announced (1889–94, 2: 10), primarily because earlier theorists had attributed to the earth and other parts of the solar system a longevity and stability known to be deceptive. He began his attack by recalling Playfair's "celebrated and often-quoted passage" (one of Lyell's favorites) in which this point of view was put forth (1802, para. 118, pp. 119–120). Though Playfair had discerned "no mark either of the commencement or the termination of the present order" in nature, Kelvin thought nothing could be further from the truth. In particular, he attacked the Laplacean astronomy to which Playfair had alluded. Various kinds of resistance to planetary motions, he thought, proved that the rotation of the earth must be slowing down. If so, it must have been rotating twice as fast a billion years ago and was probably until a few million years ago "all fluid" (1889–94, 2: 43). In any case, its physical processes throughout this time could not have been uniform. Quite certainly, therefore, a great mistake had been made, for "British popular geology at the present time" was "in direct opposition of the principles of natural philosophy" (44). Playfair and his followers, moreover, had totally disregarded the "prodigious dissipation of energy" represented by necessarily diminishing solar output, instead speaking of the existing state of things "as if it must or could have been perennial" (49). For these and other reasons, however, one must conclude that "the existing state of things on the earth, life on the earth, all geological history showing continuity of life, must be limited within some such period of past time as one hundred million years" (64). For many uniformitarians, such a limitation was unacceptable.

Replying the next year, in an address entitled "Geological Reform," Thomas Henry Huxley (1825–95) initially defended uniformitarianism, by which he meant the teachings of Hutton and Lyell. "That great though incomplete work, the *Theory of the Earth*," he commented, "seems to me to be one of the most remarkable contributions to geology which is recorded in the annals of the science. So far as the not-living world is concerned, uniformitarianism lies there, not only in germ, but in blossom and fruit" (1909 [1869], 307–308). Hutton's advanced understanding, Huxley thought, derived from vast geological knowledge (based on personal observations and travel), together with thorough training in the physical and chemical science of his day. To the latter Huxley then ascribed "Hutton's steady and persistent refusal to look to other causes than those now in operation" (308). He next quoted *Theory*, 1: 173n, 281, 371, and 200, endorsing the understanding therein while ambiguously characterizing Hutton's sentences as "oddly constructed periods . . . full of that peculiar, if unattractive, eloquence which flows from mastery of the subject" (310). Other passages established Hutton's unfortunate concern with final causes, a chimera to be avoided. Throughout his remarks, Huxley quoted Hutton rather than Lyell because the former's works were "little known, and his claims on our veneration too frequently forgotten" (312). Few present-day geologists, Huxley believed, had read Playfair's *Illustrations* and even fewer the original *Theory of the Earth*.

Though uniformitarianism was preeminently a British concept (and had even yet gained comparatively few adherents elsewhere in Europe), Huxley nonetheless criticized both Hutton and Lyell for refusing to look farther back than the earliest fossiliferous rocks. "This attempt to limit, at a particular point, the progress of inductive and deductive reasoning," he thought, had disqualified uniformitarianism as "the permanent form of geological speculation," a stature it might otherwise have attained (316). In its place Huxley now proposed Evolutionism, which was to be the history of the earth in precisely the same sense as biology was the history of life. He then went on to reconcile uniformitarianism with its supposed antagonist, catastrophism (recurring catastrophes being only a regular part of nature), and subsumed both within his own evolutionism, which accepted the sound portions of each while rejecting their arbitrary limitations.

Having thus fortified himself, Huxley next attempted to answer Kelvin more directly. Though the latter had assumed the current establishment of uniformitarianism among geologists, Huxley disagreed. "I do not suppose," he said, "that at the present day any geologist would be found to maintain absolute Uniformitarianism, to deny that the rapidity of the rotation of the earth *may* be diminishing, that the sun *may* be waxing dim, or that the earth itself *may* be cooling" (326). Whether at work or not, these factors had no demonstrable influence on geological processes as recorded and preserved in the stratified rocks. Any supposed contradiction between current geological theory and natural laws was therefore nonexistent. Even the vague limitations and curtailment of geological time proposed by Kelvin could be reconciled with contemporary geology and biology. There was, then, no need for geological reform; "we have long been reforming from within, with all needful speed" (339). Despite his impressive powers of argument, however, Huxley clearly failed to neutralize Kelvin's devastating attacks on the credibility of extended geological time.

Kelvin's rebuttal was not long in coming. Less than two weeks later (5 April 1869), he read to the Geological Society of Glasgow "Of Geological Dynamics," a three-part address beginning with a direct response to Huxley. Fully prepared to endorse the evolutionary geology of his critic, Kelvin did not object to geological theory as such but only to "the ultra-uniformitarianism of the last twenty years" (1889–94, 2: 77). He then quoted several examples to show, contrary to Huxley, that present-day writers on geology were of two contending persuasions—those who ascribed every geological result to the ordinary operations of nature and unlimited time, and those who believed that certain geological agencies were previously more intense and widespread than at present. Some of the more extreme estimations of geological time came from Charles Darwin, who required vast eons for his device of natural selection to work. A detailed reexamination of his former arguments and Huxley's inadequate responses to them then led Kelvin to reaffirm the "Catastrophism of Leibnitz, Newton, Sedgwick, Phillips, Hopkins, Forbes, Murchison, and many other true geologists, which is in no respect different as a geological doctrine from that now described by Professor Huxley under the new name 'evolutionism'" (111). An imposing presence in scientific circles, Kelvin would then repeat, elaborate, and intensify his opposition to uniformitarian geology for the rest of his life.[5]

Geikie and the Scots

Though no more immediately capable than Huxley of refuting Kelvin, Geikie chose instead to respond with nationalistic and disciplinary élan.

[5]Burchfield 1975 explicates Kelvin fully. See also Sharlin 1979, and Merrill 1969, 648–662.

On 10 March 1871, a chair specifically devoted to geology and mineralogy (honoring Roderick Murchison, who died later that year) was founded at the University of Edinburgh. As the first occupant of that chair, Geikie devoted his inaugural address to "the Scottish school of geology," which in his view had risen to prominence between 1780 and 1825. The founder of that school, James Hutton, had soon been joined by John Playfair, author of the delightful and eloquent *Illustrations*, and Sir James Hall, "the founder of Experimental Geology" (1882a, 291), a figure still better appreciated in Germany and France than in Britain. To these three, Geikie also added the names of Sir George Mackenzie, Lord Webb Seymour, and Thomas Allan.

In opposition to the biblical chronologists, it was Hutton who first declared that the geological record included no evidence of any beginning. "He, too, first clearly and persistently proclaimed the great fundamental truth of geology," Geikie disclosed, "that in seeking to interpret the past history of the earth as chronicled in the rocks, we must use the present economy of nature as our guide" (293). After quoting directly from Hutton's *Theory* (I: 160, II: 549), Geikie reiterated that "the changes of the past must be investigated in the light of similar changes now in operation" (293). This, he said, was a guiding principle of the Scottish school, and through their influence had become an axiom of modern geology. As "uniformitarianism," however, it had "unquestionably been pushed to an unwarrantable length by some of the later followers of Hutton" (293); nevertheless, a tactful compliment to Lyell followed immediately.

Hutton and his followers, Geikie continued, applied their present-day processes to phenomena both within the earth and on it. Like Steno and Moro before him, Hutton realized that earthquakes and volcanoes were not divine judgments but part of the general mechanism of our planet. Though renowned for their power, both forces were only secondary manifestations of the earth's great internal heat, the full import of which could be appreciated only after careful study of the earth's crust. Hutton devoted his adult years to that perusal. "As a result of his wanderings and reflection," Geikie observed, "he concluded that the great mass of the rocks which form the visible part of the crust of the earth was formed under the sea, as sand, gravel, and mud are laid down there now; that these ancient sediments were consolidated by subterranean heat, and, by paroxyms of the same force, were fractured, contorted, and upheaved into dry land" (295). The igneous origin of granite, basalt, and similar rocks was also established, largely through the experiments of Sir James Hall. In addition, Hutton should be regarded as "the founder of the modern doctrine of metamorphism" (296).

Attending also to sedimentary rocks, Hutton and his followers discovered many proofs of ancient convulsions and re-formations in the bent and twisted strata, culminating with the famous unconformity at Siccar Point (as described by Playfair)—the significance of which had been fully realized. Though Hutton and his friends saw evidence of former revolutions clearly, however, they had not comprehended the real extent of that great series of fossiliferous layers since elucidated by later observers. Fortunately, substantiating Hutton's theory did not require such knowledge.

For Geikie, Hutton was also the first theorist to perceive the essential orderliness of geological processes and to appreciate the immense significance of subaerial erosion. "Even some of the Huttonians themselves," Geikie observed, "refused to follow their master when he sought to explain the existing inequalities of the land by the working of the same quiet unobtrusive forces which are still plying their daily tasks around us" (302). But no incredulity or neglect could destroy the innate vitality of truth. So now, after the lapse of two generations, Hutton's views had recently been revived, especially in Britain, becoming "the war-cry of a yearly increasing crowd of earnest hard-working geologists" (302). He attributed the intermediate decline of Scottish geology to Wernerian confusion.

Geikie continued his efforts on behalf of the Scottish tradition in subsequent publications. His "Earth Sculpture and the Huttonian School of Geology" (1873) was followed by a biography of Murchison (1875). Though his predecessor had never been a uniformitarian, Geikie nonetheless included in the latter work a basic sketch of Huttonian-Wernerian warfare. "The whole of modern geology," he was now convinced, "bears witness to the influence of the Huttonian school" (I, 99). The elevation of British geology into the dignity of a science, moreover, was traceable mainly to the influence of Playfair's *Illustrations*, a book frequently praised throughout Geikie's numerous works and the one that persuaded him to become not only a geologist but a stylish geological writer as well. Speaking in London on the origins of British scenery, Geikie applied Huttonian analysis consistently to landforms throughout the island, stressing the denudation of Britain and the fluvial origin of almost all its valleys. Influenced by American examples, he thought that "The intimate relation of a system of valleys to a system of drainage lines, first clearly enunciated by Hutton and Playfair, has received ample illustrations from all parts of the world" (1905b, 143). For him, the correctness of fluvialism was no longer in question. It therefore received additional emphasis in Geikie's successful *Text-Book of Geology* (1882b) and in the significantly revised second edition (1887) of his *Scenery of Scotland*.

Despite the chronological approximation involved, Geikie used his position as president to address the 1892 meeting of the British Association for the Advancement of Science (at Edinburgh) on the centenary of Hutton's "Theory." Among his many brilliant contributions to modern thought, Hutton, he said, had "refused to invent causes or modes of operation" (1905b, 161), an insight nowhere more profoundly expressed than in his several passages reiterating the perpetual decay of exposed continents, the erosive and transporting power of rivers, and the resulting system of valleys. Hutton, moreover, "refused to admit the predominance of violent action in terrestrial changes, and on the contrary contended for the efficacy of the quiet, continuous processes which we can even now see at work around us" (163), a belief necessarily requiring an unlimited duration of past time. "He felt assured that Nature must be consistent and uniform in her working, and that only in proportion as her operations at the present time are watched and understood will the ancient history of the earth become intelligible. Thus, in his hands, the investigation of the Present became the key to interpretation of the Past. The establishment of this great truth was the first step towards the inauguration of a true science of the earth" (171). Hutton and Playfair, moreover, cited the stratified rocks to prove their shared theorem empirically.

Even so, it had to be admitted that the doctrine of uniformity had afterwards been pushed to an extreme not contemplated by its originators. Given the brevity of human experience, Geikie continued, we are by no means warranted in assuming that natural processes in the past were equivalent in force to those presently observable:

> The uniformitarian writers laid themselves open to the charge of maintaining a kind of perpetual motion in the machinery of Nature. They could find in the records of the earth's history no evidence of a beginning, no prospect of an end. They saw that many successive renovations and destructions had been effected on the earth's surface, and that this long line of vicissitudes formed a series of which the earliest were lost in antiquity, while the latest were still in progress towards an apparently illimitable future. (174)

The discoveries of William Smith, had they been adequately understood, would have corrected this overly rigid uniformitarian conception by revealing a long but unmistakable record of organic progression among both plants and animals. Nature, Geikie concluded, was not uniform but had followed "a vast and noble plan of evolution" (175) pointing clearly to a definite beginning.

Despite inadequate fossil evidence, Geikie argued, one could nonetheless investigate that beginning through astronomy and physics. In particular, the researches of Lord Kelvin had demonstrated how improbable extreme uniformitarianism was. The secular loss of heat, which demonstrably takes place both from the earth and the sun, made it quite certain that the present state of the earth could not have been its original one. "This diminution of temperature, with all its consequences," Geikie insisted, "is not a mere matter of speculation, but a physical fact. . . . It points with unmistakable directness to that beginning of things of which Hutton and his followers could find no sign" (177). Though Hutton and Playfair had both believed in periodic catastrophes, he continued (erroneously), their followers gradually abandoned any such belief, Lyell and those he influenced going rather to the opposite extreme. Such theorists failed to see that "terrestrial catastrophes even on a colossal scale might . . . be part of the present economy of this globe" (178). The Ice Age seemed a most conspicuous example; however much time it may have actually required, surely one was right to call it a catastrophe. Yet whether or not one admitted to ice ages before the latest, the agencies involved were still our present ones.

Another Huttonian accomplishment of major significance, Geikie continued, was to replace traditional acceptance of a six-thousand-year-old cosmos with awareness of one immensely older. Not only Hutton but Playfair, Lyell, and other members of the uniformitarian school all maintained the almost unlimited duration of time their theoretical position required. "It was Lord Kelvin," Geikie averred unexpectedly, "who . . . first called attention to the fundamentally erroneous nature of these conceptions" (183). By calculating the high internal heat of the earth, increasing downwards as it does, and the rate at which that heat is lost, Kelvin had been able to fix a limit to the possible age of the earth. "He showed that so far from there being no sign of a beginning, and no prospect of an end to the present economy, every lineament of the solar system bears witness to a gradual dissipation of energy from some definite starting-point" (183–184).

Further calculations by Kelvin (1862) had seemingly established also that the crust of the earth could not have formed less than twenty million years ago nor more than four hundred million. In Kelvin's opinion, the earth was perhaps one hundred million years old. As Geikie went on to argue, however, analyses of denudation and sedimentation rates produced figures of much greater magnitude. Calculating first the height of the stratigraphic column and then the widely varying estimates regarding deposition, Giekie arrived at a figure between 73 and 680 million years. No evidence whatever suggested that deposition had originally been more rapid in the past. Organic evolution and the slow changes of organisms provided another argument on behalf of extended duration; there was no reason to believe that rates had changed here either. "After careful reflection on the subject," he concluded, "I affirm that the geological record furnishes a mass of evidence which no arguments drawn from other departments of Nature can explain away, and which, it seems to me, cannot be satisfactorily interpreted save with an allowance of time much beyond the narrow limits which recent physical speculation would concede" (192). Geikie's closing remarks emphasized the evolution of landscapes, a topic owing its first impulse to the far-seeing intellects of Hutton and Playfair, both of whom were illustrious teachers of undoubted genius.

President again in 1899, Geikie chose to address the British Association for the Advancement of Science (at Dover) on the still vexing subject of geological time. Three long-held opinions relevant to this topic—cataclysmal, uniformitarian, and evolutionist—had all been challenged about thirty years earlier by a bold irruption into their camps from the side of physics. Because the dispute had continued ever since, Geikie thought a retrospective appropriate. Among the pioneers, he confirmed, "none has left his mark more deeply graven on the foundations of modern geology than James Hutton. To him, more than to any other writer of his day, do we owe the doctrine of the high antiquity of our globe" (1905b, 200). Though thoroughly aware of a vast earlier time anteceding any geological record still extant, Hutton had not attempted to speculate beyond the limits of his evidence. However much criticized by physicists, Geikie insisted, Hutton's famous conclusion that "we find no vestige of a beginning, no prospect of an end" was "absolutely true in Hutton's time and remains true today" (202). Hutton admitted that the earth had a beginning, but he could find no trace of it remaining—nor has any such relic been uncovered since. Hutton's appreciation of the vastness of geological time, moreover, enabled him to perceive the efficacy of "little causes, long continued" (*Theory*, II: 205). But he never attempted to measure nor to estimate its duration.

Playfair, however, had gone beyond Hutton in deriving his conception of geological time not only from the rocks but from the entire planetary system (1802, para. 118), which he believed indestructible by natural means. It was therefore easy to understand how the subsequent uniformitarian school, derived from the teachings of Hutton and Playfair, "came to believe that the whole of eternity was at the disposal of geologists" (1905b, 205). Later geologists, following Lyell, "became utterly reckless in their demands for time" (205), until Kelvin's paper of 1862 on the secular cooling of the earth astonished them all by proclaiming definite limits. In subsequent contributions, Kelvin and his followers reduced the broader limits previously suggested to no more than ten or twenty million years in all.

As a result of Kelvin's criticisms, geologists have revised their chronologies. Yet even physicists have in the meantime disputed the three arguments on which Kelvin depended: the history of the sun, the secular cooling of the earth, and the concept of tidal friction. Professor John Perry, for example, had only recently written in *Nature* (18 April 1895; 51:582–585) that none of the three arguments was sufficiently precise to discredit the much longer estimates of geological time proposed by paleontologists and stratigraphers. In October 1904, Geikie had the pleasure of emending his address to include the discovery of radioactivity, which extended the possible durations of both solar and subterranean heat to such an extent as virtually to destroy Kelvin's position.

It now remained only for Geikie to celebrate the essential triumph of uniformitarian geology in the revised (1905) edition of his originally briefer study, *The Founders of Geology*. Though remarkable for its attention to still earlier predecessors (chiefly classical and French), Geikie's conception of the history of geology unquestionably culminated with Hutton, whose dominant idea, that "the present is the key to the past," had become "the chief cornerstone of modern geology" (1905a, 299).[6]

Recurring Questions

With the publication of *Founders of Geology* in its revised form, Geikie superseded Lyell as the foremost historian of British geology. But the German scholar Karl von Zittel had already published an impressive *History of Geology and Palaeontology* (1901 [1899]) in which Hutton was acknowledged as "the great founder of physical and dynamical geology" (71). Both authors remain authoritative to this day, and they have been followed in their high opinions of Hutton, Playfair, Hall, and Lyell among others by many subsequent commentators. Since 1969, however, when G. L. Davies' influential history of British geomorphology (*The Earth in Decay*) appeared, there has been a considerable willingness among revisionist historians—not all of them so judicious as he—to "correct" the late nineteenth-century apotheosis of Hutton in various respects. On the whole, this fresh thinking has generated useful results, especially by exposing the superficiality of received interpretations. Often, however, corrections go too far and need themselves to be restrained by the gravitative power of facts and likelihoods. Though a survey of recent new proposals is beyond the scope of this book, certain questions have persistently recurred: What did Hutton believe? How original were his beliefs? How modern do they seem to us today? And, finally, what is Hutton's rightful place within the history of geology? In

[6]The same idea had appeared earlier in Geikie's writings, as we have already seen in part, but the elegant 1905 version became classic. For an interesting critique of Geikie and his historiography, see Oldroyd 1980.

concluding this brief survey of Hutton and his influence, I address each of these issues in turn.

What Did Hutton Believe? Hutton's geological beliefs were for him part of a larger philosophy, or even theology, of nature that one might call Agricultural Deism. According to this outlook, a central revelation of divine benevolence to man is the continuing fertility of the earth, a life-supporting fecundity made possible only by the immediate effects of the hydrological cycle and the infinitely slower but even more important effects of the petrological cycle, in which new rock and soil are continuously being created in order to replace the old. Fundamental to Hutton's point of view was a sharp-eyed appreciation for the efficacy of erosion, through which earth's soils are constantly being removed from productivity and transported to the sea. Since an overall loss of fertility was unthinkable, Hutton necessarily postulated some kind of renewing mechanism, which he found to be uplift. This uplift, in turn, was powered by some poorly known but nonetheless ubiquitous force called subterranean fire or heat, which was also responsible for consolidating loose sediments into solid rock. His completed theory therefore affirmed a majestic but extremely slow natural revolution moving like some great pump or engine through a renovative cycle that included erosion, transportation, sedimentation, consolidation, and uplift. Despite Hutton's later concern with basalt and granite, both were primarily of use to him in helping to confirm the reality of a process in which sedimentary rocks were by far the most important.

In seeking to establish his fundamental truths, Hutton was led to consider a wide variety of geological phenomena and topics, including the evolution of landscapes; the work of rivers, glaciers, and marine currents; the efficacy of present-day forces; and the immensity of geological time. With regard to each of these areas, he proved to be extremely insightful. We have already seen how Hutton worked out his theory over many years, developing, but virtually never contradicting, his earlier pronouncements and giving them increasingly wider applications. If there was ever a time when Hutton came to doubt his own conclusions, or to despair over them, we have no record of it.

How Original Were Hutton's Beliefs? Quite original, in my opinion. Though some historians have sought to derive Hutton's theory from an earlier one, perhaps Hooke's or Moro's, they have not been able to do so convincingly. Hooke, for example, emphasized earthquakes, a topic Hutton barely mentioned. Moro, similarly, became very much a Vulcanist; Hutton never was. On the evidence, there is every reason to agree with

Playfair's observation (1805b, 93) that Hutton seldom read the geological theories of others, and then only to disagree with them. Hutton's major theoretical bias came to him through the Deism and natural theology of his century. He then began with the initially disconcerting observation that precious soil was being washed away by rain and rivers and worked out the whole of his theoretical response from there: the grand paradox of loss and gain. What resemblances exist between his theory and any other are almost wholly attributable, I think, to the fact that two very talented observers were describing the same natural phenomenon.

To be sure, several of the ideas Hutton utilized had prior histories of their own. That seemingly destructive natural forces—such as volcanoes—actually serve constructive ends, for instance, was a familiar stance among the natural theologians; though Hutton *may* have derived his attitude toward volcanoes from Strabo, as is often asserted, he could just as easily have found the same idea in more contemporary sources. Hutton's originality, in this case, consisted in his applying the same kind of solution to the less immediately dramatic problem of erosion, which (as Davies has shown) was a matter of considerable debate throughout the seventeenth century. Similarly, the concept of subterranean fire, strictly defined, had been part of Stoic cosmology and was subsequently utilized by Christian theologians as a rationalization of Hell. Under Deist influence, however, the earth's heat came once again to be seen as a positive manifestation of divine intentions; Enlightenment thinkers did not, as a rule, believe in a punitive afterlife.

Deist rejections of biblical literalism and the miraculous also preceded Hutton's in calling attention to the efficacy of present-day geological forces and extended geological time. More broadly, Hutton by no means originated the concept later to be called uniformitarianism, which, as natural law, had been commonly assumed by Presocratic philosophers, Aristotle, and particularly the Stoics. It was then revived by Francis Bacon (who stressed the simplicity of nature) and substantiated by Isaac Newton. Four "Rules of Reasoning" in the latter's *Principia*, book 3, emphasized that "nature . . . is wont to be simple, and always consonant to itself." Several Enlightenment thinkers, particularly Buffon, then applied uniformitarian assumptions to selected geological problems. Hutton's contribution in this area—a very significant one—is that he was the first philosopher to base his geological theory entirely on gradualistic naturalism, having eschewed any kind of preceding cosmological myth. He worked backward from the present earth, not forward from Genesis. With the possible exception of Desmarest, Hutton greatly outdistanced any of his predecessors in his empirical arguments on behalf of perpetually acting, uniform natural processes.

As part of the eighteenth century's pagan revival, such thinkers as

Holbach, Hume, and Toulmin had adopted the classical uniformitarianism of Aristotle and Lucretius to argue, more or less openly, for the eternity of the earth. For Hutton, who claimed to believe in a finite earth (*Theory*, I: 223), uniformitarianism was primarily a methodological assumption—as it had been for Newton. Hutton's geological theorizing accepted, or at least considered, rapid continental collapse because its author remained unclear as to what held up elevated continental masses. But he certainly did not accept catastrophic uplift; any local catastrophes known to Hutton, and almost never acknowledged, were dwarfed by his ponderous regularities. He rarely claimed, moreover, to be describing anything other than the present earth. Making no far-reaching historical statements about the past, whose uniqueness in any respect he denied, Hutton argued instead that the renovative cyclical processes he had discovered would necessarily destroy any geological evidence more than a cycle or two old. However often misunderstood, his famous conclusion that we find "no vestige of a beginning, no prospect of an end" does not refer primarily to the immensity (perhaps eternity) of geological time he believed in but to his logical assumption that the surviving record of the rocks comprises only the last few perishing leaves of a constantly self-destructing but equally regenerating natural calendar whose earlier sheets are now irrecoverably disseminated as unidentifiable constituents within later ones (see *Theory*, I: 223).

Though Hutton's theory did not derive from any of his predecessors', certain interactions between those theories and his own developing one should nonetheless be considered. In all probability, the chief influence of other theories on Hutton's was negative, in that he found either their methods or their conclusions offensive and deliberately opposed them. For example, Hutton explicitly attacked the well-established eighteenth-century tradition of biblically derived geological speculations that regarded the earth as a ruin brought about by the fall of man (a calamity in which Deists generally did not believe). That tradition included numerous editions of the theory of Thomas Burnet and would culminate, for Hutton particularly, in that of John Whitehurst; both writers are mentioned by name in his own work. Other theorists, so skeptical as to be virtual atheists, accepted much of what Hutton also believed—including natural law and geological time—but then drove their speculations to what were for him intolerable depths of unbelief. As an avowed opponent of both medieval "darkness" and atheistic "skepticism" (*Theory*, I: 225), Hutton surely knew himself to be writing in part against the likes of Holbach, Hume, and Toulmin.[7]

Possible associations between Hutton and other members of the Scot-

[7]Hutton was alluding to his *Principles of Knowledge*, 1794, 1: 441, and other passages.

tish Enlightenment have not been adequately explored. It is altogether likely, for example, that both Joseph Black and John Clerk of Eldin conversed at length with Hutton throughout the years in which his theory took form. Indeed, there is very good evidence that Black was incorporating aspects of Hutton's geological thinking (or perhaps his own) into his lectures on chemistry as early as 1756; similarly, Playfair explicitly advises us that parts of Hutton's theory were almost certainly contributed by Clerk (1805b, 97). More broadly, the possibility of other, mostly philosophical, connections between Hutton and, for example, Adam Smith, Dugald Stewart, or even David Hume should not be overlooked. In all these latter cases, however, the fact remains that Hutton was specifically interested in geology whereas others, like Hume, were a good deal less so or, often, not at all.

It is of course demonstrable that Hutton utilized a variety of geological literature, most of it French, throughout his *Theory* of 1795, the chief authors being Saussure, Deluc, and Dolomieu. In addition to them, Playfair added Ferber and Bergman as being among "those on whom Dr. Hutton chiefly relied" (1802, 514), but this is less apparent. The possibility of significant indebtedness to Desmarest has occasionally been raised but never proven. Hutton's use of both Deluc and Saussure, however, remained almost entirely illustrative. Having already conceived his theory, he borrowed a selection of their data to substantiate it further (particularly after becoming bedridden, and thus no longer able to continue original fieldwork himself). Except for calling his attention to glaciers, there is no real evidence that either writer actually changed any major aspect of Hutton's already established thought.

How Modern Is Hutton's Theory? Hutton unquestionably made mistakes. Among these, four of the most important were his erroneous theories regarding the origin of flint and other stones; his dubious theory of consolidation; his unwillingness to attribute some vein deposits to water; and his failure to identify basalt with lava. As a result of the latter, he seriously underestimated the importance of volcanoes. Despite his inevitable knowledge of major seismic episodes in 1750 and 1755, he said almost nothing about earthquakes. Beyond these, a particularly significant shortcoming was his chronic inability to appreciate the geological utility of precisely identified fossils. More broadly, he did not accept the uniqueness of geological events, such as eruptions and earthquakes, or of geographical areas, or of past times. Traditional geological disasters had for him no theological significance and were altogether unimportant in the constitution of the globe. Nor was he interested in the peculiarities of individual districts, all of which were for him only parts of the same overall plan. The ideas of mapping, regional study, and systematic specimen collecting were equally foreign to him; thus, Britain's Geological Survey would not be Huttonian in origin. Except in its duration, moreover, our familiar geological timescale would have been meaningless to him. Hutton seems not to have realized that life on earth has a history, or at least chose to ignore the fact. His assumed perspective, one might say, was Neoclassical rather than Romantic, which is to emphasize that he was everywhere concerned with the basic, timeless, and enduring aspects of the earth—its formal essence—rather than with unimportant accidentals occasioned by time and place. This determined essentiality did not however prevent Hutton from analyzing specific landscapes brilliantly—indeed, his analytic imagination splendidly combined both Neoclassical and Romantic characteristics—but surviving records of his having done so are all too few. It was never his aim, moreover, to analyze any given landscape fully. Particulars were, in his mind, always subordinate to principles, as events were to processes. Once a valid theory of the earth had been established, he believed, correct interpretations of any given landscape would easily derive from it.

What Is Hutton's Rightful Place within the History of Geology? One of the strengths of Hutton's theory was that it remained for the most part at a high level of generalization. As Playfair accurately predicted, "Ages may be required to fill up the bold outline which Dr. Hutton has traced with so masterly a hand; to detach the parts more completely from the general mass; to adjust the size and position of the subordinate members; and to give to the whole piece the exact proportion and true coloring of nature" (1802, 139). Unlike previous geological theories, most of them only ingenious nonsense—to be savored as romance, then found out and discarded—Hutton's has proved to be a worthwhile foundation on which to erect a greatly elaborated but nonetheless kindred science.

Among Hutton's major contributions to the science of geology are his insistence on the basic orderliness of terrestrial processes; the centrality of heat; the duration of time; the significance of denudation, consolidation, and uplift; the efficacy of rivers; the intimate association between the earth's crust and subsurface forces; the nature and significance of plutonic and metamorphic rocks; and the ongoing nature of petrological creation. Other geological theorists have given us one or two good points, but none competes with Hutton in the number of sound principles established.

There is, then, a direct connection between Hutton's geology and our own, with the most significant work of many major figures coming in between. Clearly, we have never discarded Hutton; instead, we have built our modern science on the sturdy foundations he laid down. No prior figure in the history of geology can lay claim to such an unbroken connection with the present. Despite the merits of other plausible nominees, Hutton stands alone in the large number of profound insights he synthesized and transmitted to later workers. On the evidence, then, he is far more the precursor of our own outlook than any other theorist one can name—and therefore, as so many of his beneficiaries have proclaimed, legitimately the founder of modern geology.

Appendices

Appendix 1: Hutton and Black

The date of Hutton's first meeting with Joseph Black is unknown, but mutual interest in chemistry was almost certainly the reason. Black lectured daily on chemistry, Monday through Friday from mid-November to mid-May, at the University of Edinburgh. As of 1768, according to preserved student notes, his course (about 125 lectures in all) included several geological concepts: parallel strata, in predictable order; mountain formation by the tilting and differential erosion of strata through time; other mountain-building theories (earthquakes, volcanoes, a universal deluge, a comet); and rents and veins. He also discussed petrifaction, cave formations, and the reduction of shells through time and pressure into marble ("limestone a little hardened"). Touchstone, granite, and whinstone all contain flint (i.e., quartz), which itself is found in several forms and in a variety of strata; black or gun flints, for example, lying generally in horizontal beds, are found among strata of limestone or chalk.

Black also described the results of several experiments—apparently his own—on quartz crystals (colors alterable by heat), flint (mixed with alkalines, may be made into glass), and whin (melts remarkably when exposed to a strong fire). Limestone, moreover, hardens in fire, producing substances like marble and chalk. In discussing mineral waters, Black again dealt specifically with the presumably aqueous origin of strata and fossils. "But how are we to explain flints being formed," he asked, "for they seem not to have the least degree of solubility in water?" (165). Coal and other inflammable strata of vegetable origin were somewhat difficult to explain. All earthy and stony substances, moreover, had once been under water, but the Deluge (accepted by many) could not have produced the present, extensive deposits (McKie 1966, 55–58, 79–81, 165–166). The startling similarity between some of Black's remarks and those later published by Hutton strongly suggests that the two were already discussing geological theory in earnest. Some rudimentary form of experimental geology, moreover, was already going on (and both knew marble to be a metamorphic rock). Though Hutton's theory was certainly his own, historians should realize that both its argumentation and its substance were considerably influenced by Hutton's conversations (and perhaps experiments) with Black. Conversely, Black and other lecturers of the university may have influence from Hutton in their classroom remarks.

Appendix 2: Hutton and Toulmin

Of the several disputes regarding Hutton's originality, none has attracted more attention than the possible derivation of his thinking from that of George Hoggart Toulmin (1754–1817), who between 1780 and 1789 published four separate but textually related books affirming the eternity of the world. His possible anticipation of Hutton was first raised—in recent times, at least—by S. I. Tomkeieff, who noted the Aristotelian origin of Toulmin's concept and compared his position with Hutton's, finding some striking resemblances (1950, 398). But whether or not Hutton had actually known Toulmin's books of 1780 or 1783 Tomkeieff thought impossible to determine.

An admittedly derivative essay by Donald B. McIntyre (1963) then reaffirmed and further substantiated the alleged similarity. Juxtaposing short quotations from Toulmin (1783) and Hutton (1788), McIntyre discovered verbal echoes so compelling that he found it impossible to believe Hutton had not read Toulmin's theory prior to formulating his own. Reijer Hooykaas reinforced aspects of the comparison in two essays of 1966, one published in German and the other in English. Though tentatively accepting the possibility that Hutton *had* been influenced by Toulmin, Hooykaas emphasized fundamental differences separating the two.

As G. L. Davies (1967; 1969) next pointed out, however, Hutton's theory had in all probability been conceived well before 1780. Any textual parallels between Toulmin and Hutton, therefore, were for him explainable only by supposing that Toulmin had plagiarized some manuscript version of Hutton's theory. We know from Playfair's biography of Hutton, he added, that at least one such draft did exist, so Toulmin might easily have seen it while a medical student at Edinburgh (1776–79). Writing briefly on Toulmin in the *Dictionary of Scientific Biography* in 1976, V. A. Eyles conceded the geological similarities but also noted Toulmin's lack of field evidence, which effectively reduced his since neglected theory to nothing more than an academic exercise.

In three essays of 1978 dealing with Toulmin, Roy Porter then responded to Davies and Eyles with further information and arguments that have thus far gone largely unchallenged. Like Davies (1967), he quoted Toulmin's brief 1789 comment on Hutton's paper of the previous year:

> Wonderful and numerous, however, as the vestiges of subterraneous fire most undoubtedly are, and great as its influence is, it is still possible to extend its operations even beyond what natural facts seem absolutely to justify. Dr. Hutton, the ingenious author of a dissertation published in the Edinburgh Philosophical Transactions, while he thus perhaps too much extends the influence of fire, adopts the opinions I have so long endeavored to establish in respect of the alternate destruction and renovation of fossils, or of the substances that compose the world (204).

For Porter, "Toulmin wrote as if he and Hutton were partners in a joint enterprise," his criticism of the latter being perspicacious and showing independence of judgment (1978a, 1256). Other evidence, moreover, suggested that Toulmin was in his own right a serious student of the earth.

Whatever its author's relationship with Hutton, Porter continued, Toulmin's edition of 1789 included demonstrable plagiarisms from William Smellie's 1785 translation of Buffon's *Histoire Naturelle*. Many of the same "Huttonian" ideas present in Toulmin could also be found in the writings of John Brown, an influential medical teacher at Edinburgh, whose impact on Toulmin during his student years is beyond doubt. Perhaps, then, all concerned derived not so much from each other—whatever the precise chains of influence may have been—as from a more general body of naturalistic uniformitarian thought, already established in the eighteenth century and becoming increasingly popular.

In support of this position, Porter (like Hooykaas before him) emphasized the fundamental philosophical and religious differences separating Toulmin and Hutton, both with regard to the history of the earth and the history of man. While ignoring major aspects of Hutton's theory, for example, Toulmin postulated a series of geological catastrophes throughout human history. He also denied the existence of God and his providence while asserting the eternal existence of the earth. Such atheistic naturalism, we have seen, is entirely inconsistent with Hutton's most basic presuppositions and, together with further evidence, strongly suggests that their respective outlooks derived from competing eighteenth-century philosophies—Deism and Atheism.

Like Hooykaas and Porter, I also think that any meaningful comparisons of Toulmin and Hutton must include broadly researched discussions of eighteenth-century intellectual history. Most of what Toulmin asserted about the history of the earth, for example—such as his repeated catastrophes throughout human times—can be found in classical sources, including Plato, Aristotle, Lucretius, and Ovid. Though he cited a number of contemporary geological theorists and was evidently well read in the field, Toulmin nowhere recorded any meaningful confrontations with actual landscapes, rocks, minerals, or fossils. His methodology, in other words, was literary rather than scientific and pagan rather than Christian. Toulmin certainly regarded himself as part of the Enlightenment and held eternalist theories sincerely, but he was probably as concerned to destroy opposing views (Genesis-centered theories of the earth) as to establish his own. Characteristically, he believed that plainly stated appeals to pure reason would suffice.

I do not believe that either thinker derived from the other. Of the two, Hutton (a quarter-century older) was clearly first in the field; major portions at least of his geological theory had been formulated by the time Toulmin was born. I have also supposed that portions of his thinking appeared in the lectures of Dr. Joseph Black around that time. In all likelihood, other faculty members would also have been aware of the same ideas to some extent, particularly after Hutton moved to Edinburgh in 1767. Was, then, John Brown (1735–88) among those so influenced?

That Toulmin either read or read about Hutton's 1788 paper (he mistook the journal title) is undisputed, but no one has yet attempted to measure the considerable changes Toulmin added to his 1789 edition against possible Huttonian influence. Did Hutton ever read Toulmin, and if so when? The only real fact available to us is that Hutton never mentioned Toulmin. On the other hand, he almost never cited theoretical discussions as such. In as much as Toulmin represented opinion rather than facts, Hutton had nothing to gain by citing him. Given the mostly silent opprobrium accorded Toulmin, moreover—and undoubtedly shared by Hutton himself—acknowledgment would have been dysfunctional. That Hutton became aware of Toulmin's theories and their general import at some point, however, seems to me altogether likely.

Appendix 3: Hutton and Geological Time

Throughout Hutton's lifetime (and perhaps even now) there were four competing concepts of geological time:

1. The Scripturally derived figure of approximately six thousand years, as established by sixteenth- and seventeenth-century chronologers. According to the most commonly accepted calculation, the world began in 4004 B.C.
2. The Scriptural chronology (including the entire history of mankind) augmented by preceding geological periods (figurative "days") of much greater extent. The age of the earth was therefore unknown, but its beginning at some definite time was assumed.
3. Those periods, plus an extended history of mankind not limited by the biblical narrative.
4. An eternal earth.

Whatever they may have believed in private, both Hutton and Playfair publicly endorsed the second of these possibilities. Among those who preceded them, however, wide disagreement prevailed as to just how extensive geological time had been. Major theorists gradually extended the acceptable upper limit of that duration from around sixty-five thousand years (Buffon in public) to as much as one million (Werner). Yet no other theorist—except the widely condemned eternalists such as Holbach and Toulmin—endorsed such vague but daring lengths of geological time as Hutton did.

Appendix 4: The Authorship of Hutton's "Abstract"

Alluding to unspecified stylistic peculiarities, E. B. Bailey suggested that Hutton did not write his own abstract. "The geology is the geology of Hutton," he claimed, "but the voice is the voice of Playfair" (1967, 31). In his 1972 *Dictionary of Scientific Biography* entry on Hutton, however, V. A. Eyles rejected Bailey's contention as unnecessary. After routinely agreeing with Eyles for years, I later undertook a detailed stylistic analysis of Hutton's verbal characteristics (including syntax, grammar, and vocabulary) in his "Theory" of 1788, then did the same with his "Abstract." Though completed within a year of each other, they are in some respects highly dissimilar. A large number of words, phrases, constructions, and rhetorical techniques in the "Abstract" clearly do not belong to Hutton, though others do.

Stylistically, the relationship of Hutton's "Theory" to his "Abstract" is remarkably like that between his original "Preface" and William Robertson's elegant redaction of it. The "Abstract," moreover, often seems more akin to Robertson's version of the "Preface" than to "Theory" or any other of Hutton's works. I now believe, therefore, that the "Abstract" was also paraphrased by Robertson (it is definitely not by Playfair) from Hutton's original draft and that Hutton preferred the new version to his own.

Bibliography

The Publications and Manuscripts of James Hutton

1749. *Dissertatio Physico-Medica Inauguralis de Sanguine et Circulatione Microcosmi.* Leiden: W. Boot (submitted 12 September 1749). 34 pp. Facsimile reprint in Donovan and Prentiss 1980, with translation and commentary preceding.

1777. "Considerations on the Nature, Quality, and Distinctions of Coal and Culm, with Inquiries, Philosophical and Political, into the Present State of the Laws, and the Questions Now in Agitation Relative to the Taxes upon These Commodities." Edinburgh. 38 pp.

1785. "Abstract of a Dissertation Read in the Royal Society of Edinburgh upon the Seventh of March and Fourth of April MDCCLXXXV, Concerning the System of the Earth, Its Duration and Stability." [Edinburgh]. 30 pp. Facsimile reprints in White and Eyles 1970 and Albritton 1975; reprint in Eyles 1950. French translation as "Extrait d'une Dissertation sur le Système et Durée de la Terre," trans. Iberti. *Observations et Memoires sur la Physique, sur l'Histoire Naturelle, et sur les Arts et Metiers (Journal de Physique, de Chimie, d'Histoire Naturelle et des Arts)* 43 (1793): 3–8, with comments by the translator, pp. 8–12.

1785. "Memorial Justifying the Present Theory of the Earth from the Suspicion of Impiety." Manuscript, 5 pp. Fitzwilliam Museum, Cambridge. Published in Dean 1975.

1787. "Theory of the Earth; or an Investigation of the Laws Observable in the Composition, Dissolution, and Restoration of Land upon the Globe." Edinburgh. Separate, 96 pp.

1788. "The Theory of Rain." *TRSE* 1(2): 41–86. Read 2 February and 12 April 1784.

1788. "Theory of the Earth; or an Investigation of the Laws Observable in the Composition, Dissolution, and Restoration of the Land upon the Globe," *TRSE* 1(2): 209–304. Read 7 March and 4 April 1785. Facsimile reprint in White and Eyles 1970. German translation, *Sammlungen zur Physik und Naturgeschichte* 4 (1792): 622–725. Rev. *Monthly Review* 79 (1788): 36–38; *Analytic Review* 1 (1788): 424–425; *Critical Review* 66 (1788): 115–120; *Magazin für das Nueste aus der Physik und Naturgeschichte* 6 (4) (1790): 17–27.

1790. "Dissertation on Written Language as a Sign of Speech." *TRSE* 2(1): 5–15. Read 19 June, 17 July, and 20 November 1786.

1790. "Of Certain Natural Appearances of the Ground on the Hill of Arthur's Seat." *TRSE* 2(2): 3–11. Read before the Philosophical Society in June 1788. Rev. *Monthly Review,* n.s. 5 (1791): 197.

1790. "Answers to the Objections of M. de Luc, with Regard to the Theory of Rain." *TRSE* 2(2): 39–58. Read 3 December 1787. Rev. *Monthly Review*, n.s. 5 (1791): 197.

1790. Introductory note to Adam Smith's *Essays on Philosophical Subjects*. Edinburgh, pp. lxxxviii–lxxxix.

1792. *Dissertations on Different Subjects in Natural Philosophy*. Edinburgh: Cadell and Davies. 740 pp. Rev. *Monthly Review* 16 (1795): 246–254.

1794. *An Investigation of the Principles of Knowledge, and of the Progress of Reason, from Sense to Science and Philosophy*. 3 vols. Edinburgh: Strahan and Cadell. 649 pp., 734 pp., 755 pp.

1794. *A Dissertation upon the Philosophy of Light, Heat, and Fire*. Edinburgh: Strahan and Cadell. 326 pp.

1794. A note on Adam Smith's final days. *TRSE* 3(1): 131n.

1794. "Observations on Granite." *TRSE* 3(2): 77–85. Read 4 January 1790 and 1 August 1791.

1794. "Of the Flexibility of the Brazilian Stone." *TRSE* 3(2): 86–94. Read 7 February 1791.

1795. *Theory of the Earth, with Proofs and Illustrations*. 2 vols. Edinburgh. Printed for Messrs. Cadell, Junior, and Davies, London; and William Creech, Edinburgh. 620 pp., 567 pp. Facsimile reprints 1959, 1960, 1972, Weinheim and Codicote. Rev. J. A. Deluc, *British Critic* 8 (1796): 337–352, 466–480, 598–606.

1796–97. "Elements of Agriculture." Manuscript, 2 vols., 1045 pp. National Library of Scotland, Edinburgh.

1798. "Dissertation on the Philosophy of Light, Heat, and Fire." *TRSE* 4(i): 7–16. Reading began 7 April 1794 and continued through December.

1798. "Of the Sulphurating of Metals." *TRSE* 4(i): 27–32, 25–28 (despite mispagination, ten consecutive pages). Read 9 May 1796.

1899. *Theory of the Earth, with Proofs and Illustrations*, vol. 3. Ed. Archibald Geikie. London: The Geological Society. 292 pp. Chaps. 4–9 only, but index for all three volumes. Manuscript, 208 pp. Geological Society, London.

General Bibliography

Adams, F. D. 1895. "Hutton's Theory of the Earth." *Nature* 52:569.

Adams, F. D. 1934. Address Given at the Centenary Celebration. *TRSE* 13:209–215.

Adams, F. D. 1938. *The Birth and Development of the Geological Sciences*. Baltimore. Reprint, 1954.

Agassiz, Elizabeth Cary. 1886. *Louis Agassiz: His Life and Correspondence*. 2 vols. Boston.

Agassiz, Louis. 1838a. "On the Erratic Blocks of the Jura." *Edinburgh New Philosophical Journal* 24:176–179.

Agassiz, Louis. 1838b. "Upon Glaciers, Moraines, and Erratic Blocks" *Edinburgh New Philosophical Journal.*, 24:364–383. Read 24 July 1837, Neuchâtel.

Agassiz, Louis. 1839. "Remarks on Glaciers." *Edinburgh New Philosophical Journal.*, 27:383–390.

Agassiz, Louis. 1840. *Etudes sur les Glaciers*. Neuchatel. Facsimile reprint, 1966. Trans. A. V. Carozzi, 1967.

Albritton, Claude C. 1963a. "Philosophy of Geology: A Selected Bibliography and Index." In Albritton, ed., *Fabric of Geology*, pp. 262–363.
Albritton, Claude C., ed. 1963b. *The Fabric of Geology*. Stanford, Calif.
Albritton, Claude C., ed. 1975. *Philosophy of Geohistory*. Stroudsburg, Pa.
Albritton, Claude C. 1980. *The Abyss of Time: Changing Conceptions of the Earth's Antiquity after the Sixteenth Century*. San Francisco.
Allan, Thomas. 1812. "On the Rocks in the Vicinity of Edinburgh." *TRSE* 6:405–433. Read 4 and 19 March 1811.
Allan, Thomas. 1815a. "Remarks on the Transition Rocks of Werner." *TRSE* 7 (1):109–138.
Allan, Thomas. 1815b. "An Account of the Mineralogy of the Faroe Islands." *TRSE* 7 (1):229–268.
Anon. 1803. Review of Playfair's *Illustrations. Monthly Review* 42: 285–294.
Anon. 1803. Review of Murray's *Comparative View. Monthly Review* 42: 294–295.
Bailey, E. B. 1935. *Tectonic Essays, Mainly Alpine*. Oxford.
Bailey, E. B. 1950. "James Hutton, Founder of Modern Geology (1726–1797)." *Proceedings of the Royal Society of Edinburgh* 63(B): 357–368.
Bailey, E. B. 1962. *Charles Lyell*. London.
Bailey, E. B. 1967. *James Hutton—the Founder of Modern Geology*. London, Amsterdam, and New York.
Bakewell, Robert. 1813. *An Introduction to Geology*. London. 2d ed., 1815. Rev. *Monthly Review* 82 (1817): 164–172. Other London editions 1828, 1833, 1838. American editions (with additions by Benjamin Silliman) 1829, 1834, 1839.
Bakewell, Robert. 1823. *Travels, Comprising Observations Made during a Residence in the Tarentaise and Various Parts of the Grecian and Pennine Alps and in Switzerland and Auvergne in the Years 1820, 1821, and 1822*. London.
Beddoes, Thomas. 1791. "Observations on the Affinities between Basaltes and Granite." *Philosophical Transactions of the Royal Society of London* 81:48–70. Read 27 January 1791.
Berger, J. F. 1816. "On the Geological Features of the Northeastern Counties of Ireland." *Transactions of the Geological Society of London* 3:121–195.
Besterman, Theodore, ed. 1938. *The Publishing Firm of Cadell & Davies: Select Correspondence and Accounts, 1793–1836*. Oxford and London.
Black, Joseph. 1756. "Experiments upon Magnesia Alba, Quicklime, and Some Other Alcaline Substances." *Essays and Observations, Physical and Literary* [Philosophical Society of Edinburgh], 2:157–225. Read 5 June 1755.
[Black, Joseph.] 1770. *An Enquiry into the General Effects of Heat*. London.
Black, Joseph. 1794. "An Analysis of the Waters of Some Hot Springs in Iceland." *TRSE* 3 (2):95–126. Read 4 July 1791.
Black, Joseph. 1803. *Lectures on the Elements of Chemistry*. Edited from his manuscripts by John Robison. 2 vols. Edinburgh.
Black, Samuel. 1786. *Dissertatio Physica Inauguralis de Ascensu Vaporum Spontaneo*. Edinburgh.
Boase, Henry S. 1834. *A Treatise on Primary Geology*. London.
Boué, Ami. 1820. *Essai geologique sur l'Ecosse*. Paris.
Boué, Ami. 1822. "On the Geognosy of Germany, with Observations on the Igneous Origin of Trap. *Memoirs of the Wernerian Natural History Society* 4 (1):91–108.
Boué, Ami. 1823. "Outlines of a Geological Comparative View of the Southwest and North of France and the South of Germany." *Edinburgh Philosophical Journal* 9:128–148.
Boué, Ami. 1825. "Synoptial Table of the Formations of the Crust of the Earth and of the Chief Subordinate Masses." *Edinburgh Philosophical Journal* 13:130–145.

Brande, William Thomas. 1816. *A Descriptive Catalogue of the British Specimens Deposited in the Geological Collection of the Royal Institution*. London.

Brande, William Thomas. 1817. *Outlines of Geology*. London. Rev. *Monthly Review* 85 (1818): 214–216; 96 (1821): 291.

Breislak, Scipio. 1811. *Introduzione alla geologia*. 2 vols. Milan. Rev. *Edinburgh Review* 27 (1816): 144–163.

[Brewster, David.] 1837. Review of William Buckland's *Geology and Mineralogy Considered with Reference to Natural Theology* (1836). *Edinburgh Review* 65:1–39.

Brotsky, Peter W. 1983. "Commentaries on the Huttonian Theory of the Earth from *Transactions of the Royal Society of Edinburgh*, 1805–1815." *Earth Sciences History* 2:28–34.

Brush, Stephen. 1978. *The Temperature of History: Phases of Science and Culture in the 19th Century*. New York.

Brydone, Patrick. 1773. *A Tour through Sicily and Malta*. 2 vols. London.

Buch, Leopold von. 1813. *Travels through Norway and Lapland during the Years 1806, 1807, and 1808*. Trans. John Black, with notes by Robert Jameson. London. Orig. pub. Berlin, 1810.

Buch, Leopold von. 1821. "Ueber die Zasammunseizzung der Basaltischen Inseln und uber Erhebungs-Kratere." In K. C. Leonhard, ed., *Taschenbuch fur die Gessammte Mineralogie* 15:391–427. Read 28 May 1818.

Buckland, William. 1820. *Vindiciae Geologiae; or, The Connexion of Geology with Religion Explained*. Oxford. Read 15 May 1819.

Buckland, William. 1821a. "Descriptions of the Quartz Rock of the Lickey Hill in Worcester, and of the Strata Immediately Surrounding It, with Considerations on the Evidence of a Recent Deluge." *Transactions of the Geological Society of London* 5:506–544.

Buckland, William. 1821b. "On the Structure of the Alps and Adjoining Parts of the Continent." *Annals of Philosophy* 1:450–468.

Buckland, William. 1823. *Reliquiae Diluvianae*. London. 2nd ed., 1824.

Buckland, William. 1824. "On the Excavation of Valleys by Diluvian Action." *Transactions of the Geological Society of London*, n.s. 1:95–102.

Buckland, William. 1829. "On the Formation of the Valley of Kingsclere and Other Valleys by the Elevations of the Strata That Enclose Them." *Transactions of the Geological Society of London*, n.s. 2:119–130.

Buckland, William. 1836. *Geology and Mineralogy Considered with Reference to Natural Theology*. 2 vols. London. 2nd ed., 1837.

Buffon, Georges Louis Leclerc, Comte de. 1749. *Histoire naturelle, générale et particulière*, vol. 1. Paris.

Burke, John G. 1974. "The Earth's Central Heat, from Fourier to Kelvin." *Actes du VIIIe Congres International d'Histoire des Sciences* (for 1971):91–96.

Burnet, Thomas. 1684–90. *The Theory of the Earth*. London.

Carosi, Jan Filip. 1783. *Sur la generation du silex et du quartz*. Cracow.

Carozzi, A. V. 1990. *Histoire des sciences de la terre entre 1790 et 1815, vue à travers les documents inédits de la Société de Physique et d'Histoire Naturelle de Genève*. Geneva.

Carr, John. 1809. "On the Natural Causes Which Operate in the Formation of Valleys." *Philosophical Magazine* 33:452–459.

Challinor, John. 1954. "The Early Progress of British Geology, III: From Hutton to Playfair, 1788–1802." *Annals of Science* 10:107–148.

Challinor, John. 1971. *The History of British Geology: A Bibliographical Study*. New York.

Charpentier, Jean de. 1836. "Account of One of the Most Important Results of the Investigations of M. Venetz Regarding the Present and Earlier Conditions of the Glaciers of the Canton Vallais." *Edinburgh New Philosophical Journal* 21:210–222.

Charpentier, Jean de. 1837. "Some Conjectures Regarding the Great Revolutions Which Have So Changed the Surface of Switzerland." *Edinburgh New Philosophical Journal* 22:27–36.
Chitnis, Anand C. 1970. "The University of Edinburgh's Natural History Museum and the Huttonian-Wernerian Debate." *Annals of Science* 26:85–94.
Chitnis, Anand C. 1976. *The Scottish Enlightenment*. London.
Chorley, R. J., A. J. Dunn, and R. P. Beckinsale. 1964. *The History of the Study of Landforms, or the Development of Geomorphology, I: Geomorphology before Davis*. Frome and London.
Clark, John W., and T. M. Hughes. 1890. *The Life and Letters of the Reverend Adam Sedgwick*. 2 vols. Cambridge.
Clerk of Eldin, John. 1855. *A Series of Etchings, Chiefly of Views in Scotland . . . 1773–1779*. Edinburgh.
Clow, Archibald, and Nan L. Clow. 1947. "Dr. James Hutton and the Manufacture of Sal Ammoniac." *Nature* 159:425–427.
Cockburn, Henry. 1856. *Memorials of His Time*. Edinburgh. Reprint, 1977.
Cockburn, Henry. 1910. "The Friday Club." In Robert Chambers, ed., *The Book of the Old Edinburgh Club*, vol. 3, pp. 106–130. Edinburgh.
Conybeare, William. 1816. "Descriptive Note Referring to the Outlines of Sections Presented by a Part of the Coasts of Antrim and Derry." *Transactions of the Geological Society of London* 3:196–216.
Conybeare, William. 1829. "On the Valley of the Thames." *Proceedings of the Geological Society of London* 1:145–149.
Conybeare, William. 1830. "On Mr. Lyell's *Principles of Geology*." *Philosophical Magazine*, n.s. 8:215–219.
Conybeare, William. 1830–31. "An Examination of Those Phenomena of Geology Which Seem to Bear Most Directly on Theoretical Speculations." *Philosophical Magazine*, n.s. 8:359–362, 401–406, 9:19–23, 111–117, 188–197, 258–270.
Conybeare, William, and William Phillips. 1822. *Outlines of the Geology of England and Wales*, pt. 1. London.
Cordier, Louis. 1816. "Sur les substances minérales, dites en masse, qui servent de base aus roches volcaniques." *Journal de Physique* 84:135–161, 285–307, 352–386.
Cordier, Louis. 1827. "Essai sur la temperature de l'intérieur de la terre." *Mémoires de l'Academie des Sciences* 7:473–556.
Craig, G. Y., ed. 1978. *James Hutton's Theory of the Earth: The Lost Drawings*. Edinburgh. The drawings, intended for vol. 3, are by Clerk of Eldin. Text by Craig, D. B. McIntyre, and C. D. Waterston.
Cronstedt, Alex Frederic. 1758. *Försök til Mineralogie, eller Mineralikets Upställning*. Stockholm. English trans. Gustav von Engeström, *Essay towards a System of Mineralogy*, London, 1770.
Cozens-Hardy, B., ed. 1950. *The Diary of Sylas Neville, 1767–1788*. Oxford.
Cunningham, R. J. H. *The Geology of the Lothians*. Edinburgh, 1839. Includes App. 1, "Sir James Hall's Experiments."
Cuvier, Georges. 1812. *Recherches sur le ossemens fossiles de quadrupèdes*. 4 vols. Paris. "Discours preliminaire," 1: 1–116, became in translation Cuvier, *Essay on the Theory of the Earth*.
Cuvier, Georges. 1813. *Essay on the Theory of the Earth*. Trans. Robert Kerr, with notes by Robert Jameson. Edinburgh. 2d ed., 1815; 3d, ed., 1817; 4th ed., 1822; 5th ed., 1827, with significant changes. Facsimile of 1813, 1971.
Cuvier, Georges, and Alexandre Brongniart. 1811. *Essai sur la géographie minéralogique des environs de Paris*. Paris. Also in Cuvier, *Recherches sur le ossemens fossiles*, 1: 1–278 (separately paginated).

Daiches, David, Peter Jones, and Jean Jones, eds. 1986. *A Hotbed of Genius: The Scottish Enlightenment, 1730–1790*. Edinburgh. Includes Daiches, "The Scottish Enlightenment"; P. Jones, "David Hume", D. D. Raphael, "Adam Smith"; R. G. W. Anderson, "Joseph Black"; J. Jones, "James Hutton"; and Archie Turnbull, "Scotland and America."

Darwin, Charles. 1838. "On the Connexion of Certain Volcanic Phenomena in South America, and on the Formation of Mountain Chains and Volcanos as the Effect of the Same Power by Which Continents Are Elevated." *Transactions of the Geological Society of London*, n.s. 5 (3):601–631.

Darwin, Erasmus. 1788. "Frigorific Experiments on the Mechanical Expansion of Air." *Philosophical Transactions of the Royal Society of London* 78:43–52. Read 13 December 1787.

Darwin, Erasmus. 1806. *Poetical Works*. 3 vols. London.

Daubeny, Charles. 1826. *A Description of Active and Extinct Volcanos*. London.

Davies, G. L. 1967. "George Toulmin and the Huttonian Theory of the Earth." *Bulletin of the Geological Society of America* 78: 121–123.

Davies, G. L. 1969. *The Earth in Decay: A History of British Geomorphology, 1578 to 1878*. London.

Davies, G. L. H. 1985. "James Hutton and the Study of Landforms." *Progress in Physical Geography* 9:382–389.

[Davy, Humphry]. 1811. "Mr. Davy's Lectures on Geology." *Philosophical Magazine* 37:392–398, 465–470.

Davy, Humphry. 1830. *Consolations in Travel, or the Last Days of a Philosopher*. London.

Davy, Humphry. 1980 [1805]. *Humphry Davy on Geology: The 1805 Lectures*. Ed. Robert Siegfried and R. H. Dott. Madison.

Davy, John, ed. 1858. *Framentary Remains Literary and Scientific . . . of Sir Humphrey Davy.* London.

Dean, Dennis R. 1973. "James Hutton and His Public, 1785–1802." *Annals of Science* 30:89–105.

Dean, Dennis R. 1975. "James Hutton on Religion and Geology: The Unpublished Preface to His 'Theory of the Earth' (1788)." *Annals of Science* 32:187–193.

Dean, Dennis R. 1979. "The Word 'Geology.'" *Annals of Science* 36:35–43.

Dean, Dennis R. 1980. "Graham Island, Charles Lyell, and the Craters of Elevation Controversy." *Isis* 71:571–588.

Dean, Dennis R. 1981. "The Age of the Earth Controversy: Beginnings to Hutton." *Annals of Science* 38:435–456.

Dean, Dennis R. 1983. "John Playfair and His Books." *Annals of Science* 40:179–187.

Dean, Dennis R. 1985. "The Rise and Fall of the Deluge." *Journal of Geological Education* 33:84–93.

Dean, Dennis R. 1989. "James Hutton's Place in the History of Geomorphology." In Keith J. Tinkler, ed., *History of Geomorphology*, pp. 73–84.

De la Beche, H. T. 1829. "Notice on the Excavation of Valleys." *Philosophical Magazine*, n.s. 6:241–248.

De la Beche, H. T. 1830a. "Notes on the Formation of Extensive Conglomerate and Gravel Deposits." *Philosophical Magazine and Annals of Philosophy*, n.s. 7:161–171.

De la Beche, H. T. 1830b. *Sections and Views Illustrative of Geological Phenomena*. London.

De la Beche, H. T. 1834. *Researches in Theoretical Geology*. London.

Deluc, Jean André. 1779. *Lettres physique et morales sur l'histoire de la terre et de l'homme*. 5 vols. Paris and The Hague.

Deluc, Jean André. 1790a. "First Letter to Dr. James Hutton." *Monthly Review*, n.s. 2:206–227.

Deluc, Jean André. 1790b. "Second Letter to Dr. James Hutton." *Monthly Review*, n.s. 2:582–601.
Deluc, Jean André. 1791a. "Third Letter to Dr. James Hutton." *Monthly Review*, n.s. 3:573–586.
Deluc, Jean André. 1791b. "Fourth Letter to Dr. James Hutton." *Monthly Review*, n.s. 5:564–585.
[Deluc, Jean André.] 1796. Review of James Hutton's *Theory of the Earth. British Critic* 8:337–352, 466–480, 596–608.
Deluc, Jean André. 1809a. *Traité élémentaire de géologie*. Paris. Rev. *Monthly Review* 63 (1810): 493–500.
Deluc, Jean André. 1809b. *An Elementary Treatise on Geology*. London.
Deluc, Jean André. 1810–11. *Geological Travels*. 3 vols. London.
Deluc, Jean André. 1813. *Geological Travels in Some Parts of France, Switzerland, and Germany*. 2 vols. London.
Desmarest, Nicolas. 1774. "Mémoire sur l'origine et la nature du basalte." *Mémoires de l'Academe Royale des Sciences* (for 1771), 705–775. Read 11 May 1771. In two parts, the first incorporating his original observations of 1763. A third part was published in 1777.
Desmarest, Nicolas. 1794. "Géographie physique." *Encyclopédie méthodique* 1:732–782. Paris.
Dolomieu, Déodat G. S. T. de 1783. *Voyage aux Iles de Lipari fait en 1781*. Paris.
Dolomieu, Déodat G. S. T. de 1784. *Mémoire sur les tremblemens de terre de la Calabre, pendant l'année 1783*. Rome.
Dolomieu, Déodat G. S. T. de 1788. *Mémoire sur les Iles Ponces, et catalogue raisonné des products de l'Etna*. Paris.
Dolomieu, Déodat G. S. T. 1791–92. "Mémoire sur les pierres composees et sur les roches." *Observations sur la Physique, sur l'Histoire Naturelle et sur les Arts* 39:374–407; 40:41–62, 203–218, 372–403.
Donovan, Arthur. 1975. *Philosophical Chemistry in the Scottish Enlightenment*. Edinburgh.
Donovan, Arthur. 1977. "James Hutton and the Scottish Enlightenment—Some Preliminary Considerations." *Scotia* 1:56–68.
Donovan, Arthur. 1978. "James Hutton, Joseph Black, and the Chemical Theory of Heat." *Ambix* 25:176–190.
Donovan, Arthur, and J. Prentiss, eds. 1980. James Hutton's *Dissertatio Physico-Medica* (Leiden, 1749), with translation and commentary.
Dott, Robert H. 1969. "James Hutton and the Concept of a Dynamic Earth." In Cecil J. Schneer, ed., *Toward a History of Geology*, pp. 122–141.
Ellenberger, François. 1972a. "La métaphysique de James Hutton (1726–1797) et le drame écologique de XXe siècle." *Revue de Synthèse* 3:267–283.
Ellenberger, François. 1972b. "Les origines de la pensée huttonienne: Hutton étudiant et docteur en médecine." *Comptes Rendus Hebdomadaire des Séances de l'Académie des Sciences* 275:69–72.
Ellenberger, François. 1972c. "De Bourguet à Hutton: Une source possible des themes huttoniens: Originalité irréductible de leur mise en oeuvre." *Comptes Rendus Hebdomadaire des Séances de l'Académie des Sciences* 275:93–96.
Ellenberger, François. 1973. "La thèse de doctorat de James Hutton et la rénovation perpetuelle de monde." *Annales Guebhard* 49:497–533.
Emerson, Roger L. 1988. "The Scottish Enlightenment and the End of the Philosophical Society of Edinburgh." *British Journal for the History of Science* 21: 33–66.
Esmark, Jens. 1827. "Remarks Tending to Explain the Geological History of the Earth." *Edinburgh New Philosophical Journal* 3:107–121.
Eyles, V. A. 1937. "John Macculloch, F.R.S., and His Geological Map: An Account of the First Geological Survey of Scotland." *Annals of Science* 2:114–129.

Eyles, V. A. 1947. "James Hutton (1726–1797) and Sir Charles Lyell (1797–1875)." *Nature* 160:694–695.

Eyles, V. A. 1950. "Note on the Original Publication of Hutton's *Theory of the Earth*, and on the Subsequent Forms in Which It Was Issued." *Proceedings of the Royal Society of Edinburgh* 63(B):377–386.

Eyles, V. A. 1953. "Bibliography and the History of Science." *Journal of the Society for the Bibliography of Natural History* 3:63–71.

Eyles, V. A. 1955. "A Bibliographical Note on the Earliest Printed Version of James Hutton's 'Theory of the Earth,' Its Form and Date of Publication." *Journal of the Society for the Bibliography of Natural History* 3:105–108.

Eyles, V. A. 1961. "Sir James Hall, Bt. (1761–1832)." *Endeavour* 20:210–216.

Eyles, V. A. 1963. "The Evolution of a Chemist: Sir James Hall." *Annals of Science* 19:153–182.

Eyles, V. A. 1969. "The Extent of Geological Knowledge in the Eighteenth Century." In Cecil J. Schneer, ed., *Toward a History of Geology*, pp. 159–183.

Eyles, V. A., and Joan M. Eyles. 1951. "Some Geological Correspondence of James Hutton." *Annals of Science* 7:316–339.

Farey, John. 1809. "Observations on a Late Paper by Dr. Wm. Richardson Respecting the Basaltic District in the North of Ireland." *Philosophical Magazine* 33:257–263.

Faujas de Saint-Fond, Barthelemy. 1778. *Recherches sur les volcans eteints du Vivarais et du Velay*. Grenoble and Paris.

Faujas de Saint-Fond, Barthelemy. 1907. *A Journey through England and Scotland to the Hebrides in 1784*. Ed. Sir Archibald Geikie. 2 vols. Glasgow. Orig. pub. Paris, 1797; trans. 1799.

Ferguson, Adam. 1805. "Minutes of the Life and Character of Dr. Joseph Black." *TRSE* 5 (3):101–117. Read 3 August 1801.

Fisher, George P. 1866. *Life of Benjamim Silliman*. 2 vols. New York.

[Fitton, W. H.] 1817. Review of the *Transactions of the Geological Society of London*, vol. 3. *Edinburgh Review* 29:70–94.

[Fitton, W. H.] 1818. Review of six publications by William Smith. *Edinburgh Review* 29:310–337.

[Fitton, W. H.] 1823. Review of William Buckland's *Reliquiae Diluvianae*. *Edinburgh Review* 39:196–234.

Fitton, W. H. 1828. Presidential Address. *Proceedings of the Geological Society of London* 1:50–62. Read 15 February 1828.

Fitton, W. H. 1829. Presidential Address. *Proceedings of the Geological Society of London* 1:112–134. Read 20 February 1829.

Fitton, W. H. 1832–33. "Notes on the History of English Geology." *Philosophical Magazine*, n.s. 1:147–160, 268–275, 442–450; 2:37–57.

[Fitton, W. H.] 1839. "Huttonian Theory of the Earth" (review of Charles Lyell's *Elements of Geology*, 1838). *Edinburgh Review* 69:406–466.

Fleming, John. 1816a. "Observations on the Mineralogy of the Neighborhood of St. Andrew's in Fife." *Memoirs of the Wernerian Natural History Society* 2:145–154. Read 5 February 1813.

Fleming, John. 1816b. "On the Mineralogy of the Redhead in Angusshire." *Memoirs of the Wernerian Natural History Society* 2:339–369. Read 4 February 1815.

[Fleming, John.] 1823. Review of Cuvier's *Essay on the Theory of the Earth*, 4th ed. *New Edinburgh Review* 4:381–398.

Fleming, John. 1824. "Remarks Illustrative of the Influence of Society in the Distribution of British Animals." *Edinburgh Philosophical Journal* 11:287–305.

Fleming, John. 1825. "Remarks on the Modern Strata." *Edinburgh Philosophical Journal* 12:116–127.

Fleming, John. 1826. "The Geological Deluge, as Interpreted by Baron Cuvier and Professor Buckland, Inconsistent with the Testimony of Moses and the Phenomena of Nature." *Edinburgh Philosophical Journal* 14:205–239.
Flett, John S. 1939. "Pioneers of British Geology." *Journal and Proceedings of the Royal Society of New South Wales* 73:41–66.
Flinn, Derek. 1980. "James Hutton and Robert Jameson." *Scottish Journal of Geology* 16:251–278.
Forsyth, Robert. 1805–8. *The Beauties of Scotland.* 5 vols. Edinburgh.
Fourier, Joseph. 1819. "Mémoire sur le refroidissement séculaire du globe terrestre." *Annales de Chimie et de Physique* 13:418–437.
Fourier, Joseph. 1820. "Le refroidissement séculaire du globe terrestre." *Bulletin des Sciences par la Societe Philomatique de Paris*, pp. 58–70.
Fourier, Joseph. 1824. "Remarques générales sur les températures du globe terrestre et des espaces planétaires." *Annales de Chimie et de Physique* 27:136–167.
Fourier, Joseph. 1827. "Mémoire sur les températures du globe terrestre et des espace planétaires." *Memoire de l'Academie Royale des Sciences de l'Institut de France* 7:570–604.
Freshfield, D. W., and H. F. Mantagnier. 1920. *The Life of Horace Benedict de Saussure.* London.
Geikie, Archibald. 1865. *The Scenery of Scotland Viewed in Connexion with Its Physical Geology.* London and Cambridge. 2d ed., 1887.
Geikie, Archibald. 1868. "On Modern Denudation." *Transactions of the Geological Society of Glasgow* 3:153–190. Abstracted in *Geological Magazine* 5 (1868): 249–254. Read 26 March 1868.
Geikie, Archibald. 1873. "Introductory Address: Earth Sculpture and the Huttonian School of Geology." *Transactions of the Edinburgh Geological Society* 2:247–268. Read 6 November 1873.
Geikie, Archibald. 1875. *Life of Sir Roderick I. Murchison.* 2 vols. London.
Geikie, Archibald. 1882a [1871]. "The Scottish School of Geology." In *Geological Sketches at Home and Abroad*, pp. 286–311. London. Read 6 November 1871.
Geikie, Archibald. 1882b. *A Text-Book of Geology.* London. Later editions, 1885, 1893, 1903.
Geikie, Archibald. 1895. *Memoir of Sir Andrew Crombie Ramsay.* London.
Geikie, Archibald. 1897. *The Ancient Volcanoes of Great Britain.* 2 vols. London.
Geikie, Archibald. 1905a [1897]. *The Founders of Geology*, 2d ed. London. Reprint, 1962.
Geikie, Archibald. 1905b. *Landscape in History, and Other Essays.* London. Includes "The Origins of the Scenery of the British Islands" (pp. 130–157), an abstract of five lectures, 29 January–3 March 1884; "The Centenary of Hutton's 'Theory of the Earth'" (pp. 158–197), Presidential address, British Association for the Advancement of Science, 3 August 1892; and "Geological Time" (pp. 198–233), Presidential address, British Association for the Advancement of Science, 16 September 1899.
Geikie, Archibald. 1924. *A Long Life's Work: An Autobiography.* London.
Generelli, G. C. 1749. *Dissertazione de' crostacei, e dell' altre produzioni che sono ne monti.* Milan.
Gerstner, Patsy. 1968. "James Hutton's *Theory of the Earth* and His Theory of Matter." *Isis* 69:26–31.
Gerstner, Patsy. 1971. "The Reaction to James Hutton's Use of Heat as a Geological Agent." *British Journal for the History of Science* 5:353–362.
Gillispie, Charles C. 1951. *Genesis and Geology: A Study in the Relations of Scientific Thought, Natural Theology, and Social Opinion in Great Britain, 1790–1850.* Cambridge, Mass. Reprint, 1959.

Gillispie, Charles C., ed. 1970–80. *Dictionary of Scientific Biography*. 16 vols. New York.

Gilmer, Francis W. 1818. "The Natural Bridge of Virginia a Result of Erosion." *Transactions of the American Philosophical Society*, n.s. 1:187–192.

Gould, Stephen Jay. 1965. "Is Uniformitarianism Necessary?" *American Journal of Science* 263:223–228, 919–921.

Gould, Stephen Jay. 1987. *Time's Arrow, Time's Cycle: Myth and Metaphor in the Discovery of Geological Time*. Cambridge, Mass.

Graham, Maria. 1824. "An Account of Some Effects of the Late Earthquakes in Chili." *Transactions of the Geological Society of London*, n.s. 1:413–415.

Grant, R. 1979. "Hutton's Theory of the Earth." In L. J. Jordanova and R. S. Porter, eds., *Images of the Earth—Essays in the History of the Environmental Sciences*, pp. 23–38. Chalfont St. Giles, Bucks.

Grattan-Guinness, Ivor. 1972. *Joseph Fourier, 1768–1830*. Cambridge, Mass.

Greene, Mott T. 1982. *Geology in the Nineteenth Century: Changing Views of a Changing World*. Ithaca and London.

Greenough, G. B. 1819. *A Critical Examination of the First Principles of Geology*. London. Facsimile reprint, 1978.

Greenough, G. B. 1834. "Remarks on the Theory of the Elevation of Mountains." *Edinburgh New Philosophical Journal* 17:205–227.

Greenwood, George. 1857. *Rain and Rivers; or Hutton and Playfair against Lyell and All Comers*. London.

Grierson, James. 1826. "General Observations on Geology and Geognosy, and the Nature of These Respective Studies." *Memoirs of the Wernerian Natural History Society* 5:401–410.

Haber, Francis C. 1959. *The Age of the World: Moses to Darwin*. Baltimore.

Hall, Captain Basil. 1815. "Account of the Structure of the Table Mountain, and Other Parts of the Peninsula of the Cape." Drawn up by Professor Playfair. *TRSE* 7 (2):269–278. Read 31 May 1813.

Hall, Sir James. 1794. "Observations on the Formation of Granite." *TRSE* 3 (1):8–12. Read 4 January and 1 March 1790.

Hall, Sir James. 1805. "Experiments on Whinstone and Lava." *TRSE* 5 (1):43–75. Read 5 March and 18 June 1798. Abstracted in William Nicholson's *Journal of Natural Philosophy, Chemistry, and the Arts* 2 (October 1798): 285–288; published fully in the same, 4 (April-May 1800): 8–18, 56–65. Full publication in Nicholson's generally means that the *TRSE* separate was available.

Hall, Sir James. 1812. "Account of a Series of Experiments, Showing the Effects of Compression in Modifying the Action of Heat." *TRSE* 6:71–185. Read 3 June 1805. Abstracted in *Journal of Natural Philosophy, Chemistry, and the Arts* 9 (1804): 98–107; published fully in the same, 13 (1806): 328–343, 381–405; 14 (1806): 13–22, 113–128, 196–212, 302–318 (including an appendix not found in *TRSE*). A French translation (by M. A. Pictet, Geneva, 1807) also preceded the official version.

Hall, Sir James. 1815a. "On the Vertical Position and Convolutions of Certain Strata, and Their Relation with Granite." *TRSE* 7 (1):79–108. Read 3 February 1812.

Hall, Sir James. 1815b. "On the Revolutions of the Earth's Surface, Parts I and II." *TRSE* 7 (1):139–211. Read 16 March and 8 June 1812.

Hall, Sir James. 1826. "On the Consolidation of the Strata of the Earth." *TRSE* 10 (2):314–329. Read 4 April 1825.

Hallam, A. 1973. *A Revolution in the Earth Sciences*. Oxford.

Hamilton, Sir William. 1772. *Observations on Mount Vesuvius*. London.

Hassler, Donald M. 1986. "The Scottish Reasoning of James Hutton." *Studies in Scottish Literature* 21:35–42.

Hatch, Ronald B. 1975. "'Philosophy' and 'Science.'" *Notes and Queries* 22:24–25.
[Headrick, James.] 1804. Review of Robert Jameson's *System of Mineralogy*, vol. 1. *Edinburgh Review* 5:64–78.
[Headrick, James.] 1805. Review of Robert Jameson's *Mineralogy of Dumfriesshire. Edinburgh Review* 6:229–245.
Headrick, James. 1807. *View of the Mineralogy, Agriculture, Manufactures, and Fisheries of the Island of Arran*. Edinburgh.
Heimann, P. M., and J. E. McGuire. 1971. "Newtonian Forces and Lockean Powers: Concepts of Matter in Eighteenth-Century Thought." In Russell McCormmach, ed., *Historical Studies in the Physical Sciences* (third annual volume), pp. 233–306. Philadelphia.
Herivel, John. 1975. *Joseph Fourier: The Man and the Physicist*. Oxford.
Hobbs, William H. 1926. "James Hutton, the Pioneer of Modern Geology." *Science* 64:265.
Hoff, Karl E. A. von. 1822–41. *Geschichte der durch Ueberlieferung nachgewiesinen naturlichen Veranderungen der Erdoberflache*. 5 vols. Gotha. vol. 1, 1822; vol. 2, 1824; vol. 3, 1834; vol. 4, 1840; vol. 5, 1841.
Holbach, Baron P. H. D. d'. 1770. *Système de la nature, ou les loix du monde physique et du monde moral*. 2 vols. London.
Hooykaas, Reijer. 1959. *Natural Law and Divine Miracle: A Historical-Critical Study of the Principle of Uniformity in Geology, Biology, and Theology*. Leiden. 2d ed., 1963.
Hooykaas, Reijer. 1966a. "James Hutton und die Ewigkeit der Welt." *Gesnerus* 23:55–66.
Hooykaas, Reijer. 1966b. "Geological Uniformitarianism and Evolution." *Archives Internationales d'Histoire des Sciences* 19:3–19.
Hooykaas, Reijer. 1970. *Catastrophism in Geology*. Amsterdam.
Horner, Francis. 1853. *Memoirs and Correspondence*. 2 vols. Ed. Leonard Horner. Boston.
Howard, Philip. 1797. [*Thoughts on the Structure of the Globe and] The Scriptural History of the Earth and Mankind*. London. Rev. *Monthly Review* 25 (1798): 241–254.
Hume, David. 1779. *Dialogues Concerning Natural Religion*. London.
Huxley, Thomas H. 1909 [1869]. "Geological Reform." In *Discourses, Biological and Geological Essays*, pp. 307–342. New York.
Imbrie, John, and Katherine P. Imbrie. 1986 [1979]. *Ice Ages: Solving the Mystery*. Cambridge, Mass.
Imrie, Lt. Col. [Ninian]. 1816. "A Geological Account of the Southern District of Stirlingshire, Commonly Called the Campsie Hills, with a Few Remarks Relative to the Two Prevailing Theories as to Geology." *Memoirs of the Wernerian Natural History Society* 2:24–50. Read 1 February 1812.
Jameson, Robert. 1798. *An Outline of the Mineralogy of the Shetland Islands, and of the Island of Arran*. Edinburgh.
Jameson, Robert. 1800. *Mineralogy of the Scottish Isles*. 2 vols. Edinburgh. German translation, Leipzig, 1802. 2d Edinburgh ed. (as *Mineralogical Travels*), 1813.
Jameson, Robert. 1802a. "On Granite." *Journal of Natural Philosophy, Chemistry, and the Arts* 2:225–233.
Jameson, Robert. 1802b. "On the Supposed Existence of Mechanical Deposits and Petrifactions in the Primitive Mountains, and an Account of Petrifactions Which Have Been Discovered in the Newest Flötz Trapp Formation." *Journal of Natural Philosophy, Chemistry, and the Arts* 3:13–20.
Jameson, Robert. 1802c. "Examination of the Supposed Igneous Origin of the Rocks of the Trapp Formations." *Journal of Natural Philosophy, Chemistry, and the Arts* 3:111–119.

Jameson, Robert. 1804–8. *System of Mineralogy*. 3 vols. Edinburgh. Vols. 2 and 3 had later editions. Facsimile reprint, vol. 3 only, 1976.

Jameson, Robert. 1827. "General Observations on the Former and Present Geological Conditions of the Countries Discovered by Captains Parry and Ross." *Edinburgh New Philosophical Journal* 2:104–106.

Jameson, Robert. 1833. "Chemical Analyses of Stratified Rocks Altered by Plutonian Agency." *Edinburgh New Philosophical Journal* 15:386–389.

Jameson, Robert. 1892. "Is the Huttonian Theory of the Earth Consistent with Fact?" In Royal Medical Society of Edinburgh, *Dissertations by Eminent Members of the Royal Medical Society*, pp. 32–39. Edinburgh. Read 1796.

Jameson, Robert. 1967. "Is the Volcanic Opinion of the Formation of Basaltes Founded in Truth?" In Jessie M. Sweet and Charles D. Waterston, "Robert Jameson's Approach to the Wernerian Theory of the Earth," pp. 93–95. Read 1796.

Jefferson, Thomas. 1787. *Notes on the State of Virginia*. London. A private edition appeared in 1782.

[Jeffrey, Francis.] 1802. Review of John Playfair's *Illustrations of the Huttonian Theory* (1802). *Edinburgh Review* 1:201–216.

[Jeffrey, Francis.] 1803. Review of John Murray's *A Comparative View* (1802). *Edinburgh Review* 2:337–348.

Jeffrey, Francis. 1822. On John Playfair. In John Playfair, *Works*, vol. 1, pp. lxii–lxxvi. Also published separately.

Jones, Jean. 1982. "James Hutton and the Forth and Clyde Canal." *Annals of Science* 39:255–263.

Jones, Jean. 1983. "James Hutton: Exploration and Oceanography." *Annals of Science* 40:81–94.

Jones, Jean. 1984. "The Geological Collection of James Hutton." *Annals of Science* 41:223–244.

Jones, Jean. 1985. "James Hutton's Agricultural Research and His Life as a Farmer." *Annals of Science* 42:573–601.

Jones, Peter. 1984. "An Outline of the Philosophy of James Hutton." In V. Hope, ed., *Philosophers of the Scottish Enlightenment*, pp. 182–210. Edinburgh.

Jukes, J. Beete. 1862. "On the Mode of Formation of Some of the River Valleys in the South of Ireland." *Quarterly Journal of the Geological Society of London* 18:378–403.

Kay, John. 1837. *A Series of Original Portraits*, with anecdotes by H. Paton. 2 vols. Edinburgh.

Keir, James. 1776. "On the Crystallizations Observed on Glass." *Philosophical Transactions of the Royal Society of London* 66:530–542. Read 23 May 1776.

Kelvin, William Thomson, Lord. 1862. "On the Secular Cooling of the Earth." *Philosophical Magazine*, n.s. 25:1–14.

Kelvin, William Thomson, Lord. 1889–94. *Popular Lectures and Addresses*. 3 vols. London and New York. Includes "On the Age of the Sun's Heat" (1862), "The Doctrine of Uniformity in Geology Briefly Refuted" (1865), "On Geological Time" (1868), and "Of Geological Dynamics" (1869). All but the first are in vol. 2.

Kennedy, Robert. 1805. "A Chemical Analysis of Three Species of Whinstone and Two of Lava." *TRSE* 5 (1):76–98. Read 3 December 1798. Published fully in *Journal of Natural Philosophy, Chemistry, and the Arts* 4 (December 1800–January 1801): 407–415, 438–442.

Kidd, John. 1818. *A Geological Essay on the Imperfect Evidence in Support of a Theory of the Earth*. London. Rev. *Monthly Review* 85 (1818): 375–386.

Kirwan, Richard. 1793. "Examination of the Supposed Igneous Origin of Stony Substances." *Transactions of the Royal Irish Academy* 5:51–87. Read 3 February 1793.

Kirwan, Richard. 1794. *Elements of Mineralogy,* vol. 1. 2d ed. London. The second volume appeared in 1796.
Kirwan, Richard. 1797. "On the Primitive State of the Globe and Its Subsequent Catastrophe." *Transactions of the Royal Irish Academy* 6:233–308. Read 19 November 1796.
Kirwan, Richard. 1799. *Geological Essays.* London. Facsimile reprint, 1978.
Kirwan, Richard. 1800. "Observations on the Proofs of the Huttonian Theory of the Earth Adduced by Sir James Hall, Bart." *Journal of Natural Philosophy, Chemistry, and the Arts* 4:97–102, 153–158.
Kirwan, Richard. 1802a. "Observations on the Proofs of the Huttonian Theory Adduced by Sir James Hall, Bart." *Transactions of the Royal Irish Academy* 8:3–27. Read 8 February 1800.
Kirwan, Richard. 1802b. "An Illustration and Confirmation of Some Facts Mentioned in an Essay on the Primitive State of the Globe." *Transactions of the Royal Irish Academy* 8:29–34. Read 5 May 1800. Also in *Philosophical Magazine* 14 (1802): 14–17.
Kirwan, Richard. 1802c. "An Essay on the Declivities of Mountains." *Transactions of the Royal Irish Academy* 8:35–52. Read 28 April 1800.
Kirwan, Richard. 1802d. "A Reply to Mr. Playfair's Reflections on Mr. Kirwan's Refutation of the Huttonian Theory of the Earth." *Philosophical Magazine* 14 (1802), 3–13.
Knight, William. 1818. *Facts and Observations towards Forming a New Theory of the Earth.* Edinburgh.
Knight, William. 1900. *Lord Monboddo and Some of His Contemporaries.* London.
Lametherie, Jean Claude de. 1795. *Theorie de la terre.* 3 vols. Paris.
Laudan, Rachel. 1979. "The Problem of Consolidation in the Huttonian Tradition." *Lychnos* (1977–78): 195–206.
Laudan, Rachel. 1987. *From Mineralogy to Geology: The Foundations of a Science, 1650–1830.* Chicago and London.
Lawrence, Philip. 1977. "Heaven and Earth—The Relation of the Nebular Hypothesis to Geology." In Wolfgang Yourgrau and Allen D. Breck, eds., *Cosmology, History, and Theology,* pp. 253–281. New York and London.
Lehmann, Johann Gottlob. 1756. *Versuch einer Geschichte von Flötz-Gebürgen.* Berlin.
[Leslie, John.] 1813. Review of Leopold von Buch's *Travels through Norway and Lapland* (1813). *Edinburgh Review* 22:145–178.
Lubbock, Constance A., ed. 1933. *The Herschel Chronicle: The Life-Story of William Herschel and His Sister, Caroline Herschel.* Cambridge.
Lyell, Charles. 1825. "On a Dike of Serpentine Cutting through Sandstone." *Edinburgh Journal of Science* 3:112–126. Read May 1825.
Lyell, Charles. 1826a. "On the Strata of the Plastic Clay Formation." *Transactions of the Geological Society of London,* n.s. 2:287–292. Read 17 March 1826.
[Lyell, Charles.] 1826b. Review of the *Transactions of the Geological Society of London,* n.s. 1. *Quarterly Review* 34:507–540.
[Lyell, Charles.] 1827. Review of G. P. Scrope's *Memoir on the Geology of Central France. Quarterly Review* 36:437–483.
Lyell, Charles. 1829. "On a Recent Formation of Freshwater Limestone in Forfarshire." *Transactions of the Geological Society of London,* n.s. 2:73–96. Read 17 December 1824.
Lyell, Charles. 1830. *Principles of Geology,* vol. 1. London.
Lyell, Charles. 1832. *Principles of Geology,* vol. 2. London.
Lyell, Charles. 1833. *Principles of Geology,* vol. 3. London. Vols. 1–3 reprinted in facsimile, 1970.

Lyell, Charles. 1832–33. *Principles of Geology*, 2d ed. 3 vols. London.

Lyell, Charles. 1834. *Principles of Geology*, 3d ed. 4 vols. London.

Lyell, Charles. 1835. *Principles of Geology*, 4th ed. 4 vols. London. Other editions, 1837, 1840, 1847, 1850, 1853, 1867, 1872, 1875.

Lyell, Charles. 1838. *Elements of Geology*. London. Later editions were variously titled.

Lyell, Charles, and Roderick Murchison. 1829. "On the Excavation of Valleys, as Illustrated by the Volcanic Rocks of Central France." *Edinburgh New Philosophical Journal* 7:15–48.

Lyell, Katherine M., ed. 1881. *Life, Letters, and Journals of Sir Charles Lyell, Bart.* 2 vols. London. Facsimile reprint, 1970.

McCartney, Paul J. 1977. *Henry De la Beche: Observations on an Observer*. Cardiff.

McCosh, James. 1875. *The Scottish Philosophy*. New York.

Macculloch, John. 1814a. "On the Sublimation of Silica." *Transactions of the Geological Society of London* 2:275–276.

Macculloch, John. 1814b. "On the Junction of Trap and Sandstone at Stirling Castle." *Transactions of the Geological Society of London* 2:305–308.

Macculloch, John. 1814c. "Miscellaneous Remarks Accompanying a Catalogue of Specimens." *Transactions of the Geological Society of London* 2:388–449.

Macculloch, John. 1814d. "Remarks on Several Parts of Scotland Which Exhibit Quartz Rock." *Transactions of the Geological Society of London* 2:450–487.

Macculloch, John. 1814e. "On Staffa." *Transactions of the Geological Society of London* 2:501–509.

Macculloch, John. 1816a. "A Sketch of the Mineralogy of Skye." *Transactions of the Geological Society of London* 3:1–111.

Macculloch, John. 1816b. "A Geological Description of Glen Tilt." *Transactions of the Geological Society of London* 3:259–337.

Macculloch, John. 1819. *A Description of the Western Islands of Scotland, Including the Isle of Man*. 3 vols. London. Rev. (Leonard Horner) *Edinburgh Review* 33 (May 1820): 442–470.

Macculloch, John. 1823. "On Certain Elevations of Land Connected with the Actions of Volcanoes." *Quarterly Journal of Science, Literature, and the Arts* 14:262–295.

Macculloch, John. 1824. *The Highlands and Western Isles of Scotland*. 4 vols. Edinburgh.

[Macculloch, John.] 1826. Review of G. P. Scrope's *Considerations on Volcanos* (1825). *Westminster Review* 5:356–373.

[Macculloch, John.] 1827. Review of Charles Daubeny's *A Description of Active and Extinct Volcanos* (1826). *Edinburgh Review* 45:295–320.

Macculloch, John. 1831. *A System of Geology, with a Theory of the Earth and an Explanation of Its Connexion with the Sacred Records*. 2 vols. London.

McElroy, David B. 1969. *Scotland's Age of Improvement: A Survey of Eighteenth-Century Literary Clubs and Societies*. Pullman, Wash.

Macgregor, Murray. 1950. "Life and Times of James Hutton." In *Proceedings of the Royal Society of Edinburgh* 63(B):351–356.

McIntyre, D. B. 1963. "James Hutton and the Philosophy of Geology." In Claude C. Albritton, ed., *Fabric of Geology*, pp. 1–11.

Mackenzie, Sir George Steuart. 1811. *Travels in the Island of Iceland, during the Summer of the Year MDCCCX*. Edinburgh.

Mackenzie, Sir George Steuart. 1815. "An Account of Some Geological Facts Observed in the Faroe Islands." *TRSE* 7 (1):213–228.

McKie, Douglas. 1936. "On Thomas Cochrane's MS Notes of Black's Chemical Lectures, 1767/8." *Annals of Science* 1:101–110.

McKie, Douglas. 1965. "On Some Manuscript Copies of Black's Chemical Lectures." *Annals of Science* 21:209–255.

McKie, Douglas, ed. 1966. *Notes from Doctor Black's Lectures on Chemistry, 1767/8,* by Thomas Cochrane. Cheshire.

Maclure, William. 1817. *Observations on the Geology of the United States.* Philadelphia. Fascimile reprint, 1966.

McPhee, John. 1981. *Basin and Range.* New York.

Maillet, Benoit de. 1748. *Telliamed, ou entretiens d'un philosophe indien avec un missionnaire francois sur la diminution de la mer, la formation de la terre, l'origine de l' homme,* etc. Amsterdam. English translations: London, 1750; Urbana, Chicago, and London, 1968, ed. Albert V. Carozzi, with extensive commentary.

Mantell, Gideon. 1822. *The Fossils of the South Downs.* London.

Martin, P. J. 1828. *A Geological Memoir on a Part of Western Sussex.* London.

Mather, Kirtley F., and Shirley L. Mason. 1970 [1939]. *A Source Book in Geology, 1400–1900.* Cambridge, Mass.

Memoirs of Science and the Arts, vol. 1. 1793. Deptford.

Merrill, George P. 1969 [1924]. *The First One Hundred Years of American Geology.* New York and London.

Middleton, W. E. Knowles. 1965. *A History of the Theories of Rain.* London.

Mills, Abraham. 1790. "Some Account of the Strata and Volcanic Appearances in the North of Ireland and Western Islands of Scotland." *Philosophical Transactions of the Royal Society of London* 80:73–100. Read 21 January 1790.

Moro, Antonio Lazzaro. 1740. *De crostacei e degli altri marini corpi che si truovano su' monti.* Venice.

Morrell, J. B. 1971. "Professors Robison and Playfair, and the Theophobia Gallica: Natural Philosophy, Religion, and Politics in Edinburgh, 1789–1815." *Notes and Records of the Royal Society of London* 26:43–64.

Murray, John. 1802. *A Comparative View of the Huttonian and Neptunian Systems of Geology: In Answer to the Illustrations of the Huttonian Theory of the Earth by Professor Playfair.* Edinburgh. Facsimile reprint, 1978.

Murray, John. 1806–7. *System of Chemistry.* 4 vols. Edinburgh.

Murray, John. 1815. "On the Diffusion of Heat at the Surface of the Earth." *TRSE* 7 (2):411–434.

Oldham, Thomas. 1859. "On the Geological Structure of a Portion of the Khasi Hills, Bengal." *Memoirs of the Geological Survey of India* 1:99–210.

Oldroyd, D. R. 1974. "Some Phlogistic Mineralogical Schemes." *Annals of Science* 31:269–305.

Oldroyd, D. R. 1980. "Sir Archibald Geikie (1835–1924), Geologist, Romantic Aesthete, and Historian of Geology: The Problem of Whig Historiography of Science." *Annals of Science* 37:441–462.

Olson, Richard. 1969. "The Reception of Boscovich's Ideas in Scotland." *Isis* 60:91–103.

Olson, Richard. 1970. "Count Rumford, Sir John Leslie, and the Study of the Nature and Propagation of Heat at the Beginning of the Nineteenth Century." *Annals of Science* 26:273–304.

Olson, Richard. 1975. *Scottish Philosophy and British Physics, 1750–1880.* Princeton.

O'Rourke, J. E. 1978. "A Comparison of James Hutton's *Principles of Knowledge* and *Theory of the Earth.*" *Isis* 69:5–20.

Ospovat, Alexander M. 1967. "The Place of the *Kurze Klassifikation* in the Work of A. G. Werner." *Isis* 58:90–95.

Ospovat, Alexander M. 1974. "The Work and Influence of Abraham Gottlob Werner: A Reevaluation." *Actes du XIIIe Congres International d'Histoire des Sciences* (for 1971), 8:123–130.

Ospovat, Alexander M. 1976. "The Distortion of Werner in Lyell's *Principles of Geology.*" *British Journal for the History of Science* 9:190–198.

Page, Leroy E. 1968. "John Playfair and Huttonian Catastrophism." *Actes du XLe Congrès International d'Histoire des Sciences* (for 1965), 4:221–225.

Page, Leroy E. 1969. "Diluvialism and Its Critics in Great Britain in the Early Nineteenth Century." In Cecil J. Schneer, ed., *Toward a History of Geology*, pp. 257–271.

Pallas, Pierre Simon. 1777. *Observations sur la formation des montagnes*. St. Petersburg.

Parkinson, James. 1804. *Organic Remains of a Former World*, vol. 1. London. Additional volumes appeared in 1808 and 1811, with reprints of all. Facsimile reprint, 1970.

Pennant, Thomas. 1774. *A Tour in Scotland*. London.

Pestana, Harold R. 1979. "Rees's *Cyclopaedia* (1802–1820): A Sourcebook for the History of Geology." *Journal of the Society for the Bibliography of Natural History* 9:353–361.

Phillips, John. 1829. *Illustrations of the Geology of Yorkshire*. York. A second volume appeared in 1836, London.

Phillips, John. 1844. *Memoirs of William Smith*. London.

Phillips, William. 1818. *A Selection of Facts from the Best Authorities, Arranged so as to Form an Outline of the Geology of England and Wales*. London.

Playfair, John. 1794. Citing a remark of Hutton's. *TRSE* 3 (2):245.

Playfair, John. 1802. *Illustrations of the Huttonian Theory of the Earth*. Edinburgh. Facsimile reprints, 1956, 1964; 1975 (partial). Also in Playfair, *Works*, 1: 1–514. The French translation by C. A. Basset (Paris, 1815) includes J. Murray's *Comparative View* and a few comments.

Playfair, John. 1805a. "Investigation of Theorems Relating to the Figure of the Earth." *TRSE* 5 (1)3–30. Read 1797.

Playfair, John. 1805b. "Biographical Account of the Late Dr. James Hutton." *TRSE* 5 (3):39–99. Facsimile reprint, 1970. Also in Playfair, *Works*, 4: 33–118. Read 10 Jan 1803.

[Playfair, John.] 1806. Review of Sir James Hall on heat and compression. *Edinburgh Review* 9:19–31.

[Playfair, John.] 1811a. Review of Charles Anderson's translation (1809) of Werner on veins. *Edinburgh Review* 18:80–97.

[Playfair, John.] 1811b. Review of Cuvier on fossil bones. *Edinburgh Review* 18:214–230.

[Playfair, John.] 1811c. Review of the *Transactions of the Geological Society of London*, vol. 1. *Edinburgh Review* 19:207–229.

[Playfair, John.] 1812a. Review of Sir George Steuart Mackenzie's *Travels in Iceland* (1811). *Edinburgh Review* 19:416–435.

Playfair, John. 1812b. "On the Progress of Heat When Communicated to Spherical Bodies from Their Centers." *TRSE* 6:353–370. Read 6 March 1809.

[Playfair, John.] 1812c. Review of Cuvier and Brongniart on the Environs of Paris. *Edinburgh Review* 20:369–386.

[Playfair, John.] 1814. Review of Cuvier's *Theory of the Earth* (1813). *Edinburgh Review* 22:454–475.

Playfair, John. 1816. "A General View of the Progress of Mathematical and Physical Science since the Revival of Letters in Europe." *Encyclopedia Britannica: Supplement* 2:1–127.

Playfair, John. 1822. *Works*. Ed. James G. Playfair. 4 vols. Edinburgh.

Porter, Roy. 1977. *The Making of Geology: Earth Science in Britain, 1660–1815*. Cambridge.

Porter, Roy. 1978a. "George Hoggart Toulmin and James Hutton: A Fresh Look." *Geological Society of America Bulletin* 89:1256–1258.

Porter, Roy. 1978b. "George Hoggart Toulmin's Theory of Man and the Earth in the Light of the Development of British Geology." *Annals of Science* 35:339–352.

Porter, Roy. 1978c. "Philosophy and Politics of a Geologist: G. H. Toulmin (1754–1817)." *Journal of the History of Ideas* 39:435–450.

Porter, Roy. 1979. "Creation and Credence: The Career of Theories of the Earth in Britain, 1660–1820." In B. Barnes and S. Shapin, eds., *Natural Order*, pp. 97–124. Beverly Hills, Calif.

Porter, Roy. 1983. *The Earth Sciences: An Annotated Bibliography*. New York and London.

Prevost, Constant. 1825. "De la formation des terrains des environs de Paris." *Bulletin, Societe Philomathique de Paris*, 74–77, 88–90.

Ramsay, Andrew C. 1841. *The Geology of the Island of Arran*. Glasgow.

Ramsay, Andrew C. 1846. "On the Denudation of South Wales and the Adjacent Counties of England." *Memoirs of the Geological Survey of Great Britain* 1:297–335.

Ramsay, Andrew C. 1850. "On the Geological Phenomena That Have Produced or Modified the Scenery of North Wales." *Athenaeum*, 6 April 1850, p. 377.

Ramsay, Andrew C. 1852. "On the Superficial Accumulations and Surface Markings of North Wales." *Quarterly Journal of the Geological Society of London* 8:371–376.

Ramsay, Andrew C. 1855. "On the Occurrence of Angular, Subangular, Polished, and Striated Fragments and Boulders in the Permian Breccia of Shropshire, Worcestershire, etc." *Quarterly Journal of Geological Society of London* 11:185–205.

Ramsay, Andrew C. 1862. "The Excavation of the Valleys of the Alps." *Philosophical Magazine*, n.s. 24:377–380.

Ramsay, Andrew C. 1863. *The Physical Geology and Geography of Great Britain*. London.

Ramsay, Andrew C. 1864. "On the Erosion of Valleys and Lakes." *Philosophical Magazine*, n.s. 28:293–311.

Ramsay, William. 1918. *The Life and Letters of Joseph Black, M.D.* London.

Rappaport, Rhoda. 1964. "Problems and Sources in the History of Geology, 1749–1810." *History of Science* 3:60–78.

Raspe, Rudolph Eric. 1763. *Specimen Historiae Naturalis Globi Terraquei*. Amsterdam and Leipzig. Facsimile reprint, with translation and commentary by A. N. Iversen and A. C. Carozzi, New York, 1970.

Raspe, Rudolph Eric. 1776. *An Account of Some German Volcanos and Their Productions*. London.

Read, H. H. 1957. *The Granite Controversy*. London.

Richardson, William. 1803. "Inquiry into the Consistency of Dr. Hutton's Theory of the Earth with the Arrangement of the Strata and Other Phenomena on the Basaltic Coast of Antrim." *Transactions of the Royal Irish Academy* 9:429–487. Read 2 May 1803.

Richardson, William. 1805. "Remarks on the Basaltes of the Coast of Antrim." *TRSE* 5 (3):15–20. Read 7 March 1803.

Richardson, William. 1806. "On the Volcanic Theory." *Transactions of the Royal Irish Academy* 10:35–107. Read 10 December 1804.

Richardson, William. 1808. "A Letter on the Alterations That Have Taken Place in the Structure of Rocks on the Surface of the Basaltic Country in the Counties of Derry and Antrim. Addressed to Humphry Davy. *Philosophical Transactions of the Royal Society of London* 98:187–222.

Ritchie, James. 1952. "Natural History and the Emergence of Geology in the Scottish Universities." *Transactions of the Edinburgh Geological Society* 15:297–316.

Ritchie-Calder, Lord. 1982. "The Lunar Society of Birmingham." *Scientific American* 246:136–145.

Robinson, Eric, and Douglas McKie, eds. 1970. *Partners in Science: Letters of James Watt and Joseph Black*. London.

Royal Society of Edinburgh. 1950. "James Hutton, 1726–1797: Commemoration of the 150th Anniversary of His Death." *Proceedings of the Royal Society of Edinburgh* 63(B):351–400.

Rudwick, M. J. S. 1962. "Hutton and Werner Compared: George Greenough's Geological Tour of Scotland in 1805." *British Journal for the History of Science* 1:117–135.

Rudwick, M. J. S. 1971. "Uniformity and Progression." In D. H. D. Roller, ed., *Perspectives in the History of Science*, pp. 209–277. Norman, Okla.

Saussure, Horace Bénédict de. 1779–96. *Voyages dans les Alpes*. 4 vols. Neuchâtel. Vol. 1, 1779; vol. 2, 1786; vols. 3 and 4, 1796.

Schneer, Cecil J., ed. 1969. *Toward a History of Geology*. Cambridge, Mass., and London.

Schneer, Cecil J., ed. 1979. *Two Hundred Years of Geology in America*. Hanover, N.H.

Schofield, Robert. 1963. *The Lunar Society of Birmingham*. Oxford.

Schofield, Robert. 1970. *Mechanism and Materialism: British Natural Philosophy in an Age of Reason*. Princeton.

Scrope, George P. 1825. *Considerations on Volcanos*. London. Other editions, 1862, 1872. Rev. *Monthly Review*, n.s. 1 (1826): 24–32.

Scrope, George P. 1827. *The Geology and Extinct Volcanos of Central France*. London. 2d ed., 1858.

Scrope, George P. 1830a. "On the Gradual Excavation of the Valleys in Which the Meuse, the Moselle, and Some Other Rivers Flow." *Proceedings of the Geological Society of London* 1:170–171. Read 5 February 1830.

[Scrope, George P.] 1830b. Review of Charles Lyell's *Principles of Geology*, vol. 1. *Quarterly Review* 43:410–469.

[Scrope, George P.] 1835. Review of Charles Lyell's *Principles of Geology*, 3d ed. *Quarterly Review* 53:406–448.

Sedgwick, Adam. 1831. Presidential Address. *Proceedings of the Geological Society of London* 1:281–316. Read 18 February 1831.

Seymour, Lord Webb. 1815. "An Account of Observations Made by Lord Webb Seymour and Professor Playfair upon Some Geological Appearances in Glen Tilt and the Adjacent Country." *TRSE* 7 (2):303–375. Read 16 May 1814.

Sharlin, Harold I. 1979. *Lord Kelvin: The Dynamic Victorian*. University Park, Penn

Shea, James H., ed. 1985a. *Continental Drift*. New York.

Shea, James H., ed. 1985b. *Plate Tectonics*. New York.

Sher, Richard B. 1985. *Church and University in the Scottish Enlightenment: The Moderate Literati of Edinburgh*. Princeton.

Siegfried, Robert, and Robert H. Dott. 1976. "Humphry Davy as Geologist, 1805–1829." *British Journal for the History of Science* 9:219–227.

Silliman, Benjamin. 1810. "Sketch of the Mineralogy of the Town of New Haven." Connecticut Academy of Arts and Sciences *Memoirs* 1:83–96.

Simond, Louis. 1815. *Journal of a Tour and Residence in Great Britain during the Years 1810 and 1811*. 2 vols. Edinburgh.

Sleep, Mark C. W. 1969. "Sir William Hamilton (1730–1803): His Work and Influence in Geology." *Annals of Science* 25:319–338.

Smith, Cyril Stanley. 1969. "Porcelain and Plutonism." In Cecil J. Schneer, ed., *Toward a History of Geology*, pp. 317–338.

Smith, William. 1815. *A Memoir to the Map and Delineation of the Strata of England and Wales, with Part of Scotland*. London.

Smithson, James. 1824. "Of the Deluge." *Annals of Philosophy* 24:58–60.

Spallanzani, Lazzaro. 1798. *Travels in the Two Sicilies and Some Parts of the Apennines*. 4 vols. London. Orig. pub. 1792–97 6 vols., Pavia.

Stanley, John Thomas. 1794a. "An Account of the Hot Springs near Rykum in Iceland," addressed to Dr. Black. *TRSE* 3 (2):127–137. Read 7 November 1791.

Stanley, John Thomas. 1794b. "An Account of the Hot Springs near Haukadal in Iceland," addressed to Dr. Black. *TRSE* 3 (2):138–153. Read 30 April 1792.
Stock, J. E. 1811. *Memoirs of the Life of Thomas Beddoes, M.D.* London and Bristol.
Strange, John. 1775. "An Account of Two Giant's Causeways in the Venetian State." *Philosophical Transactions of the Royal Society of London* 65:5–47.
Sulivan, Richard Joseph. 1794. *A View of Nature, in Letters to a Traveller among the Alps*. 6 vols. London.
Sweet, Jessie M. 1967a. "The Wernerian Natural History Society in Edinburgh." *Freiberger Forschungshifte* 167:206–218.
Sweet, Jessie M. 1967b. "Robert Jameson's Irish Journal, 1797." *Annals of Science* 23:97–126.
Sweet, Jessie M., and Charles D. Waterston. 1967. "Robert Jameson's Approach to the Wernerian Theory of the Earth, 1796." *Annals of Science* 23:81–95.
Taylor, Kenneth L. 1969. "Nicolas Desmarest and Geology in the Eighteenth Century." In Cecil J. Schneer, ed., *Toward a History of Geology*, pp. 339–356.
Taylor, Kenneth L. 1971. "The Geology of Dolomieu." *Actes du XII Congres International d'Histoire des Sciences* 7:49–53.
Thompson, H. W., ed. 1927. *The Anecdotes and Egotisms of Henry Mackenzie*. Oxford and London.
Tinkler, Keith J. 1983. "On Hutton's Authorship of an 'Abstract of a Dissertation . . . Concerning the System of the Earth, Its Duration and Stability." *Geological Magazine* 120:631–634.
Tinkler, Keith J. 1985. *A Short History of Geomorphology*. London and Sydney.
Tomkeieff, S. I. 1946. "James Hutton's *Theory of the Earth*, 1795." *Proceedings of the Geological Association* 57:322–328.
Tomkeieff, S. I. 1948. "James Hutton and the Philosophy of Geology." *Transactions of the Geological Society of Edinburgh* 14 (2):253–276. Reprinted in *Proceedings of the Royal Society of Edinburgh*, 63(B):387–400.
Tomkeieff, S. I. 1962. "Unconformity—An Historical Study." *Proceedings of the Geological Association* 73:383–417.
Toulmin, George Hoggart. 1780. *The Antiquity and Duration of the World*. London.
Toulmin, George Hoggart. 1783. *The Antiquity of the World*. London.
Toulmin, George Hoggart. 1785. *The Eternity of the World*. London.
Toulmin, George Hoggart. 1789. *The Eternity of the Universe*. London.
Townsend, Joseph. 1813. *The Character of Moses Established for Veracity as an Historian, Recording Events from the Creation to the Deluge*. Bath. Rev. *Monthly Review* 74 (1814): 225–238.
Traill, Thomas Stewart. 1848. "Memoir of Dr. Thomas Charles Hope, Late Professor of Chemistry in the University of Edinburgh." *TRSE* 16: 419–434.
Treneer, Anne. 1963. *The Mercurial Chemist: A Life of Sir Humphry Davy*. London.
Tuan, Yi-Fu. 1968. *The Hydrological Cycle and the Wisdom of God*. Toronto.
Tunbridge, Paul A. 1971. "Jean André de Luc, F.R.S. (1727–1817)." *Notes and Records of the Royal Society of London* 26:15–33.
Tyrrell, G. W. 1950. "Hutton on Arran." *Proceedings of the Royal Society of Edinburgh* 63(B):369–376.
Walker, John. 1966. *Lectures on Geology*. Ed. Harold W. Scott. Chicago and London.
Watanabe, Masao. 1978. "James Hutton's 'Obscure Light'—A Discovery of Infrared Radiation Predating Herschel's." *Japanese Studies in the History of Science*, no. 17:97–104.
Watt, Gregory. 1804a. "Observations on Basalt, and on the Transition from the Vitreous to the Strong Texture Which Occurs in the Gradual Refrigeration of Melted Basalt, with Some Geological Remarks." *Philosophical Transactions of the Royal Society of London* 94:279–314. Read 10 April 1804.

[Watt, Gregory.] 1804b. Review of Scipio Breislak's *Voyage lithologique dans la campanie. Edinburgh Review* 4:26–42.

[Watt, Gregory.] 1804c. Review of Déodat de Dolomieu's *Sur la philosophie minéralogique. Edinburgh Review* 4:284–296.

Werner, Abraham Gottlob. 1774. *Von den aüsserlichen Kennzeichen der Fossilien*. Leipzig. Facsimile reprint, ed. A. V. Carozzi, Amsterdam, 1965. French translations: Dijon, 1790; Paris, 1795. English translations: Robert Jameson, *Treatise on the External Characters of Minerals*, Edinburgh, 1805; 2d ed., 1816; 3d ed., 1817; Thomas Weaver, *A Treatise on the External Characters of Fossils*, Dublin, 1805; A. V. Carozzi, *On the External Characters of Minerals*, Urbana, 1962.

Werner, Abraham Gottlob. 1787. *Kurze Klassifikation und Beschreibung der verschiedenen Gebirgsarten*. Dresden, but originally published 1786 in a local journal. Facsimile and trans., Alexander M. Ospovat, New York, 1971; a full bibliography of Werner's published works is included.

Werner, Abraham Gottlob. 1791. *Neue Theorie von der Entstehung der Gange*. Freiberg. French translation: Paris, 1802. English translation: Edinburgh, 1809.

[Whewell, William.] 1831. Review of Charles Lyell's *Principles of Geology*, vol. 1. *British Critic* 9:180–206.

[Whewell, William.] 1832. Review of Charles Lyell's *Principles of Geology*, vol. 2. *Quarterly Review* 47:103–132.

Whewell, William. 1837. *History of the Inductive Sciences, from the Earliest to the Present Time*. 3 vols. London. 2d ed., 1847; 3d ed., 1857.

[Whitaker, Thomas Dunham.] 1819. Review of Thomas Gisborne's *Testimony of Natural Theology to Christianity* (1818). *Quarterly Review* 21:41–66.

White, George W. 1973. "A History of Geology and Mineralogy as Seen by American Writers, 1803–1835: A Bibliographic Essay." *Isis* 64:197–214.

White, George W., and V. A. Eyles, eds. 1970. *James Hutton's System of the Earth, 1785; Theory of the Earth, 1788; Observations on Granite, 1794; Together with Playfair's Biography of Hutton*. Facsimiles of the original editions, introduction by V. A. Eyles, with foreword by George W. White. Darien, Conn.

Whitehurst, John. 1778. *An Inquiry into the Original State and Formation of the Earth*. London. 2d ed., 1786.

Williams, John. 1789. *The Natural History of the Mineral Kingdom*. 2 vols. Edinburgh. Rev. *Monthly Review* 6 (1791): 121–131. The second edition (1810) was noticeably less vociferous regarding Hutton.

Wilson, Leonard G. 1972. *Charles Lyell: The Years to 1841*. New Haven and London.

Witham, Henry. 1826. "Notice in Regard to the Trap Rocks in the Mountain Districts of the West and Northwest of the Counties of York, Durham, Westmoreland, Cumberland, and Northumberland." *Memoirs of the Wernerian Natural History Society* 5:475–480.

Wood, P. B. 1987. "Buffon's Reception in Scotland: The Aberdeen Connection." *Annals of Science* 44:169–190.

Zittel, Karl von. 1901. *History of Geology and Paleontology*. Trans. Maria M. Ogilvie-Gordon. London. Facsimile reprint, 1962. Orig. pub. as *Geschichte der Geologie und Paläontologie*, Munich, 1899.

Index

Library of Congress Cataloging-in-Publication Data

Dean, Dennis R.
James Hutton and the history of geology / Dennis R. Dean.
p. cm.
Includes bibliographical references and index.
ISBN 0-8014-2666-9 (alk. paper)
1. Geology. 2. Hutton, James, 1726–1797. I. Title.
QE26.D43 1992
550'.9—dc20 91-55545